Dynamic Analysis in the Social Sciences

Dynamic Analysis in the Social Sciences

Emilio J. Castilla

Sloan School of Management
Massachusetts Institute of Technology
Cambridge, MA, USA

Amsterdam – Boston – Heidelberg – London – New York – Oxford
Paris – San Diego – San Francisco – Singapore – Sydney – Tokyo
Academic Press is an imprint of Elsevier

ACADEMIC
PRESS

ELSEVIER

Academic Press is an imprint of Elsevier
84 Theobald's Road, London WC1X 8RR, UK
Radarweg 29, PO Box 211, 1000 AE Amsterdam, The Netherlands
Linacre House, Jordan Hill, Oxford OX2 8DP, UK
30 Corporate Drive, Suite 400, Burlington, MA 01803, USA
525 B Street, Suite 1900, San Diego, CA 92101-4495, USA

First edition 2007

ISBN: 978-0-12-088485-8

For information on all Academic Press publications visit our
website at books.elsevier.com

Printed and bound in USA
07 08 09 10 11   10 9 8 7 6 5 4 3 2 1

Working together to grow
libraries in developing countries

www.elsevier.com | www.bookaid.org | www.sabre.org

ELSEVIER   **BOOK AID International**   Sabre Foundation

To my parents
Emilio and Isabel

Table of Contents

List of Tables

List of Figures and Graphs

Preface

Time has become a central issue in contemporary social sciences, and its correct modeling poses a series of methodological challenges that we have begun to address with advanced longitudinal analyses. The study of social change using quantitative methodology—i.e., how and why social structures and their individuals change with time—is complex. Some of the more complicated aspects of quantitative methodology refer to the appropriate use of dynamic models that combine both time and causality. Ordinary least squares (OLS) regression and logit/probit regression are perhaps the most frequently used statistical tools in the analysis of social reality. They were designed to help analyze data sets with no temporal components. However, these standard methods are not always properly suited to address major research questions about social dynamic processes in social science research, and they can often provide inadequate modeling of the causal relationships among variables over time. To analyze change over time, there are better statistical methods, such as event history analysis and panel models. Such extensions are, however, not as simple as one may think. This book is motivated by the renewed interest in the study of social change using longitudinal data sets in current research.

Dynamic Analysis in the Social Sciences is intended for social scientists, researchers, professionals, and students interested in learning these dynamic methods and their practical applications in the analysis of social change over time. This book explains how to study change in variables over time and how to estimate multivariate models to predict such change. The first chapter presents a description of what dynamic analysis is in the social sciences and introduces an example of a longitudinal data set with several variables whose values change over time. The following two chapters cover the statistical theory behind *event history analysis* (for the modeling of events happening over time) and *cross-section time series analysis* (for the modeling of change in continuous variables). The book includes real examples and analyses of longitudinal data so that the reader can learn to use statistical software packages such as Stata, SPSS, or TDA to estimate longitudinal models. It also teaches how to design and implement a study using longitudinal data (in Chapter 4). The last chapter provides two real studies as concrete applications of the longitudinal techniques learned in this book. The book contains two useful lists of references—one with extensive comments on several basic and advanced methodology books, and one of empirical studies that estimate dynamic models that have been published in top social science journals.

I hope that the experience of exploring longitudinal data analysis and drawing conclusions about change occurring over time will make this book engaging, relevant, and useful to readers' future educational and professional lives. I assume that readers have access to a computer in order to analyze their own data sets and learn hands-on the methods discussed throughout the chapters. High school algebra and some basic statistical knowledge are the only requirements; explanations requiring the use of calculus or matrices do not appear. This book is self-explanatory and can be a practical resource to introduce this relatively new (in its usage) methodology to any reader. It can also be used as textbook for an introductory course in the methodology behind dynamic analyses. Most of the examples represent social sciences, and therefore should be interesting and understandable to social scientists as well as to scholars and professionals in other fields. The book, as a whole, leads up to an introduction to dynamic analysis. Many statistical packages could be used with this book although it was written with two particular packages in mind, Stata and SPSS, because I believe they are easy to use, fast, and flexible.

This manuscript was started in the Sociology Department at Stanford University, continued in the Management Department at the University of Pennsylvania's Wharton School, and heavily revised and completed in 2007 at the MIT Sloan School of Management—where the author currently teaches and investigates topics on work and employment using the most advanced methodology in the analysis of change over time.

Acknowledgments

Writing this book about dynamic analysis was like enacting a premonition of mine. Since I came to the United States, I have experienced changes in my life that have been both qualitative and quantitative. After I finished my bachelor's degree in economics at the Universidad de Barcelona, I left Spain for the United States to earn my PhD in sociology at Stanford. Stanford was the place where I learned how to learn. Then, I began my first job as an assistant professor of management at the Wharton School, University of Pennsylvania, where I whole-heartedly pursued the writing of this book. I am now a faculty member at the MIT Sloan School of Management. Every day I ask myself whether any model exists that could have predicted all the professional and personal accomplishments that have occurred in my life. First, I owe a great deal of gratitude to Professor Jesús M. de Miguel. I would not have come to the United States to study if not for his continuous influence, his devotion to scholarship, and his extraordinary vision. He saw in me the promise of who I have become today when nobody else could have seen it. Jesús, I thank you for introducing me to the discipline of economic and organizational sociology and for encouraging my interest in organizations, labor markets, employment, and inequality. I thank you for your energy and for your encouragement. But most importantly, I thank you for your years of training and friendship. Roberto M. Fernandez has always been supportive throughout the many years I have known him. Through good and bad times, he was there, and he provided kind words of encouragement and generous support during and after my PhD studies. My deepest thanks also goes to the Fernandez family: Carol, Julia, Alejandro, and Skippy, who have made my life in the United States always full of happy moments. Mark Granovetter, my advisor at Stanford University, has been continually helpful, both encouraging and critical at the right times. Thanks to the Granovetter family: Ellen, Sarah, and Ringo. Muchísimas gracias a todos.

I am grateful to many people for sharing their ideas and expertise over these past years, without which *Dynamic Analysis in the Social Sciences* would not have been possible. Two have had a particularly notable influence on my training in methodology, on my thinking, and on the writing of this book. First, Nancy Tuma introduced me to event history methodology during my first year in graduate school in the Department of Sociology at Stanford University. Her lectures and notes were fantastic sources for graduate students learning about this methodology and about being careful in the examination and analyses of our data sets. I also thank Nancy Tuma for hiring me as her teaching assistant for two popular PhD methodology courses at Stanford: "Sociological Methodology II: The General Linear Model" (in 1998) and "Sociological Methodology III: Advanced Models for Discrete Outcomes: Logit and Probit Models, Event History Analysis" (in 1999). As part of my teaching responsibilities, I developed documents, sections of documents, and computer samples for graduate students, some of which are the backbones of the examples and the inspiration behind my work in this book. Second, I am also extremely grateful to my wonderful boss and friend Pat Box, and to Chuck Eckman, Judy Esenwein, Ron Nakao, and the many consultants Shelley Correll, Valery Yakubovich, Elizabeth Macneany, Annelie Strath, Chris Bourg, Coye Cheshire, Stef Bailey, Kjersten Bunker, and Emeric Henry among many with whom I worked for several years at the Social Science Data and Software (SSDS) group within Stanford's Libraries and Academic Information Resources (SULAIR). SSDS provides services and support to faculty, staff, and students in the acquisition of social science data and in the selection and use of quantitative (statistical) and qualitative analysis software. As a consultant working for an amazing team empowered and motivated by Pat Box, I helped in providing some of these consulting services in a variety of ways, including one-on-one and group consulting, several methodology workshops, and in documentation preparation.[1] My involvement with this group made me think about writing a methodology book (such as this one) that would cover both theory and practice in the analysis of longitudinal data.

Various institutions have played an important role in my training and education as a scholar. This manual is the result of my research formation in many different universities around the world: the University of Barcelona (Spain), Lancaster University (United Kingdom), Harvard University, the University of California at Berkeley, the University of Michigan at Ann Arbor, and especially my prolonged stays at Stanford University, the University of Pennsylvania, and now MIT. In Spain I thank the *Fundación La Caixa* for financing my studies at Stanford University for two years. The University of Barcelona contributed to my development in economics, sociology, and methodology in the social sciences. However, Lancaster University in England is where I realized that there was an entire exciting intellectual world outside my hometown in Spain. Stanford became a magnificent and intellectually intense experience. I am finally thankful for the Clifford C. Clogg Award and Fellowship given to me by the American Sociological Association (Methodology Section) in 1998. As part of this fellowship, I was able to spend an entire summer at the Inter-university Consortium for Political and

---

[1] This Center provides excellent data and software consulting services to faculty, students, and staff. One can also benefit from the resources it makes available online by visiting http://ssds.stanford.edu. I highly recommend checking it out!

Social Research (ICPSR) at the University of Michigan, Ann Arbor. I took many methodology courses offered that summer and most importantly, I started thinking about the contents of this book, a dream that has come true today.

I feel fortunate for the help and friendship of many colleagues around the globe. In the United States, I thank Mark Granovetter, Roberto Fernandez, Ezra Zuckerman, and John W. Meyer. I am thankful to those who encouraged me and helped me as a graduate student in both the Sociology Departments and the Business School. Thanks to David Grusky, Michael Hannan, Noah Mark, Harry Makler, Francisco Ramirez, Cecilia Ridgeway, Nancy B. Tuma, Suzi Weersing, and Morris Zelditch. I thank my colleagues and friends for their personal and technical support: Shelley Correll (Cornell University), Kyra Greene, Hokyu Hwang (Stanford University), Kiyoteru Tsutsui (University of Michigan), Nancy Weinberg, Simon Weffer-Elizondo (UC Merced), Fabrizio Ferraro (IESE, Barcelona), Siobhan O'Mahony (UC Davis), and Valery Yakubovich (Wharton School). Thanks to e-mail, text-messages, and telephone, I have been able to debate these and other topics with Ruth Aguilera (University of Illinois, Urbana-Champaign), Luis Gravano (Columbia University), Anne-Marie Knott (Washington University), Mauro F. Guillén (Wharton School), Nancy Rothbard (Wharton School), Stefanie Wilk (Ohio State University), Bilian Sullivan (Hong Kong University of Science and Technology), Devah I. Pager (Princeton University), Mikolaj J. Piskorski (Harvard Business School), Elizabeth Vaquera (University of South Florida), and Mark Zbaracki (New York University). I would like to thank all of them for their support and spirit. I also want to thank my colleagues and friends at the Wharton School, University of Pennsylvania, especially Roxanne Gilmer, Gerald McDermott, Lori Rosenkopf, Daniel Levinthal, John Kimberly, John Paul Macduffie, Peter Cappelli, Michael Useem, Victoria Reinhardt, and Katherine Klein. In addition, I thank all my "new" colleages at MIT; special thanks go to Jackie Curreri for her assistance and consideration since I joined the MIT Sloan in 2005. I owe many thanks to my friends all over the world. They form that "invisible college." My friends in California: Benjamin Lawrence, William Lorié (my roommate in Escondido Village 11A), Christopher Roe, Shannon Shankle, Kristina (and her lovely one-year-old Parker) Roberts, Harry Turner, Jonathan Jewell, Frank Salisbury, and Mark Urbanek. A special thanks and remembrance to my friends in the old Europe: Ruben Hernández, David Montes, and Meritxell Pons. Lastly, I am thankful for my new friends in Boston: David Essaff, Gary Furtado, Jason Gagnon, Brian Grove, Doug Munn, John Wilhelmi, and very specially, Joseph D. Raccuglia.

At Elsevier, I benefited tremendously from the attention of Scott Bentley first and Zoe La Roche later in the publication process. Scott helped initiate this project, and oversaw its transformation from a draft into a whole new book. He helped guide it through production, keeping pace with me, the anonymous reviewers of this book, and the software development. Their patience and understanding throughout this process have made this book a very pleasant enterprise. Many anonymous reviewers offered helpful comments on earlier drafts of the book. Maurice Black, Robin Orlansky, and Joseph D. Raccuglia did an excellent job as researchers and editors of this book. Many of my graduate and undergraduate students helped me with the research and the writing of this book, especially Anjani Trivedi (at MIT) and Mauricio Achata (at Wharton).

Choosing a dedication of this book was the easiest part: To my parents, Emilio and Isabel, my brother, Sergio, my sister, Yolanda, and all my relatives in Spain for their love and support during my undergraduate education at the Universidad de Barcelona (Spain) and my graduate education at Lancaster University (United Kingdom) and Stanford University. I hope that this book helps them understand the importance of my coming to (and even staying in!) the United States.

Emilio J. Castilla
Cambridge, MA, USA, Fall 2007

Author Biography

Emilio J. Castilla is an assistant professor of Management at the MIT Sloan School of Management, where he teaches courses in human resource management and quantitative research methods. He is currently visiting the Department of Sociology at Harvard University. He joined MIT after serving on the faculty in the Management Department at the Wharton School of the University of Pennsylvania. He is a member of the Institute for Work and Employment Research at MIT and a research Fellow at both the Wharton Financial Institutions Center and at the Center for Human Resources at Wharton. He received his postgraduate degree in business analysis from the Management School in Lancaster University (UK) and his PhD (and MA) in Sociology from Stanford University. His research interests include organizational theory, economic sociology, and human resource management, with special emphasis on the social aspects of work and employment. Through his research, Professor Castilla investigates how social networks and organizations influence employment processes and outcomes over time. In his work, he formulates and tests specific hypotheses about these important issues in a variety of diverse empirical settings, using a number of unique longitudinal data sets and the most advanced quantitative methodologies.

An active scholar, Professor Castilla has presented his research all over the world. He has published chapters in several books as well as articles in a number of scholarly journals. In 2001, he was awarded the W. Richard Scott Award for Distinguished Scholarship for an article published in the *American Journal of Sociology* (with Fernandez and Moore). Other awards include the Leila Arthur Cilker Award for Excellence in Teaching (1999), the Stanford Centennial Teaching Award (1998), and the Clifford C. Clogg Methodology Award/Fellowship (American Sociological Association/Methodology Section, 1998).

Chapter 1

Longitudinal Data

One of the most prevalent issues today within the social sciences (broadly defined) is a focus on understanding and analyzing social change, i.e., the study of how and when social systems and their actors evolve over time. For a long time now, social researchers and scholars have been interested in the processes that change society. Moreover, social theories are dynamic not only in their formulation but also in their capability to explain social change. Practitioners and professionals alike have been interested in understanding the impact of policies, practices, and changes on the long-term lives of individuals, communities, groups, and organizations. Despite the importance of dynamic theories in the social sciences, the quantitative analysis of longitudinal data has been quite limited, relatively speaking. The same applies to practice, where experts and practitioners have rarely conducted dynamic quantitative analyses of longitudinal data. Yet, this type of social analysis has regained significant popularity in articles published in contemporary social science journals. Today, analysis of social change has become more important; in fact, it could be called the central underlying motivation behind contemporary social sciences research. In addition, organizations have started to create and store impressive data sets about their actors, their attitudes and actions, and their results. This growing availability of and emphasis on longitudinal data sets places new kinds of demands upon advanced statistical methodologies and, most importantly, on the resulting technical implementation of such methodologies using the available software programs.

Even though methodological techniques such as ordinary least square (OLS) regression or logit regression currently are widely employed in empirical studies and have useful applications for the static study of social processes (i.e., the study of causal relationships occurring in a given moment in time), these techniques do not sufficiently encompass and respond to this surging interest in modeling social change over time. Such models and methods for statistical analysis could be usefully expanded to incorporate the analysis of social dynamic processes. Yet, such methodological extensions to the study of social change are not as obvious as they may seem. A considerable extension of the existing statistical theories and tools is essential here in order to be able to carry out dynamic social analysis.

The basic dynamic methodology covered in this book offers two main advantages for the testing and understanding of social theories of change. The first consists of the formulation and estimation of dynamic models; i.e., models that describe the process of social change over time. The second advantage involves the development of causal models, whose primary objective is to describe how change in certain characteristics or variables directly influences change in other characteristics or variables. This means that these models focus on studying how a dependent variable, whether continuous or categorical, changes over time as a function of one or more other explanatory or independent variables, some of which may also vary or change over time. It used to be the case that social science researchers rarely employed a methodology that combined both time and causality. This book is an introduction to the dynamic analyses of social processes, examining causal relationships of variables over time. It responds to the increasing interest and need (both in academia and in practice) to understand and use dynamic methodologies to analyze dynamic processes in a simple and straightforward manner. This is accomplished by first presenting the theory behind available statistical models to understand change, and second by providing the right practical guidance to start using such techniques on a longitudinal data set using the appropriate statistical software program. The main purpose of this book is to describe these advanced methodological approaches to social change in as simple a way as possible. The reader will then be able to evaluate the validity of various social theories and their propositions regarding the causes and consequences of social change. In addition, the reader will know how to use the right commands mainly in two statistical programs, Stata and SPSS, to run the learned techniques.

The objectives of this book are multifold. First, the main statistical models and methods available for examining social change are introduced to the reader, clarified, and further developed within each chapter. Second, a necessary distinction is made, given that there are different data and variable types, between the formulation of models of change using quantitative and qualitative outcomes, especially when presenting and discussing the different methods of estimation and when using the most popular statistical software programs for longitudinal data analysis. This book is also an introduction to the use of longitudinal data in causal analysis for those who are already familiar with OLS and logistic regression types of techniques. This book is appropriate for any social scientist, practitioner, or professional interested in the formulation and basic application of empirical dynamic analysis. Included in this book are many ideas, examples, and suggestions on how to carry out empirical studies in order to examine the processes of social change. The material has been simplified and streamlined to accommodate a diverse audience and is geared especially towards those who already have acquired some basic mathematical and statistical knowledge.

1. Some Basic Concepts and Terminology

Before turning to types of longitudinal data and analyses used to predict change over time, it is important to review some basic knowledge about variables and the coding of information about a given social entity, "case," or "unit of analysis." A variable is way of measuring a property of some social unit (for example, an individual, a group of individuals, an organization, or a country) that can vary across units and over time. It is therefore possible to have data on a sample of N distinct units or cases, so that the size of the sample is N. Let the letters x and y represent two different variables on which data exist in the sample of cases. The lowercase letter "i" is used to denote a specific, hypothetical individual case in the sample. The notation x_i refers to the value that variable x takes for case i, and y_i is the value that variable y takes for case i. The subscript on the variable denotes the particular case being considered. Consequently, a basic data set can consist of two variables x and y for a sample of size N, meaning that the values of $x_1, x_2, x_3, \ldots, x_n$ and the values of $y_1, y_2, y_3, \ldots, y_n$ are known to the researcher. An easy way to store information for a sample of N units is to create a table in Microsoft Excel, with each column representing a variable and each row representing an individual case at a given period of time (with the name of the variables on the first row of the table). As the reader will see later in this section, Excel files can easily be read into the main statistical programs for their analyses—in particular, the ones to be used to analyze longitudinal data sets in this book.

	A	B	C	D	E	F	G
1	Case id	X	Y				
2	1	25000	3				
3	2	15000	5				
4	3	34000	1				
5	4	13500	4				
6	5	23400	3				
7	6	23790	5				
8	7	12343	1				
9	8	23190	6				
10	9	8900	4				
11	10	15000	6				
12							
13							

On page 3, look at an example of a hypothetical data set for a size of 10 cases (using Microsoft Excel). The first column is used to refer to the case or unit identification number (or unit ID); each number uniquely identifies each case or unit in the sample under study. Column 2 stores the value that variable x takes for each of the 10 individual cases in the sample; hence there are 10 different rows in the table (excluding the row containing the variable name). The third column does the same for variable y. So for unit 1, the value of x is 25000 (or stated using some basic terminology, $x_1 = 25000$) and the value of y is 3 ($y_1 = 3$); for unit 2, $x_2 = 15000$ and $y_2 = 5$; and so on, until unit 10 for which x_{10} equals 15000 and y_{10} is 6.

1.1.   Main Types of Variables

Before learning about univariate and multivariate statistics, it is central to understand the distinction among the different types of variables that characterize social entities and processes over time. This basic typology will be of help later in this book when deciding how to model different types of dependent variables over time. Generally speaking, there are two main types of variables: "categorical" and "continuous" variables. *Categorical* variables are also called discrete variables. They only take integer values (or whole numbers), so that each value represents a certain property, category, or group characteristic. For example, gender is a categorical variable that may take the value of 1 to refer to males and 2 to refer to females. Marital status can be another example of a categorical variable where, as an illustration, one can use 1 to refer to single individuals, 2 to refer to married individuals, 3 to refer to divorced, and 4 to refer to widowed. The value of the variable does not mean anything *per se,* except for the fact that a whole number has been given to a certain defining characteristic or trait. The term *dummy variable* (also known as indicator variable) is used to refer to a categorical variable that can only take two values, 0 or 1, and the value of 1 always indicates and points to the characteristic identified by the name of the variable. For example, it is common to include 'female' as a dummy variable in regression-type analyses to study gender differences in a social process of interest. This variable takes the value of 1 for females, 0 otherwise (i.e., males). One can imagine including a set of dummy variables in a regression analysis to see whether there are any racial differences in the mean value of a given dependent variable. African American, White, Asian, and Hispanic are four examples of dummy variables, where each variable takes the value 1 to refer to the named racial trait, 0 otherwise.

Continuous variables (also termed interval, ratio, or metric variables) are those whose range of possible values is meaningful. Such variables can be represented by a continuum of values, including whole and decimal numbers. Age and salary are examples of continuous variables. In the case of age, individual values may range from 0 up to over 100 years old. Many times, however, continuous variables are coded as discrete ordinal variables due to either the research design used when gathering the data or the lack of detailed information on the data. An illustration is the collection of respondents' incomes in a survey study. One option is to ask for a numeric value and therefore code the respondent income as a continuous variable. However, given the issues of sensitivity and privacy involved in asking individuals for income information, this variable alternatively could be collected as a discrete variable. Responses to the income question could allow the respondent to choose among a few options covering different ranges of income. Thus, for example, individuals can be classified among those who earn less than $15000 per year, those who earn between $15001 and $30000 per year, those who earn between $30001 and $40000, and so forth. This way, the question is less intrusive. Another common survey example is when age is presented in different intervals, so each variable category represents a given age range. This can be a convenient way of presenting groups of people distributed by age in five- or ten-year intervals. Typically, it is preferred to code variables as continuous variables, given that intervals aggregating a continuous variable lose some important information about the unit under analysis. Continuous variables give more precise information for a given case.

Within categorical or discrete variables, one can distinguish among "nominal," "ordinal," and "cardinal" variables. *Nominal variables* take arbitrary values, such that any arithmetic manipulation on the values (such as addition or multiplication) does not have much meaning or make much sense. These variables can only take a set of pre-determined and specified values (as decided by the researcher), so that each value merely identifies the presence or absence of certain attributes. In the case of nominal variables, no consideration is given to the quantity intrinsic to the attribute, only whether or not there is an attribute. These variables take different values that help classify some cases into certain categories or groups that frequently are arbitrarily assigned values in order to facilitate their coding and later analysis. Each value can be considered a personalized attribute identification number. A variable is *dichotomous* when it can take only two possible values. A *polytomous variable* in general has q categories ($q = 1, 2, 3, \ldots, Q$ and $Q > 2$). Examples of nominal dichotomous variables are gender (male/female) and membership in a given organization (member/non-member). Examples of nominal polytomous variables include race or ethnicity (White, African American, Asian, or Hispanic), religion (Roman Catholic, Jewish, Protestant, no religion, or other religion), and political affiliation (Democrat, Republican, or other).

In the case of *ordinal variables,* the numbers such variables take represent values that are ordered, although the numerical difference between adjacent values in the sequence is not meaningful. For example, survey researchers have often measured agreement with some statement using a 5- or 7-point Likert scale (for example, individuals' support of the legalization of gay marriage in the United States). The answer to such a statement can be coded from 1 (strong disagreement) to 5 (strong agreement); the answer could equally be coded from -100 to $+100$ or from 0 to 100. In this case, neither the extreme values nor the differences between the values have any substantive meaning; the order, however, is significant. The final type of categorical variable is called a *cardinal variable*. In this case, both the extreme values as well as the value differences have substantive importance. Examples of cardinal variables could be the number of respondent's siblings or the number of plants that a multinational company has opened in a given country, where the value 0 is not an arbitrary value, in the same manner that the difference between 1 and 2, or between 4 and 5 is not arbitrary either. Thus, the difference of 1 in both previous cases means exactly that, thus making such value difference meaningful.

Given that nominal variables are used to determine the presence of a quality or attribute for a given case, these variables are often called "qualitative" variables. And these variables are often used as dependent variables in the event history analysis that will be covered in Chapter 3. On the contrary, continuous variables along with discrete ordinal and cardinal variables are called "quantitative" variables because they all refer to the quantity of a particular quality or property of a given unit of analysis. When these variables are used to measure the dependent variable or variable whose change over time is to be investigated, the cross-sectional time series analysis is the appropriate method of analysis. This methodology will be covered in Chapter 2. Do not worry about this distinction yet, though. I will come back to these concepts later in this chapter.

1.2. *Univariate and Bivariate Statistics*

Before starting to develop any dynamic models, I would like to talk about statistics and the meaning of statistics when describing variables in data sets. A *statistic* is used to refer to a summary measure based on the values of one or more variables in a data set. A *univariate statistic* refers to a statistic for a single variable. A *bivariate statistic* refers to a statistic based on two variables. The most general statistic is the *multivariate statistic,* which is based on more than two variables. One of the most important sets of univariate statistics is the one that seeks to provide one single number or measure to summarize the "central tendency" or "location" in all of the data on a given variable. Three common measures of central tendency are the *mean* or *average,* the *median,* and the *mode.* The mean of x, usually denoted as $\bar{x}$, is the sum of all the values of the variable x in the data set, divided by the sample size N:

$$\bar{x} = (x_1 + x_2 + x_3 + \cdots + x_n)/N$$

Using the commonly used summation expression, this is similar to:

$$\bar{x} = \sum_{n=1}^{N} x_n / N.$$

Ordered statistics are those based on position in an ordered list of the data. The simplest order statistics are the *minimum* and *maximum,* that is, the first and last values in an ordered list. The other well-known order statistic is the *median.* The median is the value that lies at the halfway point of the relative frequency distribution. If N cases are listed in order from lowest to highest, the median of x is the value at position $(N + 1)/2$. The median equals the value of the middle case if n is odd, and the value of the average of the two middle values if n is even. The *mode* is the most common value of the variable x. Even though the mode may be of interest for certain purposes, it is not especially useful when measuring the central tendency of a given variable. To illustrate some of the univariate statistics, look again at the sample data set in the Excel file presented on page 3. The mean value for x is 19412.3 and the mean value for y is 3.8. The median for x is 19095 and for y is 4. The mode is 15000 for x and 3 for y.

In addition to a measure of central tendency, typically one would want to have a measure of a variable's spread or variability. The simplest measure of spread is the *range* of a variable, which is defined as the maximum value of x minus its minimum value. The most commonly used measure of spread is the *variance* (Var.) around the mean, also known as the "second moment about the mean" (the mean is the "first moment"). It is defined as the sum of the squared deviations from the mean of every value in the data, divided by the sample size N minus 1:

$$\text{Var}(x) = [(x_1 - \bar{x})^2 + (x_2 - \bar{x})^2 + (x_3 - \bar{x})^2 + \cdots + (x_n - \bar{x})^2]/(N - 1).$$

The *standard deviation* (SD) around the mean is the other most common measure of spread and is defined as the square root of the variance. It is reported more often than the variance simply because the SD has the same units of measurement as the mean. In the sample data set (page 3), x has a SD of 7657.5 (with a variance of 58637326) and y has a SD of 1.81 (with a variance of 3.29). Both the mean and variance are examples of what is typically referred to as *basic descriptive statistics* for the case of continuous variables. These descriptive statistics are often emphasized throughout this book precisely because they provide important information about the variables to be analyzed, and I highly recommend their computation, evaluation, and exploration before undertaking any advanced multivariate analyses such as those presented later in this book.

In many cases, especially when variables are not continuous, frequencies are more appropriate ways of describing the information contained in one variable than means and SDs. Frequency distributions and relative frequency distributions are very common and easy to compute with the current statistical packages. The relative frequency distribution reports the proportion of cases with each value. For example, the values of y in the example data set could be grouped as follows:

Value label	Count	Relative frequency (%)
1	2	20
2	0	0
3	2	20
4	2	20
5	2	20
6	2	20

Tables reporting frequency distributions for different groups of social entities or cases are also convenient in certain types of studies. It is possible to report the same information above using various different graphical presentations. Histograms, for example, can display frequency distributions of variables; they can show bars covering an area proportional to the frequency, proportion, or percentage of each value. Other bar graphs, stem-and-leaf plots, box and whisker plots, and pie charts are also frequently used to present these distributions graphically. Charts are visually appealing and can make it easy for a reader to make comparisons or see patterns and trends in the data. For instance, rather than having to analyze several columns of worksheet numbers, by looking at a chart,

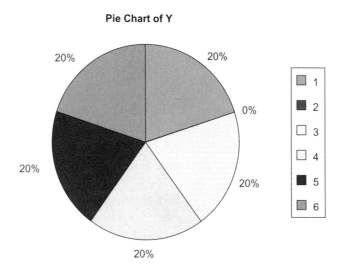

Pie Chart of Y

one can see at a glance whether the values for any given variable are falling or rising over a range of specific periods, or how the current values compare to those projected. For example, the same information from the table above can be presented as a pie chart like the one on the previous page, which was created using Excel's *Chart Wizard* features.

Excel is a very easy-to-use data storage/management program. One can create a chart on its own sheet or as an embedded object on a worksheet. To create a chart, first enter the data for the chart on the worksheet. Then select that data and use the *Chart Wizard* to step through the process of choosing the chart type and the various chart options, or use the *Chart Toolbar* to create a basic chart that can be formatted later. Excel truly makes it easy to get some of these charts in a matter of minutes.

There are several ways to begin studying the relationship between two variables by computing bivariate statistics, particularly when both variables are continuous. The commonly used method is the *scatter plot,* a graph of the two variables, x and y, for the cases in the data. In such a scatter plot, a dot, cross, or any other symbol is used to mark the (x, y) coordinate of each case in the sample. This is the way the scatter plot looks for variables x and y in the data set above presented (again constructed using the *Chart Wizard* in Excel):

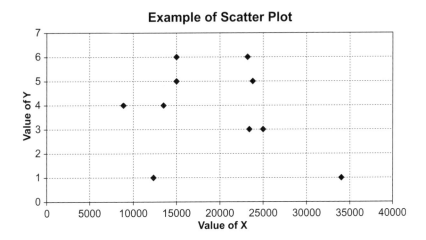

A universally used measure of the degree of relationship between x and y in the sample is called the *correlation coefficient* (also known as the Pearson correlation coefficient between x and y) usually denoted as r. r, also expressed as r_{xy}, has no units and varies between −1 and +1. When r = 0, it implies that x and y are not linearly related at all. If r = 1, then the two variables have a perfect linear relationship, with each increasing by a fixed amount for a given increase in the other. Consequently, it would be very easy to connect all the dots with a straight line. If $r = -1$, then they also have a perfect linear relationship, with one variable decreasing by a fixed amount for a given increase in the other variable. The formula for r is:

$$r_{xy} = \frac{1}{N-1} \sum_{n=1}^{N} \left(\frac{X_i - \bar{x}}{s_x} \right) \left(\frac{Y_i - \bar{y}}{s_y} \right)$$

where $\bar{x}$ and $\bar{y}$ denote the sample means of x and y, respectively, and s_x and s_y denote the sample SDs of x and y, respectively. The correlation coefficient for the example data set is -0.2675, indicating that x and y are negatively correlated with each other, which means that as variable x increases, y tends to decrease, and vice versa. The correlation coefficient does not imply causality but rather measures linear association between two variables. The graphs below show two examples of scatter plots, one for each of the extreme possible values of the correlation coefficient. The graphs are fitted for two other hypothetical data sets in which the correlation coefficients are $+1$ and -1, respectively.

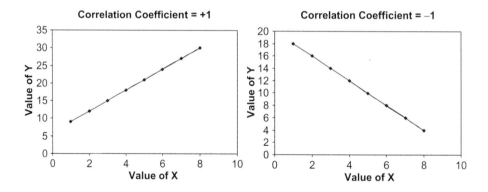

The scatter plots and the correlations assume that both the variables are continuous. However, there are many other examples of bivariate statistics in which the variables are of types other than continuous. Table 1.1 gives a summary of the different types of statistics and graphical representations available depending on the two kinds of variables one wishes to compare. The second column provides the univariate statistics available for each of the different types of variable I. Again, in the case of continuous variables, the statistics of central tendency such as mean and median (as well as the SD) are very convenient ways of summarizing information for a given sample or sub-samples of cases under analysis. For example, one can report that the average age of respondents in a sample is 27 years old (with an SD of 6 years). The best way to describe nominal variables nevertheless is by reporting

Table 1.1
Summary of univariate and bivariate statistics depending on the types of variables

VARIABLE I	UNIVARIATE STATISTICS AND GRAPHS	VARIABLE II			
		Dichotomous	Nominal (q categories)	Ordinal	Continuous (or interval or ratio)
Dichotomous	Frequencies and Histogram	Crosstabs (Chi-Square Test of Independence)			
Nominal (q categories)	Frequencies and Histogram	Crosstabs (Chi-Square Test of Independence)	Crosstabs (Chi-Square Test of Independence)		
Ordinal	Frequencies and Histogram	Nonparametric (Mann-Whitney Test)	Nonparametric (Kruskal-Wallis Test)	Crosstabs (Tau)	
Continuous (or interval or ratio)	Descriptives (mean, variance) and Frequencies	Difference of Means (T-Test)	ANOVA	ANOVA	Scatter plots, Correlation, Regression

their frequency distributions—in other words, to arrange the data in order of the values of the variable and count the number in the data with each value (as illustrated in the example above for variable *y*, on page 8). Even more typical than reporting the count is reporting the proportion or percentage of cases with each value. For example, one could describe that a sample has 40% female and 60% male respondents. In this case, the graphical presentations of frequency distributions can be informative. Histograms (bar graphs), stem-and-leaf plots, box and whisker plots, and pie charts are some common ways to present these distributions graphically. Tables reporting frequency distributions are also extremely convenient.

Table 1.1 can also help in choosing the right bivariate methods to examine the relationship between any two variables (beyond the bivariate statistics described in this chapter). Columns 3–6 provide the examples of all potential bivariate statistics comparing any two variables I and II. For example, when variable I is continuous and variable II is nominal or ordinal, one can run ANOVA (or analysis of variance). When variable I is continuous and variable II is dichotomous, one can run a difference of means test (also known as *t*-test). A basic statistics book can aid in learning more about any type of bivariate statistics (for some additional reading suggestions, consult the list of references at the end of this book).

1.3.   *First Example of a Longitudinal Data Set*

Up to this point, I have only considered an example of a *cross-sectional data set*—that is, a data set where *x* and *y* are measured for a sample of *N* units or "cross-sections" in a given moment of time. It is possible to also add a time component to any data set, by measuring *x* and *y* for a sample of *N* cases or entities at several points in time. The data set will provide information as measured by a given variable *x* or *y* for a given entity or case *i* at several points in time *t*. An example of a *longitudinal* or *temporal data set* is illustrated in the following Excel table. By temporal, I refer to the fact that the data set stores information for a sample of cases at different points of time. The term "longitudinal data" is also widely used in the social sciences to refer to the same data type. The different types of temporal or longitudinal data sets any researcher can encounter are presented and explained in more detail in section 2 of this chapter.

It is key to remember that a longitudinal data set records information about a certain number of cases or units of analysis that change over a period of time. Not every longitudinal study is amenable to the analysis of change, though. Three important data-design characteristics determine whether or not data are suited to this task:

1) Two or more waves of data. So, if one assumes that all variables in the data set are measured at discrete points in time, $t = 1, 2, 3, \ldots, T$, where T is the

end of the observation period (such T may vary across cases), then T has to be at least 2 ($T \geq 2$).

2) An outcome variable (or dependent variable) whose values change over time, called y_t or $y(t)$. Such an outcome variable y can be a discrete or a continuous variable, and is defined at a series of discrete points in time for a sample of individuals, $i = 1, 2, 3, \ldots, N$.

3) A time variable for each observation measures the passage of time in the data. Time is typically a variable that is either continuous or ordinal.

Below is an example of a longitudinal data set where the first column refers to the variable case ID, the second column refers to time, the third column refers to variable x, and the last column refers to variable y.

	A	B	C	D
1	Case Id	Time	X	Y
2	1	1	25000	3
3	1	2	15000	5
4	1	3	34000	1
5	2	1	13500	4
6	2	2	23400	3
7	2	3	23790	5
8	3	1	12343	1
9	3	2	23190	6
10	3	3	8900	4
11	4	1	15000	6
12	4	2	36749	3
13	4	3	23412	3
14	5	1	34556	4
15	5	2	13534	2
16	5	3	23456	3

In this specific example of a basic longitudinal data set, the sample size is 5 (i.e., there are 5 different cases in the sample). In addition to variables x and y, it is now worth explaining the two other important defining variables of any longitudinal data set: Column 1 refers again to the case ID, and each value that this variable takes uniquely identifies each of the five entities or cases in the sample. The second column refers to the point in time when each of the variables x and y were measured for a given case i. It is common to use the lowercase letter "t" to denote some specific time in the sample. So now the notation x_{it} refers to the value that variable x takes for case i at time t and, similarly, y_{it} is now the value that variable y takes for case i at time t. The value of x at time 2 for individual 1 is 15000 ($x_{12} = 15000$) and the value of y is 5 ($y_{12} = 5$). Another example is the value of x for individual 4 at time 3 (x_{43}) when it is 23412; and the value of y for the same individual 4 at the same time is 3 (y_{43}).

The data set organization above is an example of what has been called an *individual-period* data set, in which each case or individual has multiple records (rows) for each measurement occasion. Consequently, there is a record (or row) for each case-period observation. Another possible data arrangement is called *individual-level* data in which each case or individual has one record, and multiple variables contain the data from each measurement. Here, there are as many records as there are cases in the sample. Most software packages can easily convert longitudinal data sets from one format to the other. For example, in Stata, the *reshape* command can be used to easily accomplish this conversion. In this book, I will be working with the individual-period type of data set format (like in the Excel example shown above).

Notice that now the basic univariate and bivariate statistics presented earlier can also be computed for: (1) each individual case at different points of time, (2) all cases in the sample at a given point of time, and/or (3) the entire sample (regardless of case and/or time under consideration). So, for example, the mean value of x for individual 1 during the three time periods is 24666 ($\bar{x}_1$); and 3 is the mean value of y also for individual 1 during the same three time periods ($\bar{y}_1$). Alternatively, one could measure the mean value of x at time 1 ($\bar{x}_1$) for all cases and compare it with the mean value of y at time 2 ($\bar{y}_2$) for all cases. One could even compute the correlation coefficient between x at time 1 and y at time 2; this would be denoted as $r_{x_1 y_2}$. One could also measure the mean value of x for all cases at all times. All of these statistics are possible to compute in the case of a longitudinal data set (although some will make more sense and be more informative than others, as the reader will see in later chapters).

Additionally, with a longitudinal data set like the one described above, one can conduct some exploratory analyses that describe how cases in the data change over time. These descriptive analyses will help in examining the nature and particularities of each case or group of cases' temporal pattern of growth. Some simple graphical strategies are available. The simplest way of visualizing a variable's change over time is to examine a *growth plot* or *time plot*. This type of plot provides a temporarily sequenced record of the different values of a given variable y for a given individual. It therefore allows for the exploration of how cases or units change over time. One can easily obtain such a growth graph from any major statistical program (and even from non-statistical programs such as Excel). The figure on the next page plots change in variable y for the five hypothetical observations in the study. The coordinates (time, *y*) have been linked by a smooth line (one can choose not to smooth the time lines). This figure is just one of the many graphical options available in Excel or any statistical software program of choice:

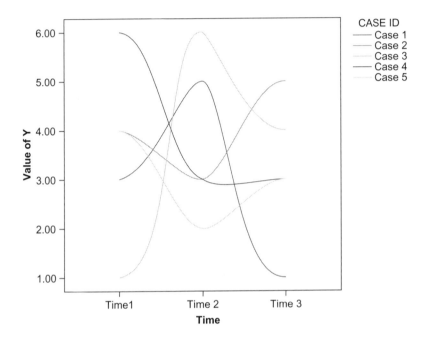

As one can see, the growth plot can be a good tool for exploring case variations in temporal patterns of dependent variables since it clearly reveals how each case changes over time. Because this plot can be confusing when there are many cases in the sample, it is highly recommended to group cases using some other variable in the sample. For example, one could display the temporal trend in salary over time for males and females, resulting in the plotting of only two curves over time.

It is always highly recommended to perform some exploratory or descriptive analysis of a longitudinal data set and its variables of interest before any more-advanced analyses are undertaken. These descriptive statistics can help in two ways. First, they can familiarize the researcher with the data set and give an idea of how each case or an average case in the sample changes over time. Second, they can show how change differs by case. Such careful and detailed exploratory analysis of a longitudinal data set will subsequently allow for better choosing of the right multivariate model.

1.4.   Reading a Data Set into SPSS or Stata

In this section, I explain how to read or input an Excel data set (such as the one created earlier in this chapter) into a convenient statistical program for analysis. I will illustrate this process for the two most popular programs: SPSS and Stata. Before doing that, I will briefly discuss each program.

SPSS is an easy-to-learn statistical analysis and data management software program (for more information go to www.spss.com). It has a graphical interface that consists of pull-down menus, dialog boxes, and windows that display and organize data, and it performs statistical and graphical tasks. One can use SPSS by pointing and clicking with the mouse. *Stata* is a very comprehensive and easy-to-learn software application that contains most of the statistical tools and models developed and used in the social sciences (for more information about this program, go to www.stata.com). It contains procedures for managing, graphing, and analyzing data, as well as certain procedures that are not available in other applications such as SPSS. One can perform most tasks using pull-down menus and windows. Both SPSS and Stata are command and window-driven. That is, they allow users to issue commands for procedures (and to store such commands in the so-called command files, also known as input files in the case of SPSS, or do-files in the case of Stata).

SPSS allows the user to enter data directly. After launching SPSS, the empty Data Editor window titled *Untitled—SPSS Data Editor* will appear. The Data Editor window consists of a grid of rows and columns. Each row can correspond to one case, and each column to a different variable. One can enter the data directly into cells in the Data Editor window and use the mouse, arrow keys, or tab key to move from one cell to another. One can also open an existing SPSS file, or open a data file saved in selected other formats (e.g., SPSS portable files, ASCII or text files, and Excel spreadsheet files). A data set from Excel can be copied and pasted directly into the Data Editor window.

To open the Excel file in SPSS 13, open the SPSS program (this is typically done by double-clicking on the SPSS icon on the desktop if a program shortcut was created when the program was installed), go to the *File* pull-down menu, select *Open,* and then *Data.* One will then be prompted by an *Open File* box where one can easily browse the computer to find the Excel file. Select Excel (*.xls) as the type of file to be read into SPSS since it will make the search much easier. Hit the *Open* button. Stata will prompt the user with the *Open Excel Data Source* box. Here, make sure that the correct Worksheet and Range from the Excel file is being read, and most importantly, given that the variable names will be on the first row of the Excel file (as in my sample data set), tick the *Read the variable names from the first row of data* box. Then hit the *OK* button. This should result in the opening of the data set in SPSS, and the program is now ready to get started with its analysis. At this point, the SPSS session should look exactly like this (if using version 13):

	Case_ID	Time	X	Y	var	var	var
1	1.00	1.00	25000.00	3.00			
2	1.00	2.00	15000.00	5.00			
3	1.00	3.00	34000.00	1.00			
4	2.00	1.00	13500.00	4.00			
5	2.00	2.00	23400.00	3.00			
6	2.00	3.00	23790.00	5.00			
7	3.00	1.00	12343.00	1.00			
8	3.00	2.00	23190.00	6.00			
9	3.00	3.00	8900.00	4.00			
10	4.00	1.00	15000.00	6.00			
11	4.00	2.00	36749.00	3.00			
12	4.00	3.00	23412.00	3.00			
13	5.00	1.00	34556.00	4.00			
14	5.00	2.00	13534.00	2.00			
15	5.00	3.00	23456.00	3.00			
16							

The data set can now be saved as an SPSS data set by going to the *File* pull-down menu and selecting *Save As*. Specify a name to give the file; the automatic added SPSS data extension is .sav. This way, one can easily open this new SPSS data set in future SPSS sessions by simply clicking on it twice (the SPSS program will automatically start). In the lower left hand side of the SPSS working screen, one can click on the *Variable View* tab. This allows the user to see (and modify) the variable names and other important variable format characteristics such as *Type of Variable, Width, Decimals, Label, Values,* and *Missing*. The *Values* option is especially useful for assigning labels to each of the values that any given variable can take (particularly when it is not so obvious what the number means). For example, one could have the variable gender where 1 = Female and 2 = Male; this option would store the meaning of 1 and 2 for the gender variable as to prevent the user from forgetting the meaning of those values in the future. The *Missing* option will allow the user to specify which values represent missing data for each of the variables in a sample.

In SPSS, one can also open text (ASCII) files containing data. To do this, go to the *File* pull-down menu and select *Read Text Data*. In the dialog box that appears, select the file and click on *Open*. After selecting the file of interest, the *Text Import Wizard* that appears will help with the reading of the data. It is necessary to tell SPSS if the file matches a predefined format, how the variables are arranged and delimited, and whether or not the variables' names are included in the file itself. A preview of the file will be shown to help the user remember its characteristics. Once the correct information has been entered, SPSS will import the data and leave it ready to be analyzed. Students, professionals, and many first-time SPSS users can benefit from using the *Statistics Coach* module. This module shows how to compute the basic statistics covered thus far and how to start graphing the variables. To run this wizard, go to the *Help* pull-down menu, and select *Statistics Coach*. Then simply follow the option that matches what is desired to accomplish.

Stata also allows the user to enter data into the Stata Data Editor window in different ways. One can enter data directly into the cells in the Data Editor window, open an existing Stata file, open an ASCII (or text) file, or open an Excel spreadsheet file. As in SPSS, to enter data directly into Stata, start the *Data Editor* window by clicking on the spreadsheet icon on the toolbar. The Data Editor window consists of a grid of rows and columns, with the rows corresponding to cases and the columns corresponding to variables. Type the values into the empty cells and use the mouse, arrow keys, or tab key to move from one cell to another.

To import data from text (ASCII) files, go to the *File* pull-down menu and select *Import*. From the submenu, choose to import data created by a spreadsheet, in fixed format, in fixed format with a dictionary, or unformatted. After selecting the format, select the file in the dialog box that appears. One must tell Stata how variables are delimited or arranged, whether the variable names are included, and so on. To open the same Excel file as before in SPSS, go to Excel and save the Excel file as a (*.csv) file (that is, a comma separated value file); this file will look exactly like an Excel file, but its format will be easier to open in Stata. After creating the Excel (.csv) data file, go to the *File* pull-down menu, select *Import,* and choose the option *ASCII data created by a spreadsheet*. Stata will then prompt the user by an *Insheet—Import ASCII data* dialog box, where one can easily browse the computer to find the .cvs Excel file. Make sure to select the Comma Separated Values (.csv) type of file. At this dialog box, double-check that the *Automatic determine delimiter* radio button option is selected. It is not necessary to enter any information in the *New variable names: (optional)* box if the first row in the Excel file contains such variable names. Hit *OK,* and the data should successfully open as a data set in Stata. One can also copy and paste a data set from Excel directly into the Stata Data Editor Window. The data should look like this (if using version 9):

As in SPSS, it is possible to save the data set as a Stata file by going to the *File* pull-down menu and selecting *Save As*. Specify a name to give the file; the newly created data file will now have the .dta extension. This way, one can easily open the new Stata data set by clicking on it twice (the Stata program will automatically start). The saved file will also contain all the values that have been entered as well as the names of the variables, the labels assigned to them, and any other formatting changes made during the Stata session.

Several programs can convert an original data set (from whichever format it currently has, such as different database management systems, spreadsheet, statistical packages, and other application packages) into an SPSS or Stata data file. Two such programs are *StatTransfer* or *DBMSCopy*, and they are both available for Windows. For more information, visit their Web sites at: www.stattransfer.com or www.dataflux.com/Product-Services/Products/dbms.asp; these Web sites may allow the downloading of the software for free for a limited time evaluation. These programs are especially useful when one needs to convert many different data sets (across different software programs) frequently.

2. Longitudinal Data

One can encounter several kinds of longitudinal (also called dynamic) data in theory and practice. In the majority of cases, the format and type of data set available will determine the methodological approach the researcher should follow for the analysis. Data sets can be classified depending on their temporal and/or cross-sectional components. By cross-sectional component, I refer to the number of units

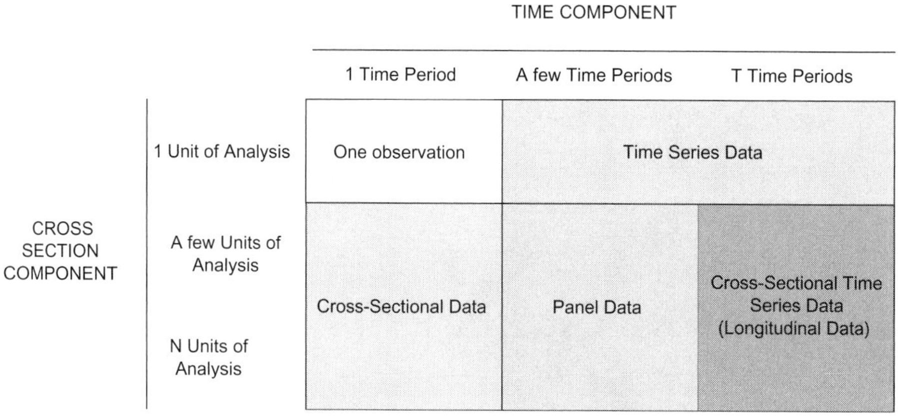

TIME COMPONENT

		1 Time Period	A few Time Periods	T Time Periods
CROSS SECTION COMPONENT	1 Unit of Analysis	One observation	Time Series Data	
	A few Units of Analysis	Cross-Sectional Data	Panel Data	Cross-Sectional Time Series Data (Longitudinal Data)
	N Units of Analysis			

Fig. 1.1 Types of data sets depending on their time and cross-section components.

of analysis (also known as cross-sections) included in the data set (for example, the number of respondents in a survey or the number of countries in a sample). By temporal component, I refer to the number of time periods included in the data set (for example, observations measured one year or observations measured each month for a period of five years). The most important types of data sets are: (1) cross-sectional data, (2) time series data, and (3) cross-sectional time series data (or longitudinal data, broadly speaking). Fig. 1.1 summarizes how different data sets can be classified depending on their temporal and cross-sectional components.

In general, a *cross-sectional data set* consists of a collection of information for entities, cases, or units of analysis—such as individuals, organizations, or countries—for a single moment of time, a time interval, or a period. Thus, the measurement of each variable for each case occurs within a narrow span of time (ideally each variable will be measured simultaneously for each case). Depending on the particular research project, a period can be defined in terms of minutes and hours (common in laboratory experiments), days, months, years, or decades (more typical in cross-national research). An example of a cross-sectional data set is the different variables published each year in the *Human Development Report* (by the United Nations Development Programme) regarding the level of economic development in many countries around the world. These variables include the population of the country, its gross national product, different levels of national investment, and the percentage of its population with secondary education. Another example is the data that one could collect by surveying a group of students, employees, or consumers in a given month about their occupations, salaries, and opinions on topics related to the research interest. These data could also store information about the respondents' basic demographic characteristics such as gender, race, country of origin, and age. The idea here is to collect a set of different variables for a group of cases or actors at a given point in time. In matrix notation, this data set could be referred to as X_{ik} and Y_{ik}, where both X and Y are two vectors of several variables, one per column, measured for a sample of N cross-sections or cases, one per row. Remember that a matrix is a rectangular array of numbers arranged in rows and columns. Therefore, the matrix X_{ik} has i rows (each row representing a given case in the sample) and k columns (each column representing one variable). So X_{55}, for example, reads as follows:

$$X_{55} = \begin{bmatrix} X_{11} & X_{12} & X_{13} & X_{14} & X_{15} \\ X_{21} & X_{22} & X_{23} & \boxed{X_{24}} & X_{25} \\ X_{31} & X_{32} & X_{33} & X_{34} & X_{35} \\ X_{41} & X_{42} & X_{43} & X_{44} & X_{45} \\ X_{51} & X_{52} & X_{53} & X_{54} & X_{55} \end{bmatrix}$$

where $i = 5$ and $k = 5$. x_{ik} generally represents any possible value in the data set. It denotes the value that variable x_k takes for unit i (and it is located in row i and column k). For example, x_{24} is the value that variable x_4 takes for case 2 in the sample collected (this entry is highlighted in the matrix). The dimension of X_{ik} is $N \times k$.

A *time series data set* is the data where many observations are taken for one case or unit, but here this one case is being measured over various time periods. Examples of this type of data include the measurement of the demographic, social, cultural, and economic information of a given country every quarter, semester, or year from 1960 to the present. These types of data are common in macroeconomic empirical studies. Here, one can also think about a data set recording all the banking transactions over the past five years. The idea here is to measure a set of different characteristics or variables for a given unit over a period of time, generating what is usually called a *time series*. In matrix notation, a time series data set could be referred to as X_{tk} and Y_{tk}, where both X and Y are two vectors of several variables, one per column, measured for one case t times. So that the matrix X_{tk} has t rows (each row representing a period of time when a given case has been observed) and k columns (each column representing one variable; the dimension of X is now $T \times k$). If $t = 5$ and $k = 5$, then X_{55} reads as follows:

$$X_{55} = \begin{bmatrix} X_{11} & X_{12} & X_{13} & X_{14} & X_{15} \\ X_{21} & X_{22} & X_{23} & X_{24} & X_{25} \\ X_{31} & \boxed{X_{32}} & X_{33} & X_{34} & X_{35} \\ X_{41} & X_{42} & X_{43} & X_{44} & X_{45} \\ X_{51} & X_{52} & X_{53} & X_{54} & X_{55} \end{bmatrix}$$

where x_{tk} generally represents any possible value in the data set. It denotes the value that variable x_k takes for the one unit of analysis at time t (and this value is located in row t and column k). For example, x_{32} is now the value that variable x_2 takes for the case under study at time 3 in the temporal sample collected for the same one unit (highlighted in the matrix above).

The third type of data set I want to cover here is *longitudinal data*. Longitudinal data are collected for a set of different variables (1) for at least two or more distinct time periods and (2) where the cases are the same (or at least comparable) from one period to the next. At the bare minimum, this longitudinal data design would permit the measurement of changes in several variables from one period of time to another. According to this definition, several types of data sets can be regarded as longitudinal. In one, data may be collected at a few distinct time periods on the same set of cases and variables in each period. This is called *prospective panel data*. A second example is when data are collected at a single period, inclusive of several past periods also counting the period that ends with the time when the data are collected; this is called *retrospective panel data*. In both the data designs, the cases and the variables are the same from one period to the next.

The two types of longitudinal data described can help generate what is known as *cross-sectional time series data* or *cross-sectional cross-time data*. Briefly, this data set provides information about the same variables for the same sample of cases at the same time periods. So, the compiling of the data occurs for the same cases or units not only in a particular temporal moment but also repeatedly at discrete time intervals thereafter. An example of this would be the same group of employees (in a given organization) being surveyed every year and asked to report information about their employment, salary, and marital status every year over a period of 10 years. This data is often called *panel data,* and consist of data collected for a set of social cases or units of analysis in only a very few points in time. For example, a data set collecting information about certain occupational achievements for a group of engineers at the age of 20, 30, and 40, yet any information about what has happened in the interim years remains uncollected (and consequently unknown). Panel data becomes longitudinal when such data is as complete as possible, that is, when the frequency of data is compiled over the smallest time intervals possible.

Fig. 1.2 plots the various types of data sets available with respect to a given variable $y(t)$ or y_t. The first one is an example of a cross-sectional data set, where one variable y is measured at time T_4 for four different cases or units of analyses. The second graph could be an example of a longitudinal data set where the temporal component is limited. Here, only two cases are plotted showing the time observation at discrete times. The third graph shows a longitudinal data set where the information available for y is available in a more continuous manner (than the one data set described in 2) over time. Here only one case is plotted over time (for simplicity of presentation), although one can easily imagine time trajectories for variable y for various cases in a given sample under study.

With regard to these various types of temporal data, cross-sectional time series data are collected for a set of cases or units i (that could be people, cities, regions, or countries) for a number of temporal units t (that could be months, years, or decades). This general example of a data set contains a maximum of $N \times T$

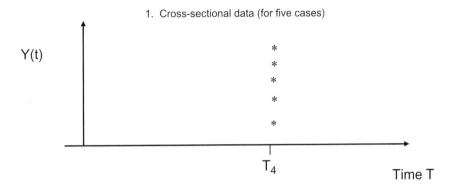

1. Cross-sectional data (for five cases)

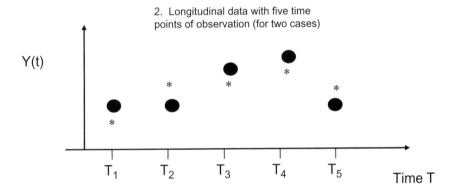

2. Longitudinal data with five time points of observation (for two cases)

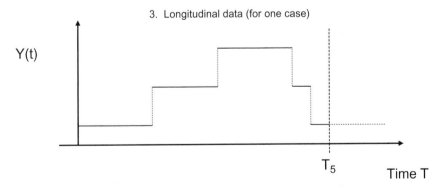

3. Longitudinal data (for one case)

Fig. 1.2 Three examples of data sets: (1) cross-sectional, (2) panel, and (3) longitudinal data set.

observations, which can only be reached when the information is available for each unit of analysis i in each of the time units t. When the number of units or cases under study is larger than the number of time units in those that are included (for example, 25 years with 100 countries), one typically says that the *cross-section component* dominates the sample. When the number of time periods is larger than the number of units of analysis (for example, 20 people during 60 months), the dominant component of the data set is the *time component.*

In matrix notation, a cross-sectional time series data set could be referred to as X_{itk} and Y_{itk}, where both X and Y are two matrices with several variables, one per column, measured for several cases i at several t times. The matrix X_{itk} has with $N \times T$ rows (representing the number of time-cases, in other words the number of times all cases have been observed) and k columns (representing the number of different variables measured for each case). Now it might be a bit more complicated to understand the structure of the matrix at first sight. For the case of 5 individuals at 4 times and 5 variables (i.e., $i = 5$, $t = 4$, and $k = 5$), X_{itk} can be written as follows:

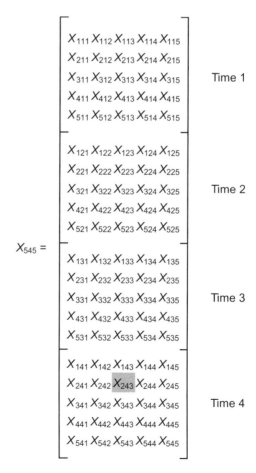

$$
X_{545} =
\begin{bmatrix}
X_{111}\ X_{112}\ X_{113}\ X_{114}\ X_{115} \\
X_{211}\ X_{212}\ X_{213}\ X_{214}\ X_{215} \\
X_{311}\ X_{312}\ X_{313}\ X_{314}\ X_{315} \\
X_{411}\ X_{412}\ X_{413}\ X_{414}\ X_{415} \\
X_{511}\ X_{512}\ X_{513}\ X_{514}\ X_{515} \\[6pt]
X_{121}\ X_{122}\ X_{123}\ X_{124}\ X_{125} \\
X_{221}\ X_{222}\ X_{223}\ X_{224}\ X_{225} \\
X_{321}\ X_{322}\ X_{323}\ X_{324}\ X_{325} \\
X_{421}\ X_{422}\ X_{423}\ X_{424}\ X_{425} \\
X_{521}\ X_{522}\ X_{523}\ X_{524}\ X_{525} \\[6pt]
X_{131}\ X_{132}\ X_{133}\ X_{134}\ X_{135} \\
X_{231}\ X_{232}\ X_{233}\ X_{234}\ X_{235} \\
X_{331}\ X_{332}\ X_{333}\ X_{334}\ X_{335} \\
X_{431}\ X_{432}\ X_{433}\ X_{434}\ X_{435} \\
X_{531}\ X_{532}\ X_{533}\ X_{534}\ X_{535} \\[6pt]
X_{141}\ X_{142}\ X_{143}\ X_{144}\ X_{145} \\
X_{241}\ X_{242}\ X_{243}\ X_{244}\ X_{245} \\
X_{341}\ X_{342}\ X_{343}\ X_{344}\ X_{345} \\
X_{441}\ X_{442}\ X_{443}\ X_{444}\ X_{445} \\
X_{541}\ X_{542}\ X_{543}\ X_{544}\ X_{545}
\end{bmatrix}
\begin{array}{l}
\text{Time 1} \\[30pt]
\text{Time 2} \\[30pt]
\text{Time 3} \\[30pt]
\text{Time 4}
\end{array}
$$

This time there is an X_{ik} sub-matrix within the X_{itk} bigger matrix; within each of those sub-matrices, x_{ik} represents the value that variable x_k takes for the one unit of analysis i at a given time t. For example, x_{243} is the value that variable x_3 takes for case 2 at time 4 (highlighted again in the matrix itself). In the example above, all the variables are collected at 4 points in time. In the most general case, X_{itk} would look as follows:

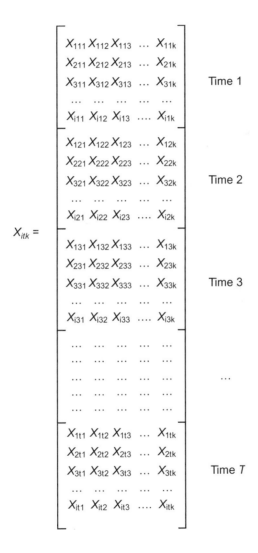

As discussed, cross-sectional data are recorded at only one point in time. However, it is very common these days for cross-sectional data to be collected in a succession of surveys with a new sample of cases each time. Another possibility for longitudinal data occurs when collecting data on the same set of variables for two or more time periods, but for different cases in each period. This is called *repeated cross-sectional data,* and the data for each period may be regarded as a separate cross-section. Because these cases are different in each period, the data have also been described as not-quite-longitudinal data sets. Caution therefore needs to be taken when applying the methods explained in this book. In general, the methods described here will not fit repeated cross-sectional data. For more information about these particular types of longitudinal data and the commonly used dynamic models, see the reading recommendations at the end of this book.

Both temporal and longitudinal analyses refer to the analysis of data collected over time. The term "dynamic analysis" has also been used to refer to this form of temporal analysis that focuses on the process of change. The dynamic analysis of data (i.e., how to study change in variables over time and how to estimate multivariate models to predict such change) is the focus of this book. The next two chapters cover the statistical theory behind *cross-sectional time series analysis* or panel analysis (for the modeling of change in continuous variables; in Chapter 2) and *event history analysis* (for the modeling of events happening over time; in Chapter 3).

2.1.   Types of Longitudinal Variables

Because this book deals primarily with the analysis of longitudinal data, I explain a useful way of classifying variables in this section. This method involves the description of the time and cross-sectional component of the different variables in the data set. The two most important variables are therefore: case ID, a number that uniquely identifies each of the different cases or units in the sample, and the time variable, a continuous/ordinal variable keeping track of when each case was observed during a period of observation T.

There are three types of variables in any longitudinal data set: (1) individual-constant variables, (2) time-constant variables; and, (3) time-varying individual variables. The *individual-constant variables* take the same value for any given case or cross-section over time, although the value differs across cases. Examples might be gender at time of birth or the socioeconomic origin of a person. Thus, this variable takes the same value for the same individual case (but such a value indeed may change across cases). *Time-constant variables* take the same value for each and every case in the sample at an instant of time, but this value changes over time. Examples include structural variables such as unemployment or inflation rates, or social security expenses for a given country (or region) that take the same value for all citizens of a given country or region. *Time-varying individual variables* are those that take diverse values for different units of study in an instant of time, and these values also vary in the short and long run.

Examples of this variable are a person's salary, a company's revenues or size, and a developing country's research and development investment. These distinctions will be important when specifying which dynamic regression models to estimate in the following chapters.

3. Causes and Causal Relationships

The objective of the next few pages is to show the utility of longitudinal statistical techniques for the purpose of investigating causes and causal explanations or relationships among different variables. Applying regression-like methods is recommended when the occurrence of changes in variable y (the dependent variable) "depends" on (or is affected by) a set of at least one other explanatory variable X (that is $x_1, x_2, x_3, \ldots, x_k$, where $k \geq 1$), also known as independent or explanatory variables. The ultimate goal of a statistical regression model is to identify the possible generalizations in explaining the attitude and the behavior of social actors. Here, I clarify some basic concepts such as "causality" and "causal relationships," especially in the case of dynamic analysis. I discuss the rules of time in any causal type of inferences and define causal relationships, which are the grounds for the development of useful statistical concepts and models in dynamic analysis.

The objective of the causal dynamic methodology is to describe the form in which variations that are observed in a set of independent variables explain variations in a given one dependent variable over time; in this sense, this is a *multivariate statistic*. The variable that represents a social process and whose variation is meant to be explained is called a *dependent variable y*. The group of variables used to explain or describe this social process is called *independent variables* or *explanatory variables X*. In the case of causal models, it is common to use the term endogenous variables Y to refer to variables that are affected by a series of other variables included in the model; and exogenous variables X to refer to those used to represent external causes of change. Therefore, an overview of the causal relationship could be represented as follows:

$$\Delta X_t \rightarrow \Delta Y_{t'} \qquad t < t'$$

This formula implies that a change within one variable x_t at the moment t causes a change in variable $y_{t'}$ at a later moment of time t'. For example, imagine the impact of an additional year of education Δx on a worker's annual salary over time $\Delta y_{t'}$. Here, I should make clear that there could be many different variables x that could affect how $y_{t'}$ changes over time (some that will be included in the vector of variables X and others that might be unobserved and consequently omitted from the models). Typically, one of the goals of the researcher is to identify and collect the set of independent variables he or she believes is driving the change in the dependent variable. In many situations, it is preferable to use the expression "causal conditions" to emphasize that the group of feasible causes behind the change of *y* is way too complex to be captured merely by a set of independent variables in a given model. For example, if one is studying how the level of education and work experience of employees affect changes in their salaries over time, the basic causal model will examine salary as the main dependent variable as a function of experience and education, two of the many independent variables one could argue affect salary growth over time. This basic model obviously ignores the possibility that there could be other exogenous variables accounting for the process of salary change that may not have been measured in a given data set. For example, previous work experience, computer knowledge, macroeconomic variables such as unemployment rates or local job supply, demographic variables such as gender or race, and other employment-related factors such as occupation and number of hours working, are among many possible explanations accounting for individual salary differences.

In addition, some of the variables *X* included in the model can be individual-constant variables, that is, variables that do not change over time. The causal relationship could now be represented as follows:

$$X \rightarrow \Delta Y_{t'}$$

In this case, the goal of the researcher is to identify differences in the pattern of change of *Y* across individual fixed attributes or characteristics. For example, if one is studying differences in employees' salary growth over time by gender, the basic causal model will include salary as the main dependent variable and gender as the main independent variable of interest.

One might argue that certain independent variables in the multivariate model could be omitted. That is why any particular set of variables *X* only explains a certain percentage of the variation of the dependent variable. The better specified the regression model is, the higher the variation explained by the set of independent variables would be. The *R-squared* or *coefficient of determination* has been defined in the basic linear regression model to learn about the predictive power of the linear regression model (OLS model). It is calculated as the ratio of explained variation (by the specified regression model with a set of independent variables) to the total variation of the dependent variable. The formula is as follows:

$$R^2 = \frac{SS_\text{M}}{SS_\text{T}}$$

where SS_M represents the model sum of squares, or *variation of predicted y* $(\hat{y}_i)$ *around the mean of y* $(\bar{y})$:

$$SS_M = \sum (\hat{y}_i - \bar{y})^2$$

and SS_T represents the total sum of squares, or variation of actual y values around the mean of y $(\bar{y})$:

$$SS_T = \sum (y_i - \bar{y})^2.$$

Consequently, the value of R^2 falls in a range from 0 to 1 and is interpreted as the "proportion of variance of y explained by x." The remaining percentage or ratio of variance in y is explained by omitted variables in the model (or non-linearities in the effect of x on y). In the simple case of a two variable (y and x) regression, R^2 equals the square of the Pearson correlation coefficient r.

Frequently in the social sciences, one is interested in studying the events that happen in the lives of certain entities or cases, e.g., individuals, teams, organizations, or countries. Examples here include migrating, getting a job, having a child (in the case of individuals), or joining a treaty or passing a law (at the national level). In this case, it is possible to establish causal relationships about how some change in variable x in the past affects the probability of change in a discrete variable y_i:

$$\Delta x_t \rightarrow \Pr(\Delta Y_{t'}) \qquad t < t'$$

This means that a change in variable x that happens at a given point in time may affect the probability of variable y also changing in the future ($t < t'$). As an example, if one is interested in studying why teenagers drop out of high school, the variable to evaluate is the probability of the student not completing high school. One could easily think of many factors or variables that could potentially increase or decrease such a probability, such as age, whether the student is working part-time, or the socioeconomic status of her parents. All these independent variables can help explain how their changes impact the probability of the student dropping out of high school, i.e., the dependent variable in this case.

In addition to the causal processes described above, any empirical studies of causal relationships using longitudinal data have to account for the temporal component or *function of time*. Often the occurrence of events and changes in some continuous variables show some temporal pattern or they can be systematic. Thus, under certain conditions, such changes in any particular variable *y* might show regular time patterns that consequently need to be modeled. First, how variables change can evolve over time. To measure change, one obviously needs a minimum of two points in time to observe how a given variable changes its value. Second, implicit in any causal temporal process is the notion that some time order exists between cause and effect relationships. The cause must precede the effect in time, according to the above formal representation: $t < t'$. As a result, there must always be an interval between the change in the variable representing the cause (independent) and the variable representing the effect (dependent). The interval may be short or long in time, but never infinite, nor zero. The cause and the effect cannot, in theory, occur simultaneously. So any empirical modeling of a causal relationship in a statistical model needs to take into consideration the existence of temporal delays between changes that are supposedly causes and the changes that are supposedly effects.

4. Longitudinal Analysis: Definition and Types

Longitudinal analysis refers generally to the study and examination of data over time. It is also sometimes called temporal or longitudinal data analysis. The term "panel data analysis" is also broadly used in the social sciences, although the term is more appropriate for the analysis of a particular type of longitudinal data, "panel data," where the number of times when such observations are recorded is small. Examples of such data can be unemployment monthly rates, annual number of people who migrate from one region to another during the past 20 years, or at the individual level, salary over time. In certain longitudinal studies, the object of analysis consists of examining the occurrence of specific life events, such as the adoption of an international treaty by the government of a country, the job promotion of an employee, or the failure to complete a work project.

   When studying any change process, one needs to be very clear about variable definitions and premises behind the methodology to study longitudinal data. First, it is imperative to define the unit of analysis or cases under study, which can be individuals, groups, organizations, societies, or any other social actors whose characteristics, behaviors, and attributes do change over time as measured by a set of dependent and independent variables. Second, changes can happen at any time to the units of analysis (this is called the temporal process). Those changes can be of a "qualitative" nature—i.e., changes that happen when cases transition from one stage to another, such as from one job *i* to beginning a new job *j*, or from being employed to being self-employed. In this case, the dependent variable to study is a discrete one, as previously defined in this chapter. However, the changes can also be of "quantitative" nature—i.e., changes that happen in magnitude or quantity in a continuous dependent variable, such as an employee's salary, a firm's sales revenues, or a country's gross domestic product. Third, it is important to acknowledge that a series of factors or explanatory variables cause or influence those changes (what is called the causal process). Such variables or factors can be time-constant or time-dependent, depending on what they are measuring for each of the cases in the sample under study.

The type of dynamic analysis of any social process developed in this book makes reference to an existing methodology that specifies, estimates, and interprets *causal* explanatory models—i.e., those in which there is one dependent variable (the one that defines the process of social change under study) as a function of a series of one or more independent variables during a period of time. A key distinction needs to be made here between those variables that cause change (independent variables) and those variables that are the outcomes or consequences to be explained (dependent variables). Again, the dependent variables are the social phenomena that the researcher is interested in explaining over time; the independent variables are those factors used to explain variation in the dependent variable under study.

In general, there are two main varieties of dynamic analysis, as defined by the nature of the dependent variable to be modeled over time (Fig. 1.3). On the one hand, there is the *longitudinal analysis of qualitative variables*. This type of analysis models nominal-discrete dependent variables, either dichotomous or polytomous. On the other hand, the *longitudinal analysis of quantitative variables* refers to those models where the dependent variable is continuous (or at least discrete and not nominal). Progress in one area has diffused slowly into the other area. Development, though, has been clear when it comes to the specification of statistical models studying categorical changes. In the next two chapters, these two types of longitudinal models will be presented and fully explored both

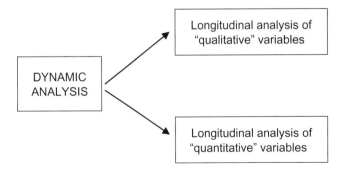

Fig. 1.3 Two types of dynamic analyses of longitudinal data sets.

theoretically and practically, using statistical software programs to estimate them using real longitudinal data sets.

A longitudinal data set records information about a certain number of cases or units of analysis that change over a period of time. Relevant to highlight here is that the longitudinal data I will refer to in this book is also called *cross-sectional time series data* for the same group of cases. I therefore refer to a set of characteristics collected for the same group of social units of analysis (whether individuals, groups, organizations, countries, or world regions) in several repetitive occasions during a certain predetermined time period. Again I do not include in the book the analysis of cross-sectional time series data of different cases each time, i.e., "repeated cross-sectional data" (for example, survey data every year from different random samples of a given population, and therefore, different unmatched units of analyses). The analysis of such data is more complicated and beyond the scope of the material covered by this book.

4.1. Why Longitudinal Analysis?

There are two main reasons why a researcher with a longitudinal data set would want to perform longitudinal analysis, rather than using the standard simple linear regression framework. The first is purely methodological: In order to avoid specification bias or error (that is, when the estimated regression coefficients differ systematically from their true values in the population) likely to occur when cross-sectional methods are applied to the modeling of longitudinal data. Second, there are often substantive reasons for wanting to analyze a phenomenon characterized by strong temporal dependencies that need to be modeled appropriately.

To further elaborate on these limitations behind the use of standard regression techniques to analyze longitudinal data sets and illustrate the importance of using longitudinal models, start by considering the standard linear regression model. Remember that the basic linear model is a way of summarizing the relationship between at least one independent variable x or vector of variables X (for simplicity here, assume only one independent variable in vector X) and y, the one dependent variable. This is accomplished by fitting a linear model with the following functional linear form:

$$y_i = \beta_0 + \beta_1 x_i + \varepsilon_i \tag{1}$$

or, in matrix notation, the model reads as follows:

$$y_i = \beta' X_i + \varepsilon_i \tag{2}$$

where β denotes the column vector containing β_0 and β_1. X_i is now the column vector that is the transpose of the ith $1 \times k$ row of X_{ik}.[2] In the case of $k = 1$:

$$X_i = \begin{bmatrix} 1 \\ X_i \end{bmatrix}.$$

In this expression (2), β_0 and β_1 are constants, β_0 is the y-intercept, and β_1 is the slope. The above equations show that x is a predictor of y. In causal terms, the model depicts X as the cause and y as its effect. As before, the i subscripts index individual cases or observations. Graphically, the equation denotes a straight line where β_0 is the height at which the line crosses the vertical Y-axis and β_1 is the slope of the line. The line rises β_1 units with each one unit increase in x_1 (if β_1 is negative, then the line actually falls as x_1 increases; when β_1 is zero, the line is horizontal). However, there can be other things accounting for change in the dependent variable y. These "other things" are represented with an error term, ε_i (epsilon), also called disturbance. The common important assumptions of this model are the following:

1) Errors have identical distributions, with zero mean and the same variance, for every value of X; or $E(\varepsilon_i) = 0$ and $\mathrm{Var}(\varepsilon_i) = \sigma^2$
2) Errors are independent, that is, unrelated to X variables or to the error of other cases; or $\mathrm{Cov}(\varepsilon_i, x) = 0$
3) Errors are normally distributed.

Here, it is also important to remind the reader that the technique called ordinary least squares (OLS) is the standard way of estimating the parameters in equation (1) and therefore guarantees a solution that "best fits" the sample data. OLS is just one of the many techniques for regression analysis, although it is by far the most commonly used. Its simplicity, generality, overall usefulness, and ideal-data properties have made it the core technique in modern statistical research. However, its theoretical advantages depend on assumptions rarely found in practice. When one departs from these basic conditions, the less one can trust OLS (for a detailed specification behind the OLS technique, that is how the values of the model coefficients are estimated, I recommend picking up any basic statistics book; consult the list of references at the end of this book for some reading suggestions). All statistical software programs are well equipped (including data management programs such as Excel) to perform OLS to fit a linear model to any data sample.

---

[2] Expression (2) is often a convenient way to refer to a single observation in the sample. In this book I will use expression (2) often — instead of the common matrix expression for the classical linear regression model: $Y = X\beta + \varepsilon$, where Y is the column vector with N observations $y_1, y_2, y_3, \ldots y_N$, X is an $N \times k$ data matrix and ε is the column vector containing the N disturbances.

Despite the fact that the standard linear model is extremely common in the social sciences, many important problems emerge when it is applied to a longitudinal data set. The first problem is the *specification bias,* which occurs when the estimated regression coefficients β differ systematically from their true values in the population. This is relevant here because in the basic regression model for a particular individual i at a particular time t, one can easily have the following true regression model in matrix form:

$$y_{it} = \beta' X_{it} + \varepsilon_{it} \tag{3}$$

with the following assumptions:

$$E(\varepsilon_{it}) = 0$$
$$\mathrm{Var}(\varepsilon_{it}) = \sigma_{it}^2 = \sigma_i^2$$
$$\mathrm{Cov}(\varepsilon_{it}, \varepsilon_{i't'}) = 0 \qquad i \neq i' \quad t \neq t'$$
$$\mathrm{Cov}(\varepsilon_{it}, x) = 0.$$

In this last equation, I assume that the variance of the error term is not only constant (or homoskedastic) but also stationary (that is, time-invariant). As I will discuss in more detail in later chapters, this is an important (but often unrealistic) constraint in the case of fitting a longitudinal data. Note that the model also assumes that the vector of coefficients β is stable over time. The above equation is a temporal model because it makes assumptions about how the dependent variable y_{it} changes over time. It assumes that y_{it} changes for two reasons. First, the dependent variable may change because the predictor variables X change. Second, the dependent variable may change because the error term ε_{it} changes over time (as the subscript t indicates). So, fitting a simple OLS regression model to longitudinal data (when the true model is model 3) can create several problems. The most obvious one is that the OLS estimates are not true values and, therefore, they are imperfect measures of the true coefficient values to be estimated. They have biases that operate in known directions; for example, their sample variances are inflated by the presence of measurement error in the outcome. One therefore needs to have some sort of model specification for how the random error terms vary (as specified in the true model 3, for example). Such available common error models will be explained in detail in the following chapters.

The basic regression equation with which I began this discussion also assumes that (1) the effects of the included variables are constant over time, (2) the variance in the error term is constant over time as well as across cases, and (3) the autocorrelation of the error terms is also constant over time. This assumption, while simplifying a researcher's life when it comes to analyzing data, is quite unrealistic. Unless the phenomenon one wants to study is in or near equilibrium, static analysis such as the one provided by the traditional OLS will give biased estimates of the true relationships among the variables of interest.

Equally important, longitudinal analysis is relevant for substantial theoretical reasons when the researcher is interested in understanding change. Decisions to quit a job, to marry, or to migrate to a different country all have a temporal dimension, and it is therefore evident that longitudinal data are essential if one is to investigate the temporal dependencies of any phenomenon or theories in any analysis. Another important concern of longitudinal research is the examination of changes, not in values or levels of variables over time, but in the *relationships* among variables over time. In this sense, static relationships are simply a subset of dynamic relationships, and, therefore, a dynamic model allows greater generalizability when it comes to understanding either change or lack of change (inertia).

Some of the methods discussed in the following chapters are ways of improving the precision and reliability of the OLS-estimated parameters in the case of a longitudinal data set. However, the need for *ad hoc* corrections of the OLS estimates has been gradually replaced by better dynamic model specification and estimation techniques (available in current widely used software programs). In what follows, I describe the two main types of longitudinal analyses covered in detail in this book, citing examples used in different social science literatures.

4.2. *Longitudinal Analysis of Qualitative Variables*

Researchers in the social sciences have long been interested in studying life events of individuals, groups of individuals or teams, organizations, or countries, just to give a few examples. The study of these occurrences or processes has been greatly developed in demographic empirical articles in order to analyze important life events such as death or birth. This has been extended to other important areas within sociology and demography such as migration, marriage, divorce, and social mobility (career change, job promotion, hiring, and turnover). The process is also used in other areas of study in social science, such as the estimation of rates of delinquency, diffusion of innovations, or rates of hospital room usage. Events of interest can refer not only to individuals but also to events happening to families, groups of people, organizations, or countries. Many studies within organizational or industrial relations research and evaluate important actions and outcomes, including union strikes, mergers and acquisitions, the organizational process of going public, and the effects of bankruptcy. In the case of international studies, scholars have been interested in studying government changes or international treaties. Technically speaking, the units under analysis in all these examples are in one specific state among a group of many possible states. The transition from one situation to another may occur at any time, and given that these changes occur over time, the typical challenge that researchers face is the use of appropriate statistical multivariate methods in order to better describe and capture the change process. More specifically, the challenge is to understand and model the factors that accelerate or decelerate the happening of events by evaluating their relative predictive importance.

Event history analysis consists of a well-developed set of statistical techniques (both exploratory and multivariate) that study events, when they happen, and which factors influence the occurrence of various types of events. This methodology is well suited for the testing or evaluation of data where the dependent variable is discrete and typically dichotomous. Consequently, the longitudinal data set to be analyzed here needs to give information about the time when those events happened for a sample of cases or units of analysis during the period of observation. Such information obviously varies for each of the cases included in the sample since the event might happen (or not happen) in a moment of time for each case or unit of analysis. The "event" is defined as a change in value for a discrete variable $y(t)$. This change can happen at any time (sometimes even multiple times, in the case of repeated events) during the period of time of study. This discrete variable can take a specific number of values, all of which are mutually exclusive. All possible values that the variable $y(t)$ can take are collectively called the "risk space." For example, a discrete variable can be the marital status of a person in all its variations: single, married, widow, divorced, or separated. Another example of a discrete variable can include the occupation of a given employee, place of residence, political party affiliation, or religious affiliation. I present the event history analysis (EHA) methodology in depth in Chapter 3 of this book.

Event history analysis has been applied frequently in demographic studies (for a thorough review of studies, see Hobcraft and Murphy [1986]), with seminal life studies in sociology (see Mayer and Tuma [1990]), as well as studies of political change and organizations. Important to mention here are the classic analyses of changes in national political structures and world regions (Bienen and Van Den Walle, 1989; Hannan and Carroll, 1981; Knoke, 1982; Tolbert and Zucker, 1983; Strang, 1990); studies of social movements (Olzak, 1989); changes in organizational forms (Barnett, 1990; Hannan and Carroll, 1989; and Halliday *et al.*, 1987); the occurrence of various types of interpersonal behaviors (Drass, 1986; Felmlee *et al.*, 1990; Griffin and Gardner, 1989; Mayer and Carroll, 1990; Robinson and Smith-Lovin, 1990); and the national adoption of laws, policies, and social innovation (Edelman, 1990; Marsden and Podolny, 1990; Soysal and Strang, 1989; Usui, 1994; Sutton, 1988). In the list of references at the end of this book, I provide an extensive list of more recently published studies using these EHA methods in their empirical analyses.

4.3. *Longitudinal Analysis of Quantitative Variables*

The researcher may be interested instead in studying the change in one metric or continuous variable over time; in other words, studying the different patterns of quantitative changes over time. Some continuous variables do change in a continuous fashion, for example, the size of organizations, as measured by the number of plants or facilities, number of workers, or the volume of sales revenues. At the individual level, important dependent variables used in longitudinal analyses in sociology have been salary or earnings per year, occupational status or prestige, or productivity. Some other continuous variables might not necessarily change, such as the prestige or status associated with a certain occupation. I present the methodology for the study of quantitative dependent variables in Chapter 2.

When social scientists study changes in certain continuous variables, they often find that those measures are recorded in a non-continuous time fashion; instead such changes are measured at discrete points in time during an observation period. A continuous recording of variation in a set of variables over time is the most adequate form for both describing and modeling change. However, the *panel data* set is probably the most frequently analyzed type of data because it is the simplest form of temporal data available to researchers. For example, in sociology there are seminal studies of the levels of collective violence (Snyder and Tilly, 1972); transformation of voting patterns (Doreian and Hummon, 1976); changes in delinquency rates (Land and Felson, 1976; Cohen and Felson, 1979; Berk *et al.*, 1981; Nelson, 1981); and variations in work organizations (Shorter and Tilly, 1970). *Time-series data* has been broadly used in macro studies, especially macroeconomic studies to estimate from consumption functions to supply and demand curves. With the current availability of longitudinal data sets (at the individual, organizational, and national level), together with the increased computing power that allows for the storing and processing of larger longitudinal databases than ever before, many social scientists have begun to examine multiple time series for different cases. And yet, this type of dynamic analysis is not used as often as one would imagine. Perhaps this is because of the complexity behind the current presentation of such models, as well as the execution of such statistical models using statistical software programs. In this book, I try to correct this by presenting both the theoretical models as well as their practical estimation using the current software in as simple a manner as possible; and I do so without losing accuracy or correctness in the presentation of such dynamic methodology. Again, more illustrations of recently published studies using these methods in their empirical analyses are provided in the list of references.

5. Organization and Characteristics of the Book

This book reviews the main perspectives employed to study change, for both qualitative and quantitative changes that occur in entities or cases (as Fig. 1.3 summarizes). Whenever possible, I try to summarize the current state of the several dynamic analysis techniques presented here. The general strategy of this book is to develop the reader's knowledge about longitudinal models so that she or he can learn different model options, formulate a model based on certain important assumptions, fit such a model to longitudinal data, and revise the model until the best fit to explain the data is found. At the end of this statistical modeling process (summarized in Fig. 1.4), I will help to interpret the model and its coefficients.

The statistical modeling I will introduce consists of a sequence of steps, as shown in Fig. 1.4. In the first stage, one will propose a model that is believed to describe the longitudinal process under study. The formulation of such a model will involve taking a number of assumptions as given. At the second stage, one will fit the model to the collected data so that the data are used to estimate the parameters or coefficients in the regression-like model as it defines the systematic relationships between the dependent variable, the independent variables, and the error term. This is achieved by estimating the unknown model parameters using different estimation techniques. At the model evaluation stage (third step), one will learn how to assess the adequacy of the fitted model, always trying to find the most parsimonious representation of the longitudinal or change process. In the statistical modeling jargon, by parsimonious, one refers to the simplest model in terms of the functional equation (with the lowest number of parameters to be estimated) that best fits the longitudinal data. At this stage, some of the model assumptions are put to the test, and one needs to evaluate several goodness-of-fit measures, analyzing the error terms, and comparing simpler models with more complex ones. If the estimated model does not seem to be the adequate one, then it will be necessary to seek improvement by going back to step one and reformulating the model. Here, model adequacy is also determined by theoretical considerations (so that parsimony alone, is neither a sufficient nor a necessary criterion to select the final model). In this book, I provide comparatively little emphasis on this fourth step to avoid complicating the basics behind each model. In the presentation of each of the available techniques, I will guide the reader in choosing the right model approach upfront, which will lead to the best model faster. In the end, after finding the appropriate longitudinal model that fits the observed data, I will help conclude the research by considering the substantive significance of the results (fifth step). This is the most exciting part of the exercise behind any research—when researchers tell the world about their findings and the recommendations drawn from their analyses. In Chapter 4, I help the reader design and undertake all the different steps behind this process of analyzing longitudinal data, which is also a longitudinal process in and of itself.

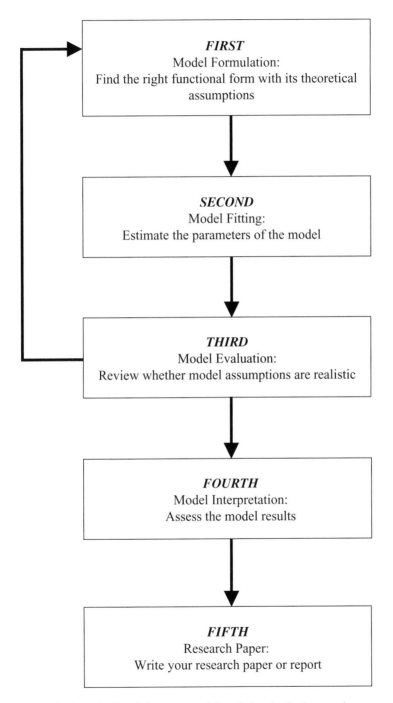

Fig. 1.4 Statistical modeling steps to follow in longitudinal research.

I will introduce the different available statistical models in steps 1 through 4, their formulation, fitting, evaluation, and interpretation in later chapters. Throughout this book, I describe the available models for the analysis of panel or cross-sectional cross-time data (when the dependent variable measures a quantitative outcome and one is interested in understanding what determines change in the continuous variable). Additionally, I present the main techniques for the examination of event history data (when the dependent variable is a qualitative outcome and one is interested in modeling the rate at which cases transition from one state of interest to another). Table 1.2 summarizes the techniques described in the book.

The rest of the book consists of four additional chapters. Chapter 2 is an introduction to the main advances in the modeling of change of quantitative variables. The methods for the analysis of panel data are reviewed, and I concentrate the reader's attention on the analysis of time series and cross-section, using fixed effects or random effects models. Chapter 3 describes the basic methodology employed for the analysis of change in qualitative variables. This methodology, broadly defined as event history analysis, has indeed revolutionized dynamic analysis in the social sciences since Tuma and Hannan's seminal theoretical manual was published in 1984. Chapter 3 can be read independently of the second chapter, depending on which methodology the reader wants to learn first. Although more attention is devoted to the statistical models *per se* and to clarify the key concepts in each type of longitudinal methodology, I also provide various practical recommendations, from how to usefully code informative data and the costs behind data and technical choices to the different statistical software programs that could be used to analyze longitudinal data. Chapter 4 presents some useful ideas (a research project "road map") for designing a research study or report and learning how to present the results of the several dynamic analyses learned in this book. I conclude in Chapter 5 with two illustrations of how dynamic analysis can be applied to investigate a social process or answer an empirical question of interest. At the end of the book, I also include a detailed bibliography to help the interested reader continue learning about dynamic methodology. Here, the reader will find a list of useful references including recommended basic and advanced methodology books, as well as examples of empirical studies that use dynamic models to study social change.

6. What You Need to Know

This book is written for those who want to analyze longitudinal data but do not know about the different basic longitudinal analysis techniques available in various statistical software programs and used in many journals, reports, and presentations. It is also useful to those who already know the basics behind such

Table 1.2
Some of the most frequently used models and techniques for the analysis of longitudinal data

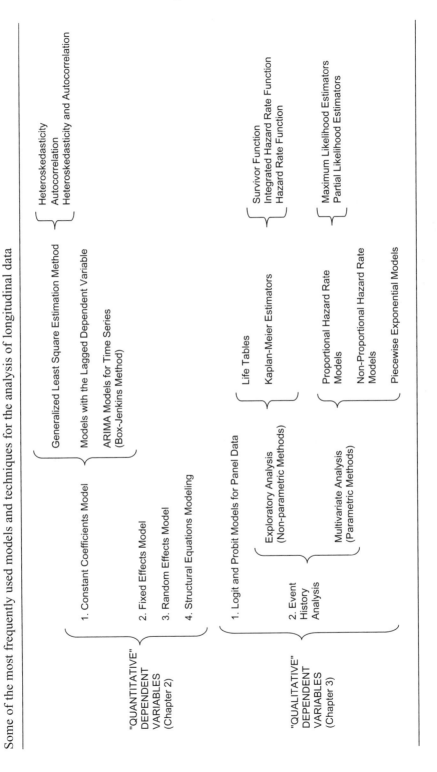

techniques but who want to learn to use the software programs available to estimate such longitudinal models. I assume some basic knowledge of statistics and probability theory. The reader should also have some theoretical and practical knowledge about multiple regression analysis and logistic regression (even though this chapter has provided the basics behind univariate and multivariate statistics). I especially recommend reviewing the basic assumptions behind linear models (in the list of references, I suggest a few readings in order to learn about multiple regression analysis as well as logit/probit regression models). If one has no understanding of the basics behind multiple regression models, it will be more challenging (although not impossible) to grasp the theory and interpretation of longitudinal regression models with temporal data.

It is not necessary to know much linear algebra, although occasionally I use the vector notation $y = \beta'X + \varepsilon$ to simplify the presentation of some of the causal models under study. Knowledge of limits, derivatives, and integers will be useful when analyzing function and procedures of estimation in dynamic models but is not necessary to follow and use the techniques covered in this book.

Finally, all the examples in the book were made using a PC with 1.59 GHz and 768 MB of memory. The software used included the most recent version of SPSS for Windows (version 13), Stata for Windows (version 9), and TDA (version 5.7) for DOS. TDA is a somewhat old-fashioned program but works in a simple and very efficient way. The Blossfeld, Hans-Peter, and Gotz Rohwer book, *Techniques of Event History Modeling: New Approaches to Causal Analysis,* provides an excellent introduction to event history models and their estimation using TDA. I also use AMOS, a program module that is easily added to SPSS, for the estimation of structural equations. Nevertheless, this book was written with the SPSS and Stata packages in mind since my experience has shown that they are two of the easiest software programs for students and practitioners to utilize (also the most widely and easily accessible software), and they allow comprehensive exploration and multivariate analysis of any data set. Throughout the book, I use many of these programs in the analysis of example data sets. This book is also a complete basic introduction to dynamic analysis, and the theoretical and empirical coverage of the material in these chapters can easily by adapted to the use of any other statistical program of choice—provided that the program allows the user to estimate the array of longitudinal techniques introduced here in this book.

Chapter 2

Longitudinal Analysis of Quantitative Variables

By longitudinal analysis of "quantitative" variables, I refer to all longitudinal models in which the dependent variable is measured as a continuous variable or an ordinal variable (with a reasonable number of ordered categories). Please re-read Chapter 1 if you still have trouble in understanding the meaning of continuous or ordinal variables. In this chapter, I examine some of the methods available for the analysis of these longitudinal data. Specifically, I describe the different longitudinal regression-like analyses of data, which include times series for a group of sections or panels (whether individuals, groups or teams, organizations, or nations). This type of data was described in the previous chapter as *cross-sectional cross-time data*.

As a quick reminder, a *time series* is a sequence of repeated measurements for a given variable during a period of time for one case or unit of analysis. For example, the gross national product of Spain from 1900 to 2006, each year, is a time series. It includes the value of the national product for the Spanish economy for each year since 1900, with 106 time observations (one for each year). In general, any time series for a given variable x_k includes the sequence of values x_{tk}, with $t = 1, 2, 3, \ldots, T$, where T is the last point in time when one observes the value of variable x_k for one given unit of analysis. On the other hand, *cross-section* refers to a collection of measurements for a group of individual cases or units at one particular point in time. Each observation therefore includes the values that a certain variable x_k take. So that x_{ik} denotes the value that x_k takes for unit i, with $i = 1, 2, 3, \ldots, N$, where N is the total number of different units in the sample. An example of cross-section is the database that includes the gross national product (x_1), the population (x_2), and the number of suicides per capita (x_3) for each country in the world in 2005.

When a vector of different variables is available for a set of social actors or units, also known as *sections* or *panels,* whose values are observed over a period of time, then the data set is called a *cross-sectional cross-time data set* (also known as a *pooled time series data set*). This type of data set is often also referred to as *panel data*. The "panel" indicates that the collection of information is available for a certain number of social actors during a certain time interval. From this, it can be easily inferred that *cross-sectional cross-time data* is a matrix of information with three key dimensions: (1) units or sections under analysis, (2) vector of both dependent and independent variables, and (3) time points. In other words, it is a time series for a group of variables for different units of analysis. When a data sample has these characteristics, it is then possible to specify a regression model with cross-section data that considers all cases in a given moment of time. Additionally, it could be interesting to perform a similar analysis for each of the time periods included in the sample and to be able to analyze possible variations of the effect of independent variables on the dependent variable in the long run. One could then evaluate whether changes exist in the functional form and the parameters of the estimated model over time. Note that in both situations, the entire sample is broken up into several sub-samples. In the first case, each sub-sample consists of all observations in a given moment of time. Thus, there will be as many sub-samples as there are periods in the sample. In the second case, there is one time series for each individual unit in the data set.

The objective of this chapter is to introduce the dynamic analysis of a continuous dependent variable considering both the time as well as the cross-sectional dimension of the longitudinal data to be analyzed. One of the many advantages of the longitudinal models that combine cross-sections with time series is that they allow distinguishing cross-sectional variation from time variation. Several models have been proposed for the analyses of these longitudinal data. I will start by presenting the most simple of statistical models. These models are frequently used in contemporary empirical studies because of how straightforward they are as well as their simplicity in terms of model specification and estimation. They are: (1) the *constant coefficient model,* (2) the *fixed effects model,* and (3) the *random effects model.* Each of these models will be presented with an example using a real longitudinal data set (described in the next section) and including the statistical software commands, which readers can use to estimate these models using their own database. Later in this chapter, I introduce some basic ARIMA models and structural equations as additional methodological tools for modeling quantitative dependent variables. To follow the content of this chapter, I recommend reviewing Chapter 1, particularly the section discussing the classical linear regression model (*Why Longitudinal Analysis?*). I also recommend further reading on the classical linear model, which can be found in the list of references at the end of this book.

1. A Practical Example

To illustrate some of the models available for the longitudinal analysis of continuous dependent variables, I work with a sample that includes national level data for the 22 country members of the Organization for Economic Cooperation and Development (OECD) for the period of time between 1960 and 1990 (for more information visit www.oecd.org). This data set is an essential tool for health researchers and policy advisors in governments, the private sector, and the academic community to carry out comparative analyses and to draw lessons from international comparisons of diverse health care systems. The data were obtained primarily from an OECD database titled *OECD Health Data 1995,* which provides relevant information about the population health and health care institutions, organizations, and practices of many OECD countries since 1960. The information studied was not available annually but was recorded every *decade* since 1960. Therefore, the data I use in this chapter include information for 22 OECD countries in 1960, 1970, 1980, and 1990.[3] So *N* equals 22 and *T* (i.e., time periods) equals 4, with a database consisting of 88 country-decade observations.

Table 2.1 shows part of the database under analysis and gives an idea of how such longitudinal data appears when stored in a computer program like Excel. It presents information for 10 of the 22 countries for each of the four decades. Each observation (or row) refers to a given individual case in a given moment of time. In this particular example, each observation is a country with information for many variables measured in a given decade. For example, one can see that the table includes four records for Spain (one for each of the four time periods available: 1960s, 1970s, 1980s, and 1990s; from now on, 1960 refers to the decade 1960–1969, 1970 refers to the decade 1970–1979, and so forth):

Spain	1	1960	69.00	0.894	0.008	54
Spain	2	1970	73.00	1.406	0.019	61
Spain	3	1980	73.00	3.887	0.147	83
Spain	4	1990	77.00	8.007	0.369	99

The first column of the data matrix on the table refers to the name of the country, and such a variable could be used as the country ID variable. The second column indicates the temporal order of the observations for each country. The third refers to the decade when the information was collected. Columns four to seven include information on four different variables measuring the following national-level characteristics, including the dependent variable, average life expectancy at birth (in years), and three independent variables, gross domestic product (GDP) per capita (in thousands of dollars), total health care expenditure per

[3] *OECD Health Data 2005*, released June 8, 2005, offers the most comprehensive source of comparable statistics on health and health systems across OECD countries.

Table 2.1
Part of a longitudinal database sorted by country name (in alphabetical order)
and by year of observation (in ascending order)

OECD Country	Observation	Year	Life Expectancy (in years) (*t*)	GDP per capita (in thousands of $) (*t* − 5)	Public Health Expenditure per capita (in thousands of $) (*t* − 5)	Percent of Population with Health Care Coverage (*t* − 5)
Australia	1	1960	70.70	1.962	0.046	100
Australia	2	1970	72.00	2.470	0.067	100
Australia	3	1980	74.00	5.942	0.322	100
Australia	4	1990	76.50	12.848	0.712	100
Austria	1	1960	68.70	1.510	0.046	78
Austria	2	1970	72.00	1.956	0.065	91
Austria	3	1980	72.00	5.138	0.262	99
Austria	4	1990	74.80	12.246	0.662	99
Belgium	1	1960	70.30	1.546	0.033	58
Belgium	2	1970	72.00	2.080	0.062	99
Belgium	3	1980	73.00	5.254	0.247	99
Belgium	4	1990	75.20	11.943	0.726	99
Canada	1	1960	71.00	1.920	0.045	68
Canada	2	1970	74.00	2.521	0.078	100
Canada	3	1980	74.00	6.022	0.332	100
Canada	4	1990	77.00	14.263	0.907	100
Denmark	1	1960	72.10	1.827	0.059	95
Denmark	2	1970	74.00	2.483	0.103	100
Denmark	3	1980	75.00	5.379	0.319	100
Denmark	4	1990	75.80	12.997	0.689	100
Spain	1	1960	69.00	0.894	0.008	54
Spain	2	1970	73.00	1.406	0.019	61
Spain	3	1980	73.00	3.887	0.147	83
Spain	4	1990	77.00	8.007	0.369	99
Sweden	1	1960	73.10	1.941	0.066	100
Sweden	2	1970	75.00	2.639	0.116	100
Sweden	3	1980	75.00	6.038	0.430	100
Sweden	4	1990	77.40	13.056	1.046	100
Switzerland	1	1960	71.20	2.822	0.057	74
Switzerland	2	1970	74.00	3.619	0.083	89
Switzerland	3	1980	75.00	7.485	0.360	97
Switzerland	4	1990	77.40	16.124	0.860	100
United Kingdom	1	1960	70.60	1.936	0.064	100
United Kingdom	2	1970	73.00	2.389	0.085	100
United Kingdom	3	1980	73.00	5.050	0.252	100
United Kingdom	4	1990	75.70	11.459	0.576	100
United States	1	1960	69.90	2.708	0.035	20
United States	2	1970	73.00	3.452	0.050	40
United States	3	1980	74.00	7.067	0.245	42
United States	4	1990	75.90	16.259	0.708	44

Note: The complete database includes all observations for the rest of OECD countries in the sample.

capita (in thousands of dollars), and percentage of population covered for health expenses. Life expectancy measures the average life span of the population in each of the countries in the sample. The GDP is the total dollar value of all final goods and services produced for consumption in a country. Total health care expenditure per capita and percentage of people covered for health expenses are two key indicators of the health care structure of OECD countries.

Thus, the first record (or row of the data matrix) provides information about the health and health care system of Spain in the 1960s (first observation available for Spain). During this period, the average Spanish life expectancy (fourth column of the matrix of data) was 69 years (this average increased by eight years in 1990), and the GDP per capita (third column; measured in thousands of dollars) was $894 (with an 11% annual rate of economic growth in the 1960–1990 period). In the 1960s, the Spanish government spent $8 per person on health care (the public health care expenditure per person was 18 times greater in 1990), and about 54% of the population was covered. By the end of the 1990s, this percentage had almost doubled, as the Spanish health care system seemed to have covered most people (i.e., 99%). All the above examples show the type of information that can be easily inferred from the records of a cross-sectional cross-time data set.

The main variable of interest or dependent variable throughout this chapter is population life expectancy (henceforth LIFEEXP, as this is the variable name used in my data set). A good way to start any longitudinal analysis is to plot the variable over time. Graphs 2.1 and 2.2 present two examples of life expectancy for different OECD countries plotted against time (in this case at four different decades). There are many possibilities here depending on what is ultimately desired in a given study. Graph 2.1, for example, shows a box-plot for each decade with a total of four boxes. With this graph one can see that the median life expectancy has increased within the OECD area since 1960, from approximately 70 to 76 years old. At the same time, there is a clear decline in variation in the levels of life expectancy across OECD countries over time. In the 1960s and 1970s, life expectancy in Portugal was quite different from that of the rest of the OECD countries (as one can clearly see in the box-plot). In later years, Portugal's level of life expectancy became quite similar to that of other OECD countries (in 1980 and 1990). The median life expectancy in OECD countries was 70 in the 1960s and 76 in the 1990s. Graph 2.2 shows the same aggregate information; this time I plot life expectancy as a function of time—the median life expectancy is plotted for each decade (i.e., the thicker black line) with the maximum and minimum life expectancy interval lines as well (above and below the median expectancy line). These two graphs were created in SPSS, and they can easily be created in Excel as well (for more detail on how to do this, see Chapter 1). The idea is to plot the dependent variable (typically over time) before any

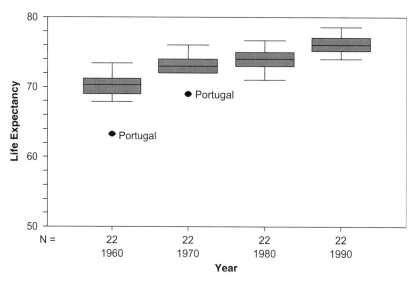

Graph 2.1   Life expectancy box-plots for the 22 OECD countries in the 1960s, 1970s, 1980s, and 1990s.

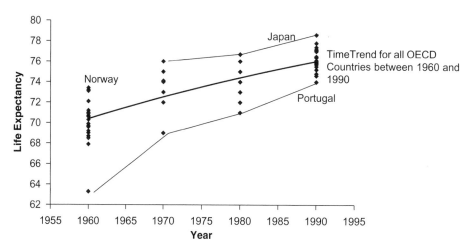

Graph 2.2   Time trend in population life expectancy for the 22 OECD countries in the 1960s, 1970s, 1980s, and 1990s.

advanced methodological techniques are used. This is an important way to begin because a simple graphical representation of the dependent variable can provide relevant descriptive information for its later longitudinal modeling.

In the following sections, I illustrate several longitudinal techniques available when the dependent variable is continuous. Using this example of longitudinal data, I investigate how variation in the average life expectancy across OECD countries is explained by national differences in the structure of the public health care systems, after controlling for the national level of development. More specifically, I examine the effects of variables relating to the structure of the national public health systems on the average life expectancy. I present a simple model that includes independent variables such as public health care expenditure per capita (in thousands of US dollars) and the level of a population's health care coverage (as the percentage of the population covered by the public sector—both lagged five years) as predictors of the contemporary life expectancy of a population of a country. For an observation in the 1990s, the level of life expectancy is regressed on the value of health care expenditure in 1985 (i.e., the measure five years earlier). The effects of those independent variables are evaluated once the GDP per capita variable is included in the model—since the impact of such independent variables controlling for the impact of national economic growth is of interest. The general equation model to be estimated, which will be further specified later in this chapter, is:

$$\textit{Life expectancy}_{it} = f(\textit{GDP pc}(t-5)_i, \textit{ health care expenditure pc}(t-5)_i, \textit{ social coverage}(t-5)_i)$$

where life expectancy in a given year t is a function of GDP per capita, health care expenditure per capita, and social coverage (five years earlier, or lagged in year $t-5$). With this model, I intend to explore the causal relationship between population life expectancy and GDP, public health care expenditure, and public coverage of health care expenses at the national level of analysis. This model can be estimated using linear regression techniques with the different model assumptions behind each of the specific statistical techniques used for its estimation. These techniques will be discussed in detail later in the chapter.

2. Multiple Regression Model with Panel Data

As explained in Chapter 1, the standard linear regression analysis is one of the most popular multivariate techniques available to analyze causal relationships. To understand the potential problems associated with the estimation of the multiple regression model with longitudinal data, in this section I examine the basic ordinary least squares (OLS) regression model as applied using a longitudinal database (like the one in Table 2.1). Consider the simple linear regression analysis in which y_{it} is a (linear) function of k independent variables (x_k, where $k = 1, 2, 3, \ldots, K$):

$$y_{it} = \beta_0 + \beta_1 x_{1it} + \beta_2 x_{2it} + \cdots + \beta_k x_{kit} + u_{it} \tag{1a}$$

or simplified:

$$y_{it} = \beta_0 + \sum_{k=1}^{K} \beta_k x_{kit} + u_{it} \tag{1b}$$

where $i = 1, 2, 3, \ldots, N$ are the number of different cases or units of analysis and $t = 1, 2, 3, \ldots, T$ are the number of different observations in time. In matrix notation, the model reads as follows (note that throughout this book, I use lowercase letters to refer to specific variables such as x_1 or x_k and uppercase letters to refer to a vector of variables such as X):

$$y_{it} = \beta_k' X_{kit} + u_{it} \tag{1c}$$

where $i = 1, 2, 3, \ldots, N$ are social units, $t = 1, 2, 3, \ldots, T$ are the number of observations over time, and $k = 1, 2, 3, \ldots, K$ are the number of different variables providing information about any given social unit i at time t.[4] The error term is u_{it} and captures the effects of all other causal factors omitted from the model. In other words, such an error term measures the observed variation of the dependent variable that is not explained by the variation observed in the k independent variables included in the regression model.

Expressions (1a), (1b), and (1c) are identical, and they will be used throughout this book. The Greek letters $\beta_0, \beta_1, \beta_2, \ldots, \beta_k$ are the model parameters (or coefficients) that one wants to estimate. β_0 is the constant term of the model, whereas the rest of the parameters are the slopes of y_{it} with respect to each one of the k independent variables included in the regression model. β_1 measures the change in the mean of y_{it} per one unit increase in x_1, when all other independent variables stay the same. Similarly, in the most general case, β_k is the coefficient on x_k and reflects change in the mean of y_{it} per one unit increase in x_k, all other independent variables being constant. In the case of a cross-sectional cross-time data set, the model (1a), (1b), or (1c) consists of a single equation with $N \times T$ observations. The parameters to estimate are K and are assumed to be equal (or constant) for all cases and for every time period in the sample; that is to say, such coefficients are time and case/unit constant.

Many times, especially when comparing coefficients' size and determining whether one independent variable's effect is larger than another's, standardized regression coefficients are provided. Standardized regression coefficients or

---

[4] Henceforth, I use β_k or simply β to denote the column vector containing the model coefficients $\beta_0, \beta_1, \beta_2, \ldots,$ β_k. X_{kit} or X_{it} is the column vector with k rows (one for each independent variable) for individual i at time t; that is, the transpose of the ith $1 \times k$ row of the matrix X_{itk} (of dimension $(N \times T) \times K$).

"beta weights" can be obtained from unstandardized coefficients by using the following formula:

$$b_k^* = b_k(s_k/s_y)$$

where b_k^* is the unstandardized regression coefficient of a given independent variable x_k, measuring the effect of variable x_k on the dependent variable y. s_k is the standard deviation of x_k and s_y is the standard deviation of y. In the most general case, using standardized coefficients implies examining standard deviation changes in the dependent variable per one standard deviation increase in any of the independent variables in the model. So that b_k^* reflects change in the average of y_{it}, in standard deviation terms, per one standard deviation increase in x_k, other things being equal.[5] While unstandardized coefficients are easier to interpret, standardized ones are appropriate for making comparisons across variables' effects (the standardized coefficients are not appropriate for comparisons across different subsamples or samples because they depend partly on the variance of the variables included in the model).

Most statistical programs perform the OLS calculations by using matrix algebra. The least squares criterion is the procedure used to minimize the sum of squared residuals in equations (1a), (1b), or (1c), $\Sigma(u_{it})^2$. Software programs are capable of finding the estimates of the model coefficients in equation (1) that minimize the sum of squared residuals over all observations. I will bypass the details of this type of estimation and instead focus on hypothesis testing using regression models. Three statistics are relevant and consequently important to understand: the *t*-test, the *F*-test, and the R^2 statistic. The first one is the *individual coefficients t-test*, which tells the significance level of each of the regression coefficients in the sample. To test the null hypothesis that an individual population coefficient equals zero:

$$H_0: \beta_k = 0$$

one would have to examine the *t*-statistic:

$$t = b_k/s_k$$

where b_k is the sample regression coefficient (i.e., the estimate of β_k using the data sample under study) and s_k is the estimated standard error. The alternative to the above null hypothesis is that the population coefficient is not zero, or:

$$H_1: \beta_k \neq 0.$$

---

[5] The standardized regression coefficients can be obtained from a regression where all variables included in the regression model are in standard-score form. We can easily transform any variable in a standard score by subtracting its mean and dividing it by its standard deviation as follows: $z_i = (x_i - \bar{x})/s_x$, where z_i is the standard-score form of x_i.

The t-test formula, $t = b_k/s_k$, indicates how far the sample regression coefficient b_k is from the hypothesized population parameter in estimated standard errors. If H_0 is true, then $\beta_k = 0$, and this t-statistic follows a theoretical t-distribution with $N - k$ degrees of freedom (where N is the total number of observations in the sample and k is the number of model parameters including the constant term or β_0). A large value of t implies a low probability, allowing for the rejection of H_0 and the acceptance instead of the alternative hypothesis that H_1: $\beta_k \neq 0$. In general, any coefficient for which the obtained p-value is less than 0.05 is said to be *statistically significant*.[6] The t-test formula is a two-tailed test because the alternative hypothesis is that the coefficient for the kth independent variable is different from zero (or in other words, the alternative hypothesis is not specifying any direction in the effect of a given independent variable x_k on y). Computer statistical software programs routinely calculate the two-tailed test results; and later I will detail where to read these statistics from the Stata software output. Some theories may actually suggest or hypothesize a positive or negative impact of an independent variable on a dependent variable *a priori*. So for example, if it has been proposed that the effect of the independent variable is positive then one should test:

$$H_0: \beta_k \leq 0$$
$$H_1: \beta_k > 0.$$

Only positive values of the estimated coefficient would support this directional hypothesis H_1. Alternatively, if one believes that the population coefficient is negative, then the following should be tested:

$$H_0: \beta_k \geq 0$$
$$H_1: \beta_k < 0.$$

To obtain a one-tailed t-test probability, simply divide any regression table's two-tail probability (or p-value) in half. If the estimated coefficient lies in the direction specified by H_1, the one-tailed p-value is equal to the usual two-sided p-value (typically reported by any statistical software when estimating OLS) by 2. If the estimated coefficient lies in the direction specified by H_0, then H_0 cannot be rejected.

In the most general case, the t-statistic allows one to test hypotheses regarding individual values for β_k. To test the null hypothesis that an individual population

---

[6] The probability cutoff point is arbitrary. Typical cutoff points are 0.05, 0.01, and 0.001 and one must always be very clear about which cutoff points are used in the study. Theoretically, if our decision rule is to reject H_0 when $p < 0.01$, one can say that we have a one percent chance of being wrong (called Type I error). Although such lower cutoff point (of 0.01) also raises the likelihood of the opposite error to occur, that is, failing to reject H_0 when in fact it is false (Type II error). Again, the 0.001, 0.01, and 0.05 are typical cutoff points.

coefficient equals a concrete value q:

$$H_0: \beta_k = q$$

one would have to examine the *t*-statistic:

$$t = (b_k - q)/s_k$$

where b_k is the sample regression coefficient (i.e., the estimate of β_k using the data sample under study), q is the concrete value that is being hypothesized, and s_k is the estimated standard error. In this case, the alternative to the above null hypothesis is that the population coefficient is not equal to q, or:

$$H_1: \beta_k \neq q.$$

Again, one would reject H_0 and instead support H_1 if the *t*-test two-sided probability is less than 0.05 (as a typical decision rule).

It is also common to construct *confidence intervals* to assess the importance of the effect of any given independent variable on the dependent variable of interest. Construct confidence intervals by adding and subtracting the *t*-value times the estimated standard error of the coefficient. So the confidence intervals goes from $b_k - $ *t*-value $\times s_k$ to $b_k + $ *t*-value $\times s_k$. Or, as follows:

$$[b_k - \text{\textit{t}-value} \times s_k, b_k + \text{\textit{t}-value} \times s_k]$$

$$\text{or} \quad b_k - \text{\textit{t}-value} \times s_k \leq \beta_k \geq b_k + \text{\textit{t}-value} \times s_k.$$

Typically, one must choose a *t*-value or *t* for the desired level of confidence from the theoretical *t*-distribution with $N - k$ degrees of freedom. The most frequently reported 95% confidence interval uses the *t*-value of approximately 1.96 (and researchers round up to the value of 2). For the 99% confidence interval, the *t*-value equals 2.576. One can also display the same information graphically as follows:

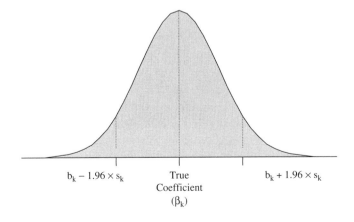

$b_k - 1.96 \times s_k$ True Coefficient (β_k) $b_k + 1.96 \times s_k$

So that forming a 95% confidence interval for an estimated coefficient of 10 (with a standard deviation of 3) is calculated. The value of the lower extreme of the interval is $10 - 1.96 \times 3 = 4.12$ and the upper extreme is $10 + 1.96 \times 3 = 15.88$. So that based on the sample, one can be 95% confident that the true parameter coefficient lies approximately between 4 and 16, or in the formula expression, $4 \leq \beta_k \geq 16$.

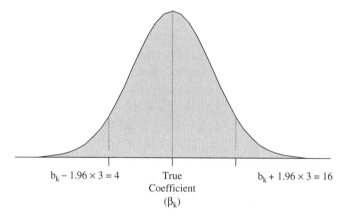

$b_k - 1.96 \times 3 = 4$ True $b_k + 1.96 \times 3 = 16$
Coefficient
(β_k)

The second important test in the multiple linear regression model is the *over-all F-test*. This test covers the significance of all slope regression coefficients in the model (excluding the constant). The overall *F*-test is considered an overall goodness-of-fit model test because it measures whether the proposed regression model with k independent variables explains variation in the dependent variable better (and consequently better predicts the values of the dependent variable) than its own average value. The null hypothesis is the following:

$$H_0: \beta_1 = \beta_2 = \cdots = \beta_k = 0.$$

This *F*-statistic follows an *F*-distribution with k (i.e., number of independent variables) and $N - (k + 1)$ degrees of freedom (where N is the total number of observations in the sample). A low enough *p*-value (for example, $p < 0.01$) indicates that one can reject the null hypothesis that all regression coefficients are equal to zero (significant at the 0.01 level). This means that *at least one* of such coefficients is different from zero—but not necessarily that each and every coefficient is different from zero.

Finally, the R^2 is an overall goodness-of-fit model statistic in the basic linear regression model. As mentioned in Chapter 1, the R^2 is the ratio of explained variation (by the specified regression model with a set of independent variables) to the total variation of the dependent variable. The formula is the following:

$$R^2 = \frac{SS_M}{SS_T}$$

where SS_M represents the model sum of squares, or variation of predicted y ($\hat{y}_i$) around the mean of y ($\bar{y}$):

$$SS_M = \sum (\hat{y}_i - \bar{y})^2$$

and SS_T represents the total sum of squares, or variation of actual y values around the mean of y ($\bar{y}$):

$$SS_T = \sum (y_i - \bar{y})^2.$$

The value of R^2 falls in a range from 0 to 1 and is interpreted as the "proportion of variance of y explained by x." The remaining percentage or ratio of variance in y is explained by omitted variables in the model (or non-linearities in the effect of x on y). In many occasions, it is preferable to report what is called the adjusted R^2, whose exact formula is the following:

$$\text{Adjusted } R^2 = R^2 - \frac{k-1}{N-k}(1 - R^2) = 1 - \frac{SS_R / (N-k-1)}{SS_T / (N-1)}$$

where SS_T represents the total sum of squares, or variation of actual y values around the mean of y and SS_R represents the total sum of squared residuals or $\sum (u_{it})^2$. The adjusted R^2 is calculated in order to account for the complexity of the regression model (in terms of the number of parameters to be estimated) relative to the complexity of the data (in terms of number of cases). The adjusted R^2 is recommended when increasing the number of parameters to be estimated.

I used the Stata software program to estimate the OLS regression model that fits the OECD health data. The general OLS equation model to be estimated is the following:

$$Life\ expectancy_{it} = \beta_0 + \beta_1 GDP\,pc(t-5)_i + \beta_2 health\ care\ expenditure\,pc(t-5)_i$$
$$+ \beta_3 social\ coverage(t-5)_i.$$

With this model, I explore the causal relationship between population life expectancy and GDP per capita, public health care expenditure per capita, and public coverage of health care expenses at the national level of analysis. To estimate the coefficients for the three independent variables in Stata, I use the command `regress` immediately followed by the name of the dependent variable, followed by the list of independent variables (or their computer variable names). So I typed "`regress lifeexp gdppc pubpc hlthcov`" in the Stata command box once the OECD data set had been read by Stata. LIFEEXP is the Stata variable name for the variable life expectancy, GDPPC is for GDP per capita, PUBPC is for health care expenditure per capita, and HLTHCOV is for social coverage. The summary of the main OLS regression results are reported in the following table:

Independent Variables	OLS Model	
	B	St. Error
Constant	67.179***	0.773
GDP per capita (in thousands of $)	0.390**	0.126
Health care expenditure per capita (in thousands of $)	−1.464	1.965
Public coverage of health care expenses (% of population)	0.043***	0.009
F-statistic:		
Compare with the baseline model	59.98***	
Degrees of freedom	(3 and 84)	

There are two relevant columns to look at in order to see the impact of the different independent variables on life expectancy. The first column of the table, headed typically by "Coef." or "B," gives the estimated model parameters of interest (in this case the unstandardized partial coefficients). This column alone helps in the writing of the estimated OLS model as follows:

$$Life\ expectancy_{it} = 67.179 + 0.39 \times GDP\ pc(t-5)_i + (-1.46)$$
$$\times\ health\ care\ expenditure\ pc(t-5)_i + 0.04$$
$$\times\ social\ coverage(t-5)_i.$$

The interpretation of the least squares coefficients is straightforward. For example, $\beta_1 = 0.39$ indicates that a $1000 increase in GDP per capita is associated, on average, with a 0.4 year increase in life expectancy. $\beta_2 = -1.46$ indicates that a $1000 increase in the per capita health care expenditure is associated, on average, with a 1.46 year decrease in national life expectancy. Finally, $\beta_3 = 0.04$ indicates that a one unit increase in the percent of public coverage of health care expenses of a country's population results, on average, in a 0.04-year increase in population life expectancy. The intercept or constant is the average value of life expectancy when all independent variables are equal to zero. The intercept is usually of little direct interest because the fitted value when all independent variables are equal to zero is not substantively important. Here, however, the intercept denotes that countries with GDP per capita, health care expenditure, and social protection all set to zero have on average a 67.2-year life expectancy.

The second column of the table "St. Error," reports the standard error of the coefficient. One could compute the *t*-statistic for the significance of independent variable "Public Coverage" by dividing the coefficient 0.043 by the standard error 0.009. The *t*-value is 4.78 indicating the significance of this model parameter (that is, testing whether the model parameter is equal to zero). This large *t*-value implies a low probability value, allowing for the rejection of H_0 and the acceptance of the alternative hypothesis or H_1: That the coefficient associated to public coverage is significantly different than zero (this coefficient is significant at the 0.001 level). In article tables reporting final analyses, it is typical to see asterisks marking which coefficients are significant. For example, the effect of GDP on life expectancy is positive and significant at the 0.01 level (because the ** indicates that the *t*-statistic *p*-value is equal to or less than 0.01). The same applies to public coverage of health care expenses, which has a positive and significant effect on population life expectancy (the *** indicates that the *p*-value is equal to or less than 0.001). The variable health care expenditure per capita is not a significant predictor of life expectancy (at the 0.05 level); the *t*-value for this variable is 0.74 (i.e., -1.46 divided by 1.965), which implies a probability value higher than 0.05. Finally, the constant term indicates that the average life expectancy for countries with zero GDP per capita, zero health care coverage, and zero public expenditure is 67.2 years (the *** indicate that the *p*-value associated to the constant term is equal to or less than 0.001).

From the output included on the table, one could compute the 95% confidence interval for the coefficient of GDP per capita. Using the formulas presented earlier for computing confidence intervals, one can be 95% confident that the true parameter coefficient of GDP per capita on life expectancy lies between 0.143 (i.e., $0.39 - 1.96 \times 0.126$) and 0.637 (i.e., $0.39 + 1.96 \times 0.126$). Similarly, one can be 99% confident that the true parameter for GDP per capita lies between 0.065 (i.e., $0.39 - 2.576 \times 0.126$) and 0.714 (i.e., $0.39 + 2.576 \times 0.126$).

The adjusted R^2 of the model and *F*-test help to evaluate the linear predicting power of any OLS model. In this case, the R^2 term is 0.67, indicating that the specified model explains 67% of the (linear) variance in population life expectancy. The *F*-test is almost 60, with 3 ($k - 1$) and 84 ($N - k$) degrees of freedom. With those two degrees of freedom, this *F*-statistic results in a *p*-value well below 0.001 leading to the rejection of the null hypothesis that all regression coefficients are equal to zero (or H_0: $\beta_1 = \beta_2 = \beta_3 = 0$). Remember that this means *at least one* of these coefficients is different from zero—but not necessarily that each and every one of the three is different from zero (as one can see in the OLS table the coefficient for public expenditure on health is not significantly different from zero). All these basic OLS regression statistics are calculated and reported by the Stata program, including coefficient estimates, standard errors, *t*-values, and *p*-values for each of the coefficients. The `regress` command also displays the overall goodness-of-fit measures for any OLS regression model, such as the R^2 and adjusted R^2, as well as the overall F-test.

3. Limitations of the OLS Regression Model with Panel Data

The simplicity in the formulation and estimation of such a linear regression model by OLS clearly comes with some important drawbacks (especially when it comes to modeling change and longitudinal processes). The OLS formulation ignores the temporal dimension of any data set and consequently raises important estimation problems when applied to a longitudinal data set where the number of observations is $N \times T$. The most intuitive limitation that arises is that the model estimates a vector of coefficients that is assumed to be equal for all the N cases and time invariant, that is, the coefficients are constant over time. Another shortcoming concerns the estimation procedure.

The OLS regression model as specified and estimated in equations (1a), (1b), and (1c) assumes that:

$$E[u_{it}] = 0 \text{ for all } i \text{ or social units} \tag{a}$$

$$\text{Var}[u_{it}] = \sigma^2 \text{ for all social unit } i, \text{ and all points of time } t \tag{b}$$

$$\text{Cov}[u_{it}, u_{js}] = 0 \text{ for all agent } i \neq j, \text{ and all points of time } t \neq s \tag{c}$$

$$\text{Cov}[u_{it}, X_{kit}] = 0 \text{ for all } i \text{ and } t. \tag{d}$$

Expressions (a) through (d) are the most important OLS assumptions. Each will be discussed here. First, OLS regression parameter estimates assume that the errors follow a normal distribution with a mean of zero (assumption (a)) and some variance σ^2 that is the same for each of the units of analyses i—what is called the *homoskedasticity assumption*—and it is time invariant or constant (assumption (b)). In addition, it assumes that these errors are not correlated across sections or cases (assumption (c)) nor are they correlated with any of the independent variables x_k included in the model (assumption (d)). In the econometrics literature, the error term is said to behave as a white noise process or to be "well-behaved." The violation of such assumptions is likely to occur when estimating the OLS regression model with any cross-sectional cross-time data set because observations may not to be independent from each other: These observations can be correlated across panels and/or over time. When that is the case, the OLS estimates of the regression model with a longitudinal data set are not necessarily the most efficient estimates nor are they unbiased with the minimum variance. Consequently, they are not the Best Linear Unbiased Estimators (known as BLUE). Also important to note is that any of the hypotheses tested using either t-tests or overall F-tests assume that the errors are normally distributed with all the properties, (a) through (d), presented above. Thus, the hypothesis testing as proposed in the previous section might not be accurate when using a cross-sectional cross-time data set.

Therefore, when estimating the regression model as in equation (1) by OLS using a longitudinal data set caution is advised. Because the working assumptions behind OLS do not necessarily apply to the longitudinal data set, relevant (sectional and temporal) information about the longitudinal data is transferred to the error structure. So that the error term in equation (1) can now be divided into three distinctive components:

$$u_{it} = \alpha_i + \phi_t + \varepsilon_{it}. \tag{2}$$

This time, the error has an individual component that is invariable over time (α_i) although this component varies by individuals or cases. In addition, a temporal component is invariable across individuals (ϕ_t), although it does vary over time. Finally, there is a component ε_{it} that varies across panels and over time, which represents the effect of all the omitted variables. Given this linear structure of the error term, u_{it} is no longer random. Consequently, the analysis of $N \times T$ observations fitting equation (1a), (1b), or (1c) described above implies that the errors are correlated over time for the same unit of analysis. At the same time, such error terms can be correlated among many cases at a given time. This correlation problem gets even more serious when some correlation exists in the error terms both over time and across cases or sections. In such an instance, the true error term structure is represented properly in equation (2) only.

Using some statistical jargon now, there is *heteroskedasticity* when the variance of the errors varies across cases or sections in the sample. There is *auto-correlation* or *serial correlation* when the error terms corresponding to the same case or section under analysis are not independent over time. When this is the case, the error terms are correlated over time for each case or unit of analysis. Heteroskedasticity and serial correlation are common problems when analyzing longitudinal data sets using OLS. As such, the problem of heteroskedasticity is typically associated with the analysis of cross-sectional data, where the observations are obtained from a population of people, companies, or countries. The autocorrelation problem appears most frequently in the case of analysis of time series, when data is collected for the same unit or case in successive time periods.[7]

---

[7] For those with knowledge of matrix algebra, in the extreme case of pure heteroskedasticity, the matrix $E(UU')$ is still diagonal, but the elements of the diagonal are not equal. In the case of autocorrelation, matrix $E(UU')$ is not diagonal and has distinct elements outside the diagonal indicating that pairs of error terms are correlated at different points of time.

The classic OLS regression assumes both *homoskedasticity* (assumption (b)) and *no serial correlation* (assumption (c)).[8] These two assumptions suggest that there is no relationship among the values of a variable at different moments in time (for a given individual or case), for different individuals or cases (in a given moment of time), or for different cases at different moments in time. These assumptions are not realistic in the case of longitudinal data. Consider an intuitive example: One can easily imagine how the economic growth of Portugal in 1997 is probably related to its economic growth during the previous year and that probably, it will also determine its future economic growth. One could also argue that the economic growth of Portugal in 1996 is probably related to the growth of other countries of the European Union in 1996, especially the growth of neighboring national economies such as France and Spain. It is this temporal and cross-sectional dependence that demands implementing the appropriate longitudinal methodology.

When the OLS regression model is estimated for panel data with $N \times T$ observations, the estimators continue to be unbiased, although they are no longer the unbiased linear estimators with the minimum variance. This could mean that the estimators might not be as accurate as desired. A relatively quick and easy way to solve this problem is to consider a model in which all the effects of the independent variables (that is, the coefficients β_{kit}) differ for each individual case i and/or change at every moment in time t. This procedure of estimating different models for different cases, for example, is equivalent to estimating a different model with the data corresponding to a single social actor for all different periods of time—for example, subdividing the $N \times T$ sample of observations into N sub-samples (one for each of the cases) and repeating its analysis. This procedure allows for examining whether cases in the sample behave differently over time (notice that Y is now a vector of variables, one for each individual i, in the sample):

$$Y_{it} = \beta_{0i} + \beta_{1i}x_{1it} + \beta_{2i}x_{2it} + \cdots + \beta_{ki}x_{kit} + u_{it} \tag{3a}$$

or, in matrix notation:

$$Y_{it} = \beta_i' X_{it} + u_{it} \tag{3b}$$

where $i = 1, 2, 3, \ldots, N$ and $t = 1, 2, 3, \ldots, T$. Note that the vector of parameters β now varies for each social agent i (as expressed with subscript i appearing in each one of the coefficients), but such a vector of coefficients does not vary over time. Equations (3a) and (3b) consist of estimating a system

---

[8] Serial correlation is obviously not an issue for cross-sectional data (given that such data do not have any temporal dimension).

of N equations with as many equations as social units or cases there are in the sample:

$$y_{1t} = \beta_1'X_{1t} + u_{1t} \qquad \text{for case } i = 1$$
$$y_{2t} = \beta_2'X_{2t} + u_{2t} \qquad \text{for case } i = 2$$
$$y_{3t} = \beta_3'X_{3t} + u_{3t} \qquad \text{for case } i = 3$$
$$\cdots$$
$$y_{Nt} = \beta_N'X_{Nt} + u_{Nt} \qquad \text{for case } i = N.$$

Alternatively, one could be interested in analyzing possible variations of the linear regression model over time. Thus, one could estimate different equations in time, generating a system of equations where there is now one equation for each time period. Such a system of equations is a variation of equations (3a) and (3b):

$$Y_{it} = \beta_{0t} + \beta_{1t}x_{1it} + \beta_{2t}x_{2it} + \cdots + \beta_{kt}x_{kit} + u_{it} \qquad (4a)$$

or, in matrix notation:

$$Y_{it} = \beta_t'X_{it} + u_{it} \qquad (4b)$$

where $i = 1, 2, 3, \ldots, N$ and $t = 1, 2, 3, \ldots, T$. In this case, there is a system of T equations (one equation for every time period t). The model allows for the estimation of a different set of parameters for every period of time t. Note that in this case the parameters to be estimated allow change over time, although those parameters do not change across units or cases at a given moment of time:

$$y_{i1} = \beta_1'X_{i1} + u_{i1} \qquad \text{for } t = 1$$
$$y_{i2} = \beta_2'X_{i2} + u_{i2} \qquad \text{for } t = 2$$
$$y_{i3} = \beta_3'X_{i3} + u_{i3} \qquad \text{for } t = 3$$
$$\cdots$$
$$y_{iT} = \beta_T'X_{iT} + u_{iT} \qquad \text{for } t = T.$$

Both solutions presented above divide the original longitudinal data into several sub-samples for analysis. In the first case, there are as many sub-samples as individual cases considered in the sample. With the second solution, the sample is divided into different time sub-samples. Consequently, note that both solutions make limited use of the longitudinal aspect of the sample under study. Alternative available techniques are advised for estimating dynamic models in these cases.

4.   Models for the Analyses of Panel Data

In the previous section, I presented alternative regression models that could be used to properly analyze longitudinal data sets with continuous dependent variables. On the one hand, some models assume that there are no differences in the effects of the explanatory variables on the dependent variable either by individuals or through the passage of time; consequently, a *common model* for the set of $N \times T$ observations should be estimated (by applying OLS standard regression to the panel data). Conversely, other regression models assume that the effects differ for each individual and/or at each moment in time. As a result, *different models* should be estimated for each individual and/or each time period. To assume that the regression coefficients are identical for all individuals in the sample as well as over time is extremely restrictive and difficult to believe. By doing so, one also completely ignores the longitudinal nature of the data. At the same time, assuming that the vector of coefficients is different for each social actor or case is excessively general. There is not much to gain in efficiency with the *pooling* of time series data given how time-consuming and expensive it can be to obtain this type of data. For this reason, social researchers have chosen to estimate intermediate longitudinal regression models in their empirical studies.

In the rest of this chapter, I discuss, in detail, several ways of modeling the relation between a continuous dependent variable y_{it} and a set of independent variables x_{kit} in the case of longitudinal data. The goal of these models is to obtain reliable and efficient estimators, that is to say, unbiased estimators with minimum variance. Each model makes a series of explicit assumptions about the existing relation between the explanatory variables and/or the nature of the error of the regression equation. These assumptions are to be taken seriously, and the choice of model ultimately depends on the structure and characteristics of the particular data under study. Depending on the social process being investigated, one could possibly estimate models with a given structure of errors. The objective of these regression models could also be the identification of a model in which the effect of an explanatory variable x_k on a dependent variable y_{it} is the same for all the social units or cases of the sample. The objective can also be to investigate whether this effect is constant over time.

The first model presented in this chapter is the so-called *constant coefficients model* in which the coefficients for the effect of any variable x_k on the dependent variable y_{it} are assumed to be constant for all the units or cases in the sample. In this type of model, certain correlation structures of errors are defined and estimated as part of the methodological procedure. The *fixed effects model,* or *least squared dummy variable model* capture the existing variation in the sample due to the presence of different cases or units with the inclusion of a set of $N - 1$ dichotomous variables in the regression equation. These variables are also referred to as *dummy variables* d_i: One dummy variable is included for each social actor in the sample, excluding the social actor or agent of reference. It is possible to account for differences over time by using dummy variables, as well. In this latter case, one will include a set of $T - 1$ dichotomous variables c_t. By including these c_t dummy variables, one is estimating the effects of the different independent variables on the dependent variable, controlling for possible shifts in the constant term over time. A more complicated version of this model consists of incorporating a series of $N - 1$ dichotomous variables to control for the individual effect of each unit of analysis in the sample on the dependent variable *and* a series of $T - 1$ dichotomous variables to control for the effect of time. This model can be easily extended to include interactions between social agents and time periods as well.

The third type of model is called the *random effects model*. This model assumes that the variation across sections or individuals (and/or over time) is random and is therefore captured and specified explicitly in the error term of the equation. This model is estimated using generalized least squares (GLS), where a given correlation structure of the error term is incorporated in the estimation process behind the longitudinal regression model. In addition, the *structural equation model* (SEM) is proposed when the number of time periods observed in the sample is small. In this type of model, instead of trying to estimate a single regression model with a given error structure, one estimates simultaneously a series of structural equations in which the effect of time as well as the variance and covariance of all model variables (due to the presence of different cases at different time periods) are explicitly modeled. These models allow for estimating different model coefficients for different groups of individuals and/or time units. They also facilitate the testing of hypotheses comparing the magnitude of coefficients across different equations (also called *cross-equation parameter testing*).

Each of these statistical models will be further described and illustrated in the next sections of this chapter. Each section includes some of the theory behind and examples of regression models estimated using the longitudinal OECD health data set sample. I also include the key commands used (as well the results or output files obtained) when I run such models using SPSS and Stata. One could easily apply the lessons from the following sections to analyze one's own longitudinal data set where the dependent variable is continuous. Because each sub-section builds on the previous one, I recommend not skipping sub-sections within this chapter.

5. Constant Coefficients Model

The constant coefficients model is the most straightforward of all dynamic models. This model assumes that regression coefficients measuring the effects of any given independent variables x_k on the dependent variable of interest y_{it} are the same for all cases in the sample at all times. Recall the classical regression equation (1) in which y_{it} is a linear function of k explanatory variables:

$$y_{it} = \beta_0 + \sum_{k=1}^{K} \beta_k x_{kit} + u_{it} \tag{1a}$$

with $i = 1, 2, 3, \ldots, N$ cases or sections in the sample under study, and $t = 1, 2, 3, \ldots, T$ observation points in time. In matrix notation, this can be written as follows:

$$y_{it} = \beta_k' X_{it} + u_{it}. \tag{1b}$$

There are k parameters to be estimated, and these k parameters are considered equal or constant for all cases in the sample regardless of the time period. The estimation of such an equation by the OLS technique assumes that the variance of the random disturbances or errors is a constant σ^2, which is identical for each individual or case—again, this is called the assumption of *homoskedasticity*. In addition, it assumes that these errors are not correlated over time, so that the error term for individual i at time t is not related to the error term for the same individual i at a different time s. This is called the assumption of *no serial correlation*. In other words, the OLS methodology assumes the following error structure:

$$\text{Var}[u_{it}] = \sigma^2 \text{ for all cases } i, \text{ and for all points of time } t$$

$$\text{Cov}[u_{it}, u_{js}] = 0 \text{ for all cases } i \neq j, \text{ and for all points of time } t \neq s$$

or to express these two assumptions in matrix notation, if Σ_t is the variance–covariance matrix of the random errors (also known as the correlation structure of the errors at time t), such a matrix has the following structure:

$$
\Sigma_t =
\begin{bmatrix}
\sigma^2 & 0 & 0 & \cdots & 0 \\
0 & \sigma^2 & 0 & \cdots & 0 \\
\cdots & \cdots & \cdots & \cdots & \cdots \\
0 & 0 & 0 & \cdots & \cdots \\
0 & 0 & 0 & \cdots & \sigma^2
\end{bmatrix}
$$

where the values or elements on the diagonal are the variances, and the ones off-diagonal are the covariances so that the element on column j and row i represents the covariance between the error terms for individuals j and i. The dimension of Σ_t is $N \times N$. Note that the OLS methodology assumes that the error matrix is diagonal—with constant variance on the diagonal of the matrix and covariance of zero in the off-diagonal cells of the matrix. Also note that this matrix is assumed to be constant over time, i.e., regardless of the time t at which it is measured. Given these standard assumptions of the linear regression model, the OLS estimators are both unbiased and efficient. The estimators are unbiased because these regression coefficient estimators are precise and close to the true population parameters. They are efficient because the OLS estimators have the minimum variance.

In a cross-sectional cross-time data set, the assumptions behind the OLS estimation are clearly violated because there is likely to be some relationship between any two periods of time within the same social unit or panel. Additionally, there might be some relationship between any two panels in a given point of time. Because the OLS regression equation does not specify such relationships in the model, such lack of specification is captured only in the error, which can affect the regression estimates in the case of a cross-sectional cross-time data set. Thus, the longitudinal data create the opportunity for several potentially important violations. In the case of heteroskedasticity, or *cross-sectional hetero-skedasticity,* the variance of the random errors now differs by case or cross-section in the sample. That is to say, in mathematical notation, that now $\text{Var}[u_{it}] = \sigma_i^2$, with σ_i^2 being different from σ_j^2 (for any case i different from j). One can write Σ_t at time t, as follows:

$$
\Sigma_t =
\begin{bmatrix}
\sigma_1^2 & 0 & 0 & \cdots & 0 \\
0 & \sigma_2^2 & 0 & \cdots & 0 \\
\cdots & \cdots & \cdots & \cdots & \cdots \\
0 & 0 & 0 & \cdots & \cdots \\
0 & 0 & 0 & \cdots & \sigma_n^2
\end{bmatrix}.
$$

Another situation occurs when, independently of what is assumed about the variances of errors at different periods of time, the errors are also correlated across cases or panels at a period of time. This is called *cross-sectional corre-lation* (or correlation across panels). That means $\text{Cov}[u_{it}, u_{jt}]$ is now different from zero and equal to σ_{ij}^2; and consequently, the errors are correlated across different cases at time t. So expressing both cross-sectional heteroskedasticity and cross-sectional correlation of the errors in matrix notation, one can write Σ_t as follows at time t:

$$\Sigma_t = \begin{bmatrix} \sigma_1^2 & \sigma_{12}^2 & \sigma_{13}^2 & \cdots & \sigma_{1n}^2 \\ \sigma_{21}^2 & \sigma_2^2 & \sigma_{23}^2 & \cdots & \sigma_{2n}^2 \\ \cdots & \cdots & \cdots & \cdots & \cdots \\ \sigma_{n-11}^2 & \sigma_{n-12}^2 & \sigma_{n-13}^2 & \cdots & \cdots \\ \sigma_{n1}^2 & \sigma_{n2}^2 & \sigma_{n3}^2 & \cdots & \sigma_n^2 \end{bmatrix}.$$

Another common complication with longitudinal data occurs when the random errors are correlated over time for a given case or cross-section. Here, it is possible that errors in two consecutive time periods are highly correlated. Another possibility is for all errors to be correlated regardless of the time difference. This situation is known as the problem of *autocorrelation,* or serial correlation, where errors for each case exhibit time or serial dependence and consequently are correlated at successive points in time. Autocorrelation can be written as $\text{Cov}[u_{it-r},$ $u_{it}]$ being different from 0 for at least one lag r (where $r > 0$). From here, the most general case of autocorrelation emerges when there is correlation in the errors from different cross-sections and different time points, so that $\text{Cov}[u_{is}, u_{it}] = \sigma_{ijst}^2$ now changes for different combinations of case i and j and different times s and t. What this means in matrix notation is that there is a different Σ_t for each time point t, up to T different covariance–variance matrices, Σ.

As one can imagine, heteroskedastic and/or autocorrelated errors are quite common when it comes to the estimation of the linear regression model using longitudinal data. Heteroskedasticity by itself does not make the OLS estimators biased, although it causes the standard errors of the regression coefficients to be inefficient—i.e., they do not have the minimum possible variance. Autocorrelation may also cause OLS estimates to be inefficient. In addition, autocorrelation can cause the standard errors of the regression to be biased. So to sum up, because of the presence of heteroskedasticity and/or autocorrelation, the estimate of the variance of the error is incorrect. Consequently, any of the statistics based on the error variance such as the R^2, the F-test, or the t-tests for the coefficients are unreliable and the risk of not rejecting false hypotheses about the statistical significance of certain parameters may now be considerably higher. To solve these problems, the most widespread procedure consists of specifying the variance–covariance structure of such random errors and considering such a structure when estimating the main regression equation. Here, it is important that the time and/or cross-section of the errors are captured in the estimation. Examples of frequently used models and techniques follow, including heteroskedastic models, autoregressive models, models with the lagged dependent variable, models with instrumental variables, and the basic ARIMA models.

5.1. Heteroskedastic Models

As explained in the previous section, in the presence of heteroskedastic or auto-correlated errors in the model, the OLS estimators continue to be unbiased, but they are no longer unbiased linear estimators with the minimum variance. For efficiency reasons, the constant coefficients model should be estimated using the *Generalized Least Square method* (GLS). This estimation method explicitly assumes some variance–covariance error structure when estimating the regression coefficients. By explicitly, I mean that certain parameters of such correlation error structures are estimated using this method. Under heteroskedasticity and serial autocorrelation, the GLS estimators are linear, unbiased, and have minimum variance (for more information about GLS, consult the suggested readings in the list of references of this book). The GLS method is, without a doubt, a general case and sometimes not easy to estimate, especially when there are many variance and covariance parameters to be estimated. This typically happens when one wants to remedy both cross-sectional and serial correlation together. If this is the case, the error structure is as follows:

$$\text{Var}[u_{it}] = \sigma_{it}^2$$

$$\text{Cov}[u_{it}, u_{js}] = \sigma_{ijts}^2$$

where the variance of the term errors is different for each individual or case and it is also allowed to vary over time. The covariance matrix is different from zero and differs by individuals i and j, and at times t and s.

This is the most general variance–covariance error matrix in which the variance of random errors changes by case or individual i and over time t. In addition, the covariance is different from zero, and its particular value depends on the individual cases i and j. The sub-indexes t and s indicate that the covariance also depends on the two time points when such variance is calculated. All this together means that it is necessary to introduce an additional number of parameters to be estimated in the model. These variance–covariance parameters to be estimated are included in the model in addition to the regression coefficients of the main regression equation. Using the GLS method, one calculates the generalized least squared estimators. In the case of N individuals observed at T times, the maximum number of newly specified parameters in the variance–covariance error structure is calculated by the following formula:

$$\frac{(N \times T)[(N \times T) + 1]}{2}.$$

For example, under the assumption of heteroskedasticity and autocorrelation, in a sample of 20 individual cases or cross-sections at five different time points, the number of parameters to be estimated is 5050. Clearly this is an impossible mission and the model is not identified. There is no solution to such an estimation problem, because one cannot estimate more parameters than there are observations in a sample. Due to the impossibility of estimating so many parameters given the sample size, it is common to think about imposing simple assumptions about the temporal and/or cross-sectional behavior of the random errors in the model. I will present some of the most popular assumptions used in the estimation of coefficients by the GLS method. Which specific error assumptions one chooses to make depends entirely on the data structure and the social process under analysis. In other words, the parameter constraints imposed on the behavior of the error term should be based on both the data and fit of the model, but also the model should ultimately make theoretical sense to both the researcher and the audience intended to be motivated by the study.

I will address the problem of *cross-sectional heteroskedasticity* first (also known as heteroskedasticity across panels or sections). Here, the model to be estimated assumes that the variance of each of the individuals or panels in the sample differs. This assumption is frequent when analyzing macro data at the country, state, or city level, for example, where there exists great variation across individual cases. In the heteroskedastic model, the two main working assumptions are:

$$\mathrm{Var}[u_{it}] = \sigma_i^2$$

$\mathrm{Cov}[u_{it}, u_{js}] = 0$ for all agents $i \neq j$, and for all points of time $t \neq s$,

or in matrix notation, the variance–covariance structure is given by:

$$\Sigma_t = \begin{bmatrix} \sigma_1^2 & 0 & 0 & \cdots & 0 \\ 0 & \sigma_2^2 & 0 & \cdots & 0 \\ \cdots & \cdots & \cdots & \cdots & \cdots \\ 0 & 0 & 0 & \cdots & \cdots \\ 0 & 0 & 0 & \cdots & \sigma_n^2 \end{bmatrix}$$

where the variances of the random errors differ by individual case i, but the model still assumes that there is no correlation in the error from different cases or cross-sections at the same time points or from different cross-sections and different time points. In addition to $k + 1$ parameters to be estimated in the main regression equation (i.e., one for each independent variable plus the constant term), one must estimate N different error variances, one for each individual in the sample.

If using the Stata software program to estimate the GLS regression coefficients, use the command xtgls immediately followed by the name of the dependent variable, followed by the list of independent variables (their computer variable names). One will have the option of specifying additional estimation subcommands or options; such subcommands go after the list of independent variable names, separated by a comma.[9] To specify the heteroskedastic model, the subcommand one must include is `panels(hetero)`. With this subcommand, one assumes the structure of error variance and covariance presented above. Type: "`xtgls lifeexp gdppc pubpc hlthcov, i(cntry) panels(hetero)`" in the Stata Command box once the data set has been read by Stata (to learn how to read any data set into Stata, consult Chapter 1, or the *Stata Manual*). The `i( )` option is used to specify the variable that uniquely identifies each individual case in the sample. In this case, such variable is named CNTRY in Stata representing the country ID variable where each number is used to uniquely identify each OECD country in the sample. So the Stata Command box appears as follows:

■ **Stata Command**

`xtgls lifeexp gdppc pubpc hlthcov, i(cntry) panels(hetero)`

Press ENTER for Stata to calculate and estimate the heteroskedastic model. The window will look something like this (by default, Stata shows the command lines in white and the output of commands in green and yellow):

```
. xtgls lifeexp gdppc pubpc hlthcov, i(cntry) panels(hetero)

Cross-sectional time-series FGLS regression

Coefficients:  generalized least squares
Panels:        heteroskedastic
Correlation:   no autocorrelation

Estimated covariances      =        22      Number of obs      =        88
Estimated autocorrelations =         0      Number of groups   =        22
Estimated coefficients     =         4      Time periods       =         4
                                            Wald chi2(3)       =    320.18
Log likelihood             = -142.7974      Prob > chi2        =    0.0000

     lifeexp |      Coef.    Std. Err.       z    P>|z|     [95% Conf. Interval]
     gdppc   |   .3185234    .1015405      3.14   0.002     .1195077    .5175391
     pubpc   |  -.5083589   1.543434      -0.33   0.742    -3.533434   2.516717
     hlthcov |   .0378577    .0078995      4.79   0.000     .0223749    .0533406
      _cons  |     67.742    .7292319     92.89   0.000     66.31273   69.17126
```

---

[9] For more detailed information on how to use Stata to estimate longitudinal models (beyond the introductory nature of this book), please consult the *Stata Reference Manuals*. Release 9 of Stata includes one book titled *Longitudinal/Panel Data*, which contains all the commands covered in this section of the book. There are also many excellent resources available online on how to get started with Stata. Also see list of references at the end of this book.

From this point on, instead of showing a screen shot of Stata, I will include the command lines, used when estimating several models in Stata, in **bold italics**, followed by the output provided by the software program in non-italics. So the very same heteroskedastic model estimated above (on page 69) would be shown like this:

xtgls lifeexp gdppc pubpc hlthcov, i(cntry) panels(hetero)

```
Cross-sectional time-series FGLS regression

Coefficients:     generalized least squares
Panels:           heteroskedastic
Correlation:      no autocorrelation

Estimated covariances      =         22     Number of obs     =       88
Estimated autocorrelations =          0     Number of groups  =       22
Estimated coefficients     =          4     Time periods      =        4
                                            Wald chi2(3)      = 320.18
Log likelihood             = -142.7974      Prob > chi2       = 0.0000
-----------------------------------------------------------------------
 lifeexp |    Coef.   Std. Err.     z    P > |z|    [95% Conf. Interval]
---------+-------------------------------------------------------------
   gdppc |  .3185234  .1015405    3.14   0.002     .1195077    .5175391
   pubpc | -.5083589  1.543434   -0.33   0.742    -3.533434    2.516717
 hlthcov |  .0378577  .0078995    4.79   0.000     .0223749    .0533406
   _cons |    67.742  .7292319   92.89   0.000     66.31273    69.17126
-----------------------------------------------------------------------
```

where LIFEEXP is the dependent variable that measures population life expectancy; GDPPC is GDP per capita; PUBPC is public health care expenditure per capita; HLTHCOV is health coverage (as a percentage of the population covered for health costs). Again, CNTRY is the country ID variable where each number is used to uniquely identify one OECD country in the sample. It is important to emphasize that the i(cntry) is essential in the command line. The i(cntry) option allows Stata to recognize each of the individuals or panels in the sample; in general, it specifies the variable name that contains the individual or panel information for each observation in the sample. In the case of heteroskedastic models, one could also specify the t(year) option. Such an option provides the time dimension of the data set and specifies the variable name that contains the time period to which each of the observations in the sample belongs to. In the case of models assuming cross-sectional heteroskedasticity and/or cross-sectional correlation (and assuming no serial correlation or auto correlation), the results are the same whether or not one includes the t(year) command in the command line. In the case of models assuming serial correlation though, both the i(cntry) and t(year) options are required in the xtgls command line.

When discussing some of the main results in this longitudinal model, there are two relevant columns to look at in order to see the impact of the different independent variables on life expectancy. The second column of the Stata output, headed by "Coef.," gives the estimated model parameters of interest. The $P > |z|$ column (fifth column) gives the significance of these model parameters. By looking at the output, one can see that the effect of GDP per capita on life expectancy is positive and significant at the 0.01 level. The same applies to health coverage, which has a positive and significant effect on population life expectancy (also significant at the 0.001 level). The variable public health care expenditure per capita does not seem to be a significant predictor of life expectancy (the significance is 0.742). The constant term indicates that the average life expectancy for countries with zero GDP per capita, zero health care expenditure per capita, and zero health care coverage, is 67.7 years. Note that instead of the t-value, Stata now provides z-values when reporting the significance level for each of the model coefficients. This is because when estimating the regression model by GLS (instead of OLS), the z-distribution is the one to be used in this case.

Before I move on to other variance–covariance error matrix structures, let me point out important parts of the Stata output file. First, always check that the number of observations ($N \times T$) as reported by Stata is consistent with the data set. Do the same check for the number of groups (or individuals or units of analysis) and the number of time periods. Looking at the example Stata regression output (on the upper right-hand side), there are 88 observations, with 22 countries observed at 4 points in time:

```
No. of obs        =  88
No. of groups     =  22
No. of time periods =   4
```

In this case, an easy way of checking that the coefficients have been estimated by the GLS method, assuming that panels (or sections) are heteroskedastic, and assuming no autocorrelation, is to look at the following section of output:

```
Cross-sectional time-series FGLS regression

Coefficients:    generalized least squares
Panels:          heteroskedastic
Correlation:     no autocorrelation
```

The above four additional lines in the Stata output file are important in verifying the type of estimated model.

One should always look at the overall goodness-of-fit of the model to make sure that nothing went wrong with the estimation, and most importantly, that the model is significant. In the case of GLS estimation models, look at the *overall χ^2-statistic*. This test allows for testing the significance of all the regression coefficients (excluding the constant term) in the model (similar to the overall *F*-test in OLS). This is why this test is also considered as an overall goodness-of-fit model test. The null hypothesis is the following:

$$H_0: \beta_1 = \beta_2 = \cdots = \beta_k = 0.$$

Theoretically, this goodness-of-fit statistic is distributed as a χ^2 or chi-square distribution with k (i.e., number of independent variables excluding the constant term or β_0) degrees of freedom. A low enough *p*-value (for example, $p < 0.01$) indicates that one can reject the null hypothesis that all regression coefficients are equal to zero (at the 0.01 significance level). In this case, the χ^2-test is 320.18 with three degrees of freedom (one for each of the independent variables to be estimated excluding the constant term). Such a test is significant at the 0.0001 level as the `Prob > chi2` in the output indicates:

```
Wald chi2(3)   =    320.18
Log likelihood = -142.7974
Prob > chi2    =    0.0000
```

One can also test the constraints that the coefficients of PUBPC and HLTCOV (representing the two main health care characteristics) are all zero. One way of accomplishing this is by using the `test` command (one of the many post-estimation commands available in Stata):

test pubpc hlthcov

```
( 1) pubpc   = 0
( 2) hlthcov = 0

        chi2( 2) =   23.46
      Prob > chi2 = 0.0000
```

A more precise test would be to refit the model, apply the proposed constraints, and then calculate the *likelihood-ratio test,* as follows. For the first step, one needs to save the estimates of the current model by using the command `estimates store` followed by the chosen name for the regression equation (in the command line below I use the name "`full`" and I will use this model name later in order to compare models):

xtgls lifeexp gdppc pubpc hlthcov, i(cntry) panels(hetero)
estimates store full

Next, one should fit the constraint model, which in this case is the model omitting the PUBPC and HLTHCOV variables:

```
xtgls lifeexp gdppc, i(cntry) panels(hetero)

Cross-sectional time-series FGLS regression
Coefficients:   generalized least squares
Panels:         heteroskedastic
Correlation:    no autocorrelation

Estimated covariances      =        22    Number of obs      =       88
Estimated autocorrelations =         0    Number of groups   =       22
Estimated coefficients     =         2    Time periods       =        4
                                          Wald chi2(1)       =   357.36
Log likelihood             = -146.2368    Prob > chi2        =   0.0000
------------------------------------------------------------------------
lifeexp |     Coef.   Std. Err.     z    P > |z|    [95% Conf. Interval]
--------+---------------------------------------------------------------
gdppc   |   .3375747  .0178574   18.90   0.000     .3025749    .3725746
_cons   |  70.71668   .1671459  423.08   0.000    70.38908    71.04428
------------------------------------------------------------------------
```

Then, the command `lrtest` can compare the latest estimated model with the previously saved model using the `estimates store` command (remember that it is called "`full`"):

```
lrtest full .

(log-likelihoods of null models cannot be compared)
Likelihood-ratio test                 LR chi2(2)   =     6.88
(Assumption: . nested in full)    Prob > chi2  =  0.0321
```

Comparing results, `test` reported that the two variables were jointly significant at the 0.001 level; however, the `lrtest` reports they are significant at the 0.05 level. Given the quadratic approximation made by `test`, one could argue that `lrtest`'s results are more accurate than using the `test` command. Note that the `lrtest` command compares the last fitted model "." (i.e., the dot specifies the results of the most recent estimated model) with the model stored as "`full`." In other words, in this case, "`full`" is the unconstrained model and "." is the constrained model—this is because the constrained model is equal to the full model when both PUBPC and HLTHCOV are constrained to be equal to zero. In the windows version of Stata, the names in (`Assumption: . nested in full`) are actually links, and one can click on any of the two names, either "." or "`full`," in order to re-display the results of that particular model. Note that I also got the message "log-likelihoods of null models cannot be compared." One can safely ignore this warning message. The "null models" (also known as baseline models) referred to in the message correspond to the constant-only

models (i.e., the models that only estimate the constant term). Some commands compute and temporarily store the likelihood function for the constant-
only model, but `xtgls` does not. When `lrtest` does not find the likelihood
value for the corresponding null model, it produces that warning message.
However, the likelihood ratio test for the full model is correctly calculated and,
therefore, the results testing the constraints that PUBPC=HLTHCOV=0 can
be trusted.

Following a similar procedure, one could easily test for the presence of panel-
level heteroskedasticity. First, estimate the full heteroskedastic model; then type
the `estimates store` command to save the likelihood (just as before). This
time I named the full heteroskedastic model as "heteromodel." Notice that
I used the `quietly` command so that the estimates of the full heteroskedastic
model are not displayed again:

```
quietly xtgls lifeexp gdppc pubpc hlthcov, i(cntry) panels(hetero)
estimates store heteromodel
```

One can now fit the model without heteroskedasticity by typing the same `xtgls`
command as before without the `panels` option as shown below:

```
xtgls lifeexp gdppc pubpc hlthcov, i(cntry)
```

```
Cross-sectional time-series FGLS regression

Coefficients:   generalized least squares
Panels:         homoskedastic
Correlation:    no autocorrelation

Estimated covariances      =          1    Number of obs    =       88
Estimated autocorrelations =          0    Number of groups =       22
Estimated coefficients     =          4    Time periods     =        4
                                           Wald chi2(3)     =   188.52
Log likelihood             = -159.3932     Prob > chi2      =   0.0000
-------------------------------------------------------------------------
 lifeexp |     Coef.    Std. Err.      z     P > |z|    [95% Conf. Interval]
---------+---------------------------------------------------------------
   gdppc |    .390826    .1239565    3.15   0.002     .1478758    .6337762
   pubpc |  -1.464312   1.919982   -0.76   0.446    -5.227406    2.298783
 hlthcov |   .0438394    .0083358    5.26   0.000     .0275016    .0601772
   _cons |   67.17923    .7556205   88.91   0.000     65.69824    68.66022
-------------------------------------------------------------------------
```

Notice that the homoskedastic model is now being fitted (i.e., the one assuming the variance of the errors is the same for all panels or individuals in the sample). There is a trick here in order to compare the heteroskedastic with the
homoskedastic model. Typically, `lrtest` infers the number of constraints when
fitting nested models by looking at the number of parameters estimated. For

`xtgls`, however, the panel-level covariances are not included among the parameters estimated. Therefore, one must specify with `lrtest` how many constraints are to be imposed. In Stata, the number of panels or groups is stored in $e(N, g)$. So given that in the second model, I am constraining all these parameters to be single value, the number of constraints (i.e., $N - 1$) can be computed, stored, and displayed by typing the two following command lines:

```
local df=e(N_g)-1
display `df'
```

The presence of panel-level heteroskedasticity test is then obtained by typing:

```
lrtest heteromodel . , df(`df')
```

```
(log-likelihoods of null models cannot be compared)

Likelihood-ratio test                LR chi2(21)      =    33.19
(Assumption: . nested in heteromodel)
                              Prob > chi2      =  0.0441
```

In this case, the *p*-value is below 0.05 and the null hypothesis that the errors are homoskedastic can be rejected. In other words, the heteroskedastic model seems to fit the data better than the homoskedastic one.

Now, consider another possible specification of the error variance–covariance matrix. A more complex structure than the previous one is one in which the error terms are heteroskedastic with *cross-sectional correlation*. Again, this is a situation in which the errors are assumed to be heteroskedastic and also correlated across individuals or panels. In such a case, the variance structure is specified as follows:

$$\text{Var}[u_{it}] = \sigma_i^2$$

$$\text{Cov}[u_{it}, u_{jt}] = \text{Cov}[u_{jt}, u_{it}] = \sigma_{ij}^2 \text{ for all agents } i \neq j,$$

and for all points of time *t* or in matrix notation:

$$\Sigma_t = \begin{bmatrix} \sigma_1^2 & \sigma_{21}^2 & \sigma_{31}^2 & \cdots & \sigma_{n1}^2 \\ \sigma_{21}^2 & \sigma_2^2 & \sigma_{23}^2 & \cdots & \sigma_{n2}^2 \\ \cdots & \cdots & \cdots & \cdots & \sigma_{n3}^2 \\ \sigma_{n-11}^2 & \sigma_{n-12}^2 & \sigma_{n-13}^2 & \cdots & \cdots \\ \sigma_{n1}^2 & \sigma_{n2}^2 & \sigma_{n3}^2 & \cdots & \sigma_n^2 \end{bmatrix}.$$

Note that now because of the $\text{Cov}[u_{it}, u_{jt}] = \text{Cov}[u_{jt}, u_{it}] = \sigma_{ij}^2$, the error matrix is symmetric, which considerably reduces the number of covariance parameters to estimate in the model (almost by half). So, in addition to the estimation of the $k + 1$ parameters in the main regression equation, I estimate N different variances (one variance parameter for each individual or panel) and also $N(N - 1)/2$ covariances. In total, I am proposing to estimate $N(N + 1)/2$ new parameters. To specify such a model in Stata, one must include the subcommand `panels(correlated)` this time. In the example of the 22 OECD countries used throughout this chapter (88 observations in total), it does not make sense to specify such a model, because the number of parameters to estimate for this procedure would be 253 (and this will mean that there are more parameters than observations and consequently impossible to estimate). This procedure is applicable only when there are a few social units or individuals in the sample.

For the purpose of illustrating an example of cross-sectional correlation error specification, take the sub-sample of OECD countries in Southern Europe, at the time mainly Spain, France, Greece, Italy, and Portugal (five countries or "groups" using Stata terminology). The literal copy of the Stata output is as follows (again the line of commands is included in ***bold italics***):

xtgls lifeexp gdppc pubpc hlthcov if soeurope==1, i(cntry)
t(year) panels(correlated)

```
Cross-sectional time-series FGLS regression

Coefficients:  generalized least squares
Panels:        heteroskedastic with cross-sectional correlation
Correlation:   no autocorrelation

No. est. covariances      =      15     No. of obs          =      20
No. est. autocorrelations =       0     No. of groups       =       5
No. est. coefficients     =       4     No. of time periods =       4
                                        chi2(3)             = 560.60
Log Likelihood            = 61.5999     Pr > chi2           = 0.0000
-------------------------------------------------------------------------
lifeexp |    Coef.     Std. Err.     z    P > |z|   [95% Conf. Interval]
--------+----------------------------------------------------------------
  gdppc |   .5016911    .480748    1.044  0.297   -.4405575    1.44394
  pubpc |  -1.924937   6.767688   -0.284  0.776  -15.18936    11.33949
hlthcov |   .0596876   .0041236   14.475  0.000    .0516055     .0677697
  _cons |  66.02496    .7882638   83.760  0.000    64.48       67.56993
-------------------------------------------------------------------------
```

Notice that to estimate this cross-sectional correlation model, I have included the `if soeurope==1` expression right after the list of independent variables. This `if` expression qualifier, in particular, restricts the scope of any Stata command to those observations with values of the variable SOEUROPE equal to 1. SOEUROPE is a dummy variable that takes the value of 1 if the OECD country is in Southern Europe, otherwise it takes the value of 0. The model is now estimated only for the sample of five OECD countries in Southern Europe.

In this case of heteroskedasticity with cross-sectional correlation, I also specify the `t(year)` option in the command line (even when the results are the same whether or not I include the `t(year)` option).

Notice that the estimated effects of the independent variables in this model are very similar to the estimated effects in the previous model (the one including all 22 OECD countries), with the only difference being that GDPPC now seems to be insignificant. Reducing the sample to five OECD countries (i.e., the ones in Southern Europe) could be the reason that the slope coefficients are now insignificant. In other words, I might have lost some statistical power by reducing the number of cases in the sample under analysis (once again recall that only five OECD countries were analyzed in order to illustrate this heteroskedastic model with cross-sectional correlation).

In Stata, the estimated covariance structure is stored in `e(sigma)`. So if interested in looking at the estimated values of such covariance structure or Σ, one could easily retrieve such an estimated matrix by typing the following command:

`matrix list e(sigma)`

```
symmetric e(sigma)[5,5]
               _ee          _ee2          _ee3          _ee4          _ee5
_ee       .18505516
_ee2     -.00611586      4.265434
_ee3      .56260436     -.52043609      1.8054306
_ee4      .87271199     -2.3133049      3.0335721      5.6921635
_ee5      .27479754      2.9022233       .4645791     -.35543571      2.4201383
```

The dimension of the variance-covariance matrix is 5×5; this is so now because the analyzed sample includes only the five OECD countries located in the South of Europe. Also notice that Σ is symmetric, so that $\text{Cov}\,[\mu_i\,\mu_j] = \text{Cov}\,[\mu_i\,\mu_j]$. This is again an important assumption because it reduces the number of parameters to be estimated.

5.2. Autoregressive Models

As one can imagine, several additional error structures can be specified when estimating constant coefficient models. A more general structure (than the ones presented up to now) can allow for *serial correlation* (or autocorrelation within panels). One popular error structure assumed in empirical studies is the serial correlation that follows what is called an autoregressive process of first order or AR(1). This process occurs when the errors at time t are correlated with the errors in the previous time period $t - 1$, as specified below:

$$u_{it} = \rho_i\, u_{it-1} + \varepsilon_{it} \tag{2a}$$

where $i = 1, 2, 3, \ldots, N$ refers to each of the individuals or panels and $t = 1,$ 2, 3, $\ldots, T$ time periods. The residual ε_{it} is well behaved.[10] This formula shows that the errors at time t can be predicted by the value of such errors at time $t - 1$; all the errors now follow an autoregressive process of first order, where the autocorrelation parameter ρ_i might vary for each individual or panel i in the sample. Such a parameter ρ_i simply measures the effect of the error at time $t - 1$ on the contemporary error at time t.

In the simplest of these autoregressive models, the serial correlation parameter is equal for all individuals or panels i in the sample, and the formula can be rewritten as:

$$u_{it} = \rho u_{it-1} + \varepsilon_{it} \tag{2b}$$

where the parameter ρ is now constant across individuals and over time. In Stata, the xtgls command permits the user to specify this option to control for serial correlation, simply by adding the option corr(ar1) to the xtgls command. This option now introduces the assumption that each individual has a structure of errors that follows an autoregressive process of first order where the autocorrelation parameter ρ is the same for all individuals. Computationally, this means that only one additional parameter needs to be estimated, the auto-correlation parameter, in addition to the regression coefficients in the main model. This allows many more degrees of freedom when estimating the model in comparison with the heteroskedastic models in the section before. Using the example data for 22 OECD countries, the Stata command line and output looks as follows:

```
xtgls lifeexp gdppc pubpc hlthcov, i(cntry) t(year) panels(hetero)
corr(ar1)

Cross-sectional time-series FGLS regression

Coefficients:    generalized least squares
Panels:          heteroskedastic
Correlation:     common AR(1) coefficient for all
                 panels (0.2269)

No. est. covariances      =        22   No. of obs          =        88
No. est. autocorrelations =         1   No. of groups       =        22
No. est. coefficients     =         4   No. of time periods =         4
                                        chi2(3)             =    298.88
Log Likelihood            = -103.2376   Pr > chi2           =    0.0000
```

---

[10] ε_{it} is a white noise process. This is the same as saying that its expectation is 0 for all social units i; that Var$[\varepsilon_{it}] = \sigma^2$ for all social units i, and for all points of time t; and that Cov$[\varepsilon_{it}, \varepsilon_{js}] = 0$ for all agents $i \neq j$, and for all points of time $t \neq s$.

```
-------------------------------------------------------------------------------
lifeexp |     Coef.    Std. Err.     z    P > |z|   [95% Conf.  Interval]
--------+----------------------------------------------------------------------
  gdppc |    .5028704   .1532503    3.281   0.001    .2025053    .8032355
  pubpc |   -2.417245   2.343006   -1.032   0.302   -7.009452    2.174962
 hlthcov |   .0557448   .0079505    7.012   0.000    .0401621    .0713274
  _cons |    66.24101   .7429641   89.158   0.000    64.78483     67.6972
-------------------------------------------------------------------------------
```

This time, in addition to the variables used in the previous Stata command line, `t(year)` is required and therefore added to the `xtgls` command with YEAR being the time variable taking the four numeric values 1960, 1970, 1980, and 1990. It is now assumed that there is serial correlation, meaning that for each social unit the error terms are correlated over time. As I mentioned earlier in this section, in the case of models assuming serial correlation, both the `i(cntry)` and `t(year)` options are required.

One can see that the estimated parameters have much of the same magnitude and significance as the previously estimated heteroskedastic models (notice that the command line also includes the `panels(hetero)` option; one could have estimated the homoskedastic model with serial correlation by omitting the `panels(hetero)` option in the command line). By looking at the "Correlation" line in the Stata output file, it is seen that ρ is 0.2269 (also called the AR(1) coefficient or the autocorrelation parameter). Such a parameter is the same for all "panels" or countries in the sample:

```
Correlation:    common AR(1) coefficient for all panels (0.2269)
```

If it is not clear which structure of the error to specify in the model, a good approach is always to begin with the simplest assumption—that is, the autoregressive process of first order (following the methodology of analysis of time series). One can also start with an autoregressive process of superior order, a moving average process or MA. I have started with the autoregressive process of first order because it is the most commonly applied error model in empirical work. Also, it is easy to estimate using most statistical software programs, including SPSS and Stata. The `xtgls` command in Stata allows the user to specify three options when estimating these models. The first option is a structure with no autocorrelation with the `corr(independent)` (this is the default). The second one is a structure with serial correlation in which the correlation parameter is common for all panels or `corr(ar1)` option (as specified in formula (2b)). The third option is a structure with serial correlation in which the correlation parameter is unique for each panel or `corr(psar1)` (as specified in formula (2a)). One can also investigate other error specifications, especially if suspecting some other type of behavior from the error term. After estimating any regression-like model, one can always display the estimated covariance–variance matrix using the display command in Stata, which can be useful in assessing the fit of the model.

In Stata, if one specifies the command corr(psar1) instead, then one would assume that each social case, unit, or panel follows an autoregressive process of different first order (option three presented in the previous paragraph). So the GLS model is now estimating ρ_i where ρ is different for each individual case i. Thus, N parameters are now being estimated.[11] Below is the example in Stata using the same OECD data set:

```
xtgls lifeexp gdppc pubpc hlthcov, i(cntry) t(year) panels(hetero)
corr(psar1)

Cross-sectional time-series FGLS regression

Coefficients:  generalized least squares
Panels:        heteroskedastic
Correlation:   panel-specific AR(1)

No. est. covariances       =          22    No. of obs          =          82
No. est. autocorrelations  =           5    No. of groups       =          22
No. est. coefficients      =           4    No. of time periods =           4
                                            chi2(3)             =      470.37
Log Likelihood             =   -95.36228    Pr > chi2           =      0.0000

------------------------------------------------------------------------------
 lifeexp |     Coef.    Std. Err.     z     P > |z|   [95% Conf. Interval]
---------+--------------------------------------------------------------------
   gdppc |   .5839556   .1314624    4.442    0.000    .3262941    .8416171
   pubpc |  -3.728153   2.051813   -1.817    0.069   -7.749633    .2933275
 hlthcov |   .0630744   .0059341   10.629    0.000    .0514439    .074705
   _cons |   65.64805   .5679779  115.582    0.000    64.53483   66.76127
------------------------------------------------------------------------------
```

where YEAR is the variable that identifies the time period for each observation. In the case of estimating a model with autocorrelated errors with either corr(ar1) or corr(psar1), the data *must* be scattered equally in time. Note that the following line in the Stata output file indicates that there is now a panel-specific AR(1), that is, one for each country in the sample:

```
Correlation:    panel-specific AR(1)
```

Notice also that PUBPC or public expenditure now seems to have a negative and significant effect on life expectancy although the coefficient is barely significant at the 0.1 level. After any estimation command, one can type ereturn list to have a look at the results that Stata temporarily stores for

---

[11] To learn more about the command xtgls, read the *Stata Reference Manual*.

further computations. The panel-specific AR(1) parameters are stored as `e(rho)` in Stata. So to retrieve the value of such parameters and after fitting the model with `xtgls`, type `display e(rho)` to look at the different "psar1" coefficients:

```
display e(rho)
```

```
.5204901885415697 .4685629443188115 1.082596807239912 −.3486365233156286
.1952773103918368 .58164685286083 98 −.4501217184544233 .5228931409360409
.7117570489720546 −.4819955697158464 −.5059168021688094 −.3024991 270542374
−.1085349961208232 .9
```

Another way to obtain the same results using a different estimation method, the maximum likelihood estimation (ML), is to add the `mle` subcommand after the comma as follows:

```
xtreg lifeexp gdppc pubpc hlthcov, mle
```

Some of the serial autocorrelation in the error terms can be approximated, although not always, with an autoregressive process of first order or AR(1). Nevertheless, given the large amount of distortion in the estimation of standard deviations of the coefficients when using the OLS method in the case of serial autocorrelation, it is advisable to use the maximum likelihood estimation (ML) of these models in which the autoregressive process is estimated together with the main equation of interest. The steps behind this ML procedure, also referred to as *autoregressive models*, are simple to explain. First, the statistical procedure fits a regression model by OLS. Later, the estimated errors are used to find some autocorrelation pattern among the error terms. If u_t is correlated with u_{t-1}, then the process that better describes the behavior of the error is an autoregressive process of first order; if u_t is correlated only with u_{t-2}, the process is an autoregressive of restricted second order, and so on. In general, the autoregressive process can be of order p:

$$u_{it} = \sum_{p=1}^{p} \rho_{pi} u_{it-p} + \varepsilon_{it} \tag{3}$$

where $i = 1, 2, 3, \ldots, N$ individuals or panels and $t = 1, 2, 3, \ldots, T$ time points, and where ε_{it} is well-behaved or white noise. The equation above describes the random error as an autoregressive process of order p, where the parameter of autocorrelation between u_{it} and u_{it-p}, varies depending on the time lag p between any two errors, and depending on whom the individual or panel i is.

Once the autoregressive process is properly identified, the autoregressive regression models incorporate it in the estimation (using the ML estimation method) of the regression model of interest. These methods of maximum probability for autoregressive models can be calculated with Stata and many other

standard statistical software programs. Thus, the command `xtgee` with the
option `corr( )` allows the user to estimate structures of serial autoregressive
error correlation that are more complex than the ones assumed thus far by the
command `xtgls` (GEE refers to generalized estimation equations). The command `xtgee` assumes a common structure of serial correlation for all individuals
(or panels)—that is, it assumes that ρ is the same for all individual cases i. For
example, one can assume an autoregressive model of first order similar to the
one previously estimated with `xtgls,corr(ar1)` (this model is reported in
Table 2.2, columns 5 and 6):

```
xtgee lifeexp gdppc pubpc hlthcov, i(cntry) t(year) corr(ar1)
```

```
    Iteration 1: tolerance = .02919665
    Iteration 2: tolerance = .00242164
    Iteration 3: tolerance = .000573
    Iteration 4: tolerance = .00010089
    Iteration 5: tolerance = .00001702
    Iteration 6: tolerance = 2.853e-06
    Iteration 7: tolerance = 4.775e-07
```

```
General estimating equation for panel data       Number of obs     =         88
Group and time vars:                   id year   Number of groups  =         22
Link:                                  identity   Obs/group, min    =          4
Family:                                Gaussian                avg  =       4.00
Correlation:                             AR(1)                  max  =          4
                                                  chi2(3)           =     155.71
Scale parameter:                       2.540717   Prob > chi2       =     0.0000
Pearson chi2(84):                        213.42   Deviance          =     213.42
Dispersion (Pearson):                  2.540717   Dispersion        = 2.540717
------------------------------------------------------------------------------
 lifeexp |      Coef.    Std. Err.     z    P>|z|    [95% Conf. Interval]
---------+--------------------------------------------------------------------
   gdppc |    .6123653   .1892261    3.236  0.001    .241489     .9832416
   pubpc |   -4.019716   3.009071   -1.336  0.182   -9.917385   1.877954
 hlthcov |    .0587939   .0098313    5.980  0.000    .0395248    .0780629
   _cons |    65.91751   .9005741   73.195  0.000    64.15242    67.6826
------------------------------------------------------------------------------
```

One of the additional advantages of using the command `xtgee` is that the
command `xtcorr` (typed right after the `xtgee` command) displays the estimates of the error correlation matrix at different time points—assuming the
same correlation matrix for all individuals in the sample (in Stata, this is the
estimated *within-id correlation matrix*). The dimension of such matrix is $T \times T$;
this is the matrix containing the variances and covariances among errors at
different points in time for each case or panel (and assuming the same structure
for all panels or cases in the sample). In this last case, after estimating the

Table 2.2
Regression coefficients for several longitudinal models compared with the OLS model coefficients

Independent Variables[a]	OLS Model		Year and Country Fixed Effects Model[b]		Autoregressive Models			
					AR(1)		AR(2)	
	B	St. Error	B	St. Error	B	St. Error	B	St. Error
Constant	67.179***	0.773	66.834***	1.111	65.918***	0.901	65.66***	0.774
GDP per capita (in thousands of $)	0.390**	0.126	0.247	0.151	0.612***	0.189	0.482***	0.170
Health care expenditure per capita (in thousands of $)	−1.464	1.965	−2.454	1.640	−4.02	3.009	−2.412	2.651
Public coverage of health care expenses (% of population)	0.043***	0.009	0.041**	0.010	0.059***	0.010	0.066***	0.009
F-statistic tests								
Compared with baseline model	59.98***		28.10***					
Degrees of freedom	(3 and 84)		(27 and 60)					
χ^2-statistics								
Compare with baseline model[c]					155.71***		238.35***	
Degrees of freedom					3		3	

$N = 88$ observations (countries per year).

Note: The levels of significance are: $*p < 0.05$; $**p < 0.01$; $***p < 0.001$ (two-tail tests).

[a] All independent variables of interest are delayed 5 years.

[b] For presentation purposes, the coefficients for the year and country fixed effects are omitted from the table (although they are included in the model estimation).

[c] The model labeled baseline is a model with no independent variables (only the constant term).

xtgee model, this is the 4 × 4 within-id correlation matrix that I obtained simply by typing the xtcorr command:

xtcorr

```
Estimated within-id correlation matrix R:

          c1        c2        c3        c4
r1     1.0000
r2     0.4087    1.0000
r3     0.1670    0.4087    1.0000
r4     0.0683    0.1670    0.4087    1.0000
```

After careful examination of the error correlation matrix, it is clear that the error correlation declines as the time delays/lags increase (in this case, $t = 4$). This makes sense because it reflects a situation where the most recent past values of any variable generally have a stronger impact on the current values of that variable compared with less recent past values.

One could also produce a simple test for autocorrelation in panel-data models. This time, given that the GLS with an autocorrelation procedure does not produce the maximum likelihood estimates, the likelihood-ratio test procedure cannot be used as it was used in the case of models with heteroskedasticity. However, Wooldridge (2002, pp. 282–283) derived a simple test for autocorrelation in panel-data models. Drukker (2003) provides simulation results showing that the test has good size and power properties in reasonably sized samples. A user-written program, called xtserial was written by Drukker himself to perform this test in Stata. To install this user-written program, type:

findit xtserial
net sj 3-2 st0039
net install st0039

To use xtserial, first run the tsset command to inform Stata about the longitudinal nature of the data set, and then use the xtserial command to specify the dependent and independent variables. In this case, because CNTRY is the country ID variable and YEAR is the time variable, the Stata commands are as shown below, followed by the output:

tsset cntry year
```
        panel variable:   cntry, 1 to 22
         time variable:   year, 1 to 4
```

xtserial lifeexp gdppc pubpc hlthcov
```
  Wooldridge test for autocorrelation in panel data
  H0: no first-order autocorrelation
       F( 1, 21) = 0.144
       Prob > F = 0.7081
```

A significant test statistic indicates the presence of first-order serial autocorrelation, that is, when Prob $> F$ is lower or equal than 0.01 or 0.05, depending on the desired level of significance. The F-statistic does not seem to detect any first-order autocorrelation structure in the OECD longitudinal data.

In the case of an autoregressive model of the second order, the command line in Stata is the following (the results of this command line are reported in Table 2.2, columns 7 and 8):

```
xtgee lifeexp gdppc pubpc hlthcov, i(cntry) t(year) corr(ar2)
```

Remember that the command `xtgee` with the option `corr( )` allows estimating structures of serial autoregressive error correlation more complex than the ones assumed by the command `xtgls`. Notice that typing `corr(ar2)` specifies the second-order component of the autoregressive model. One could also specify the most complete correlation structure of errors in different points in time. To do this, do not specify any error structure, and consequently, all the coefficients of the error correlation matrix with different time lags are then estimated by the statistical software subroutine used. This is the most general type of autoregressive model. In Stata, this is accomplished with the option `corr(unstructured)` in the `xtgee` command, obtaining the following:

```
xtgee lifeexp gdppc pubpc hlthcov, i(cntry) t(year) corr(unstructured)
```

```
Iteration 1: tolerance = .1667933
Iteration 2: tolerance = .03887066
Iteration 3: tolerance = .05103919
Iteration 4: tolerance = .0479788
Iteration 5: tolerance = .04230803
[many other program iterations are eliminated from this output file]
Iteration 59: tolerance = 1.324e-06
Iteration 60: tolerance = 1.084e-06
Iteration 61: tolerance = 8.880e-07
```

```
General estimating equation for panel data    Number of obs     =        88
Group and time vars:                id year   Number of groups  =        22
Link:                               identity   Obs/group, min   =         4
Family:                             Gaussian             avg    =      4.00
Correlation:                    unstructured             max    =         4
                                               chi2(3)           =    433.75
Scale parameter:                    2.644272   Prob > chi2       =    0.0000
Pearson chi2(84):                     222.12   Deviance          =    222.12
Dispersion (Pearson):               2.644272   Dispersion        = 2.644272
```

```
-----------------------------------------------------------------------------
lifeexp |    Coef.    Std. Err.     z     P>|z|    [95% Conf. Interval]
--------+--------------------------------------------------------------------
 gdppc  |  .5082155   .1355443    3.749   0.000    .2425536    .7738774
 pubpc  | -2.725373   2.075234   -1.313   0.189   -6.792757    1.342011
hlthcov |  .0469739   .0057951    8.106   0.000    .0356158    .0583321
 _cons  |  66.99605   .4736671  141.441   0.000    66.06768    67.92442
-----------------------------------------------------------------------------
```

Notice now that the "Correlation" line of the Stata output indicates that the correlation matrix is set to be unstructured and consequently a different parameter is estimated for each of the cells in the within-id variance–covariance matrix.[12] As before, the estimated error correlation structure can be retrieved by typing the xtcorr command as follows:

xtcorr

```
Estimated within-id correlation matrix R:

          c1        c2        c3        c4
r1    1.0000
r2   -0.1035    1.0000
r3    0.3273    0.7730    1.0000
r4    0.0957    0.6975    0.5692    1.000
```

The within-id correlation matrix could also have been displayed using the estat wcorrelation command right after using the xtgee command to estimate the model. This correlation matrix is quite different from the previous one in which an autoregressive process of first order was assumed (on page 82). Notice that all the possible error correlations over time are taken into consideration when estimating the main equation model. It shows that the serial correlation of the errors decrease as the time lag increases, although residuals separated by small lags seem to be more correlated than, for example, an AR(1) would imply. This makes sense, although it might not always be the case.

Table 2.2 shows the results of several autoregressive models I estimated using the OECD health care data set. All such models are compared to the basic (non-dynamic or static) linear regression model estimated by OLS and the year and country fixed effects model (that I describe later in detail in Section 6 of this chapter). If the specified autoregressive model is the correct one, then the estimation problems associated with the estimation by OLS disappear simply because the right serial (temporal) and cross-sectional behavior of the error terms is accounted for. The techniques presented here will provide better estimators, that is, model parameters that are more precise (better standard deviations) and more reliable overall goodness-of-fit model statistics and independent variable significance tests. Consequently, these models will help to perform better testing of research hypotheses regarding the impact of some independent variables of interest on the dependent variable over time. For additional information about GLS and generalized estimating equations (GEE), see the additional reading suggestions in the list of references.

---

[12] Again, this is the correlation matrix containing the variances and covariances among errors at different points in time within each case or panel, assuming the same correlation matrix for all panels in the sample.

5.3. Models with the Lagged Dependent Variable

One of the main problems of the autoregressive models is the difficulty in interpreting each of the ρ_i parameters of autocorrelation in a substantive way. This has been one of the main reasons why some researchers have long preferred regression models that include lags of the dependent variable as independent variables in the regression equation. These models are the so-called *dependent variable lagged models* or *linear models with the lagged dependent variable*. According to the most basic specification of these models, the lagged dependent variable y_{it-1} (lagged one time unit) is included in the main equation as an independent variable. Sometimes more lags of the dependent variable are included—especially after assuming that the temporal correlation is an issue and therefore needs to be modeled in the main regression equation. In the simplest case, when including the first time lag of the dependent variable, the model to estimate is the following:

$$y_{it} = \beta_0 + \eta y_{it-1} + \sum_{k=1}^{K} \beta_k x_{kit} + u_{it} \tag{4a}$$

where again $k = 1, 2, 3, \ldots, K$ is the number of different independent variables of interest, $i = 1, 2, 3, \ldots, N$ is the number of different individuals or panels in the sample observed at $t = 1, 2, 3, \ldots, T$ time points. More complicated versions of model (4a) also include several lagged independent variables, x_{kit-1}.

In the models with the lagged dependent variable as summarized in (4a), y_{it}, the dependent variable whose variation will be modeled, is also known as the *endogenous variable*. y_{it-1} is the lagged endogenous variable (lagged one period of time); x_{kit} is any of the explanatory or exogenous variables whose lags can also be included in the causal model if there is some methodological as well as substantive reason for doing so; and finally, u_{it} is the error term in the model. This regression equation is usually used to overcome the problem of error correlation associated with the OLS estimation technique of a longitudinal data set. In addition, when y_{it-1} is included in the model, one is controlling for the effect that the previous values of the endogenous dependent variable has on its contemporary value. This is a way of estimating a growth model from time $t - 1$ up to time t. Under these lagged dependent variable models, the effect of each of the independent variables on the dependent variable of interest can be estimated more accurately and precisely than in the classical OLS regression model.

The coefficient associated with the lagged dependent variable y_{it-1} is denoted by η. This coefficient is also called the "rate of discount," that is to say, the rate at which previous values of the dependent variable y_{it} influence its current values. These models are also called *growth models*. This is because if the rate of discount is assumed to be one and the lagged dependent variable is moved to the right-hand side of the equation, one would model growth in the dependent variable, or:

$$y_{it} - y_{it-1} = \sum_{k=1}^{K} \beta_k x_{kit} + u_{it}. \tag{4b}$$

This lagged dependent variable model has been extremely popular in longitudinal empirical studies, especially panel data studies. In my opinion, these models have been estimated in practice without worrying too much about the serious problems associated with the use of this technique. First, as with the OLS parameter estimators, models with some lag dependent variables are usually unstable, meaning that the model coefficients estimates are quite sensitive to the individual cases included in the sample under analysis. Thus, such model parameters may take different values depending on the sub-sample that is being analyzed. Second, despite the inclusion of y_{it-1} in the model, this is not necessarily solving the problem associated with the serial autocorrelation of the random error. To be able to do so, it is still necessary to specify an autoregressive process that accounts for the potential heteroskedastic autocorrelated behavior of the random disturbances. Finally, the estimation of parameters in models with lags of the endogenous dependent variable introduces an important source of bias, which may complicate the estimation of the model. Such a complication usually happens because one of the main assumptions behind the basic OLS regression model is being violated—i.e., the absence of some significant correlation between each of the explanatory variables and the error term. In these models, the lagged dependent variable, as it appears in the right-hand side of the regression equation, could be naturally correlated with the error term. Putting y_{it-1} in the right part of the equation, since it is correlated with the error term, can make the estimation of the model considerably more difficult. Mainly because the effect of y_{it-1} is endogenous and is correlated with the error term, the matrix of variances and covariances is singular, and consequently its inversion is impossible.[13]

A commonly used dynamic panel-data model that permits past values of the dependent variable to affect its current value is the model estimated in Stata using the `xtabond` command. Such a command fits a dynamic panel-data model using the Arellano–Bond estimator, which allows the efficient use of lagged levels of the dependent variable and lagged levels of the predetermined independent variables. For additional information about this model in Stata, look for the `xtabond` command in the *Stata Reference Manual*.

---

[13] For more information, see Greene [1997].

5.4. Models with Instrumental Variables

Many times, it makes theoretical sense to include lags of the dependent variable in the regression models (especially in the case of the modeling growth of the dependent variable). In these situations, and in order to overcome the complications associated with these lagged dependent variable models (described in the previous section), many researchers have decided to go with the estimation of such models in two stages, the so-called *two-stages regression model*. These models have also been commonly denominated as models with *instrumental variables*. The parameter estimators in two stages consist of considering a variable $I_{it}*$ or instrument, first in the equation estimation. The instrument is a linear combination of a vector of independent variables Z_{it}:

$$I_{it}* = a + \sum_{r=1}^{R} b_r z_{kit} + u_{it}. \tag{5}$$

The main reason to use this instrument I is because it is highly correlated with y_{it-1}, so one can include the instrument afterwards instead of the lag dependent variable (which is highly correlated with the errors) when estimating the coefficients in the main equation. In other words, the instrument is meant to solve some of the problems associated with the inclusion of the lagged dependent variable as in independent variable in the linear regression model (identified in the previous section). Now the previous equation model is specified as follows:

$$y_{it} = \beta_0 + \eta I_{it}* + \sum_{k=1}^{K} \beta_k x_{kit} + u_{it} \tag{6}$$

where $I_{it}*$ is the instrument as predicted by the equation estimated at stage one or equation (5). At the second stage of this estimation process, the second equation is estimated (6), where the instrument is now one of the independent variables. To solve the problems associated with the model successfully, including the lagged dependent variable as an independent variable, the instrument variable should meet two key conditions. First, the instrument variable must be orthogonal to the error term of the equation, u_{it}, that is, it should not be correlated with the error term at all. Second, the instrument must be a variable that is highly correlated with the lagged endogenous variable for this variable to be a good instrument. If the instrument variable satisfies both of these two conditions, a unique solution for the regression model can be found and this method is then highly recommended.

In Stata, one can estimate models with instrumental variables using the `xtivreg` command. This command offers five different estimators when fitting longitudinal models in which one of the right-hand side independent variables is an instrument variable. These estimators are two-stage least squares. The following is the general form of this command:

```
xtivreg lifeexp hlthcov (pubpc = gdppc), i(cntry)

G2SLS random-effects IV regression      Number of obs      =        88
Group variable: cntry                   Number of groups   =        22

R-sq: within  = 0.7866                  Obs per group: min =         4
      between = 0.2739                                avg =       4.0
      overall = 0.6385                                max =         4

                                        Wald chi2(2)       =    240.18
corr(u_i, X) =       0 (assumed)        Prob > chi2        =    0.0000
------------------------------------------------------------------------
 lifeexp |    Coef.    Std. Err.      z     P > |z|  [95% Conf. Interval]
---------+--------------------------------------------------------------
   pubpc |   4.416291   .3956054    11.16   0.000    3.640919   5.191664
 hlthcov |   .0564502   .0092477     6.10   0.000    .0383251   .0745754
   _cons |   66.60325   .810031     82.22   0.000    65.01562   68.19088
---------+--------------------------------------------------------------
 sigma_u |  .90274157
 sigma_e |  1.1991853
     rho |  .36171629  (fraction of variance due to u_i)
------------------------------------------------------------------------
Instrumented:  pubpc
Instruments:   hlthcov gdppc
```

where, for the purposes of illustrating how this command works in Stata, I have modeled PUBPC as a function of GDPPC. PUBPC is the instrument variable in this example, which is why it is included in the xtivreg command as (pubpc = gdppc) in brackets. Notice how both independent variables are extremely significant in predicting life expectancy. This time PUBPC is positive and significant.

A clear advantage of these longitudinal models using instrumental variables is that their estimation is easy with OLS. Understandably, the estimated parameters are better the higher the correlation between the instrumental variable and the lagged endogenous variable. There exist, nevertheless, a few major shortcomings worth mentioning here. One obvious difficulty is finding an instrument that is statistically appropriate and that also has a substantive reasonable interpretation. For theoretical reasons, sometimes it does not make sense to include an instrument that does not have any theoretical foundation or reason to be included in the model equation. Another problem concerns the nature of the selected instrument variable. The instrument must be a linear combination of some explanatory variables that do not appear in the main regression model (6), so a different vector of independent variables, in this case Z_{kit}, is used to predict such an instrument variable. When the explanatory variables that compose the instrument also appear as explanatory variables in the main equation, the use of such an instrument could generate some multicolinearity, violating one of the other main assumptions behind the basic OLS regression equation. This occurs when one of the independent variables included as explanatory in the main model is a linear combination of a subset of other independent variables. Moreover, in practical terms, finding variable candidates to predict such an instrument, which are not already included in the main equation, is typically quite challenging. It is unusual to have a sufficient number (at least one) of different explanatory variables to be able to generate such an instrument and to avoid the risk of multicolinearity.

The longitudinal regression models with at least one lagged endogenous variable or with an instrument variable are often used (and recommended) as an alternative procedure to the autoregressive models described above. This is so even when there is no guarantee that the procedure of instrumental variables corrects for the presence of autocorrelation. The reason is that those models are far more intuitive with some appealing theoretical and substantive interpretations. Furthermore, if such an instrument is appropriately identified, these models are preferred over any others explained in this chapter because of their simplicity and yet, substantive importance when it comes to modeling growth in the dependent variable. When such models are estimated, I highly recommend accounting for the presence of certain error correlations in the estimation of such models (as I suggested doing using Stata earlier in this chapter).

5.5. The Basic ARIMA Models

Based on the work of Box–Jenkins (1976), a certain type of model, called an ARIMA model, has recently become popular in the analysis of longitudinal data. To explain the behavior of a dependent variable, its own past observations are used in the model before incorporating the effect of any other independent or explanatory variables. The dependent variable in these models is usually referred to as an *endogenous variable*, a common name for a variable influenced by other exogenous or independent variables in this type of methodology. Although advances in the incorporation of such techniques are being developed, ARIMA models have been used almost *exclusively* for the analysis of time series— remember that this is when the data set contains several time observations for one social unit or case. The basic principle behind any of the ARIMA models (and the Box–Jenkins methodology in general) is that the effect of a set of explanatory variables *x* on *y* must be considered *after* one has been able to control for the previous past effects of the dependent (endogenous) variable on itself. By doing this, one examines to what extent some covariates explain variation in the endogenous variable that is not explained by variation in the previous values of the endogenous variable. The Box–Jenkins methodology highlights the necessity of finding out and accounting for the process that autogenerates (and explains) the variation of the endogenous variable over time.

An important requirement for this technique to be appropriate is that all the variables to be included in the regression model must be *stationary*. Using Box–Jenkins' 1976 definition, a stationary series has a constant mean level and a constant dispersion or variance around that mean level. A more intuitive definition of such stationary is that the variable does not have any time trend—i.e., the variable has no systematic change in the average of the values of the variable over time (a fixed mean) and no cyclical variations of such a variable over time (a constant variance).

Consequently, before including any independent variables in the multivariate regression model (also called, following the Box–Jenkins terminology, the "transfer function"), such variables must attain stationary. The purpose of transforming such time series into being stationary is to eliminate any time trend so that they have a constant average and a constant variation. If the variable is not stationary, it should be transformed. Its time trend can be eliminated by differentiating the variable, that is to say, by taking the difference between the current and the past value of such a variable. For example, the first difference of y_t (or ∇y_t) is $y_t - y_{t-1}$. The second difference (or $\nabla^2 y_t$) is $\nabla y_t - \nabla y_{t-1}$. Normally, any linear average trend over time of any variable y_t disappears by taking its first difference, whereas the quadratic trend disappears with the second difference. Using Box–Jenkins' terminology, the non-stationary variables that can be transformed into stationary by means of their differences of order d are known as homogenous processes of order d. If the variable does not show evidence of being stationary because its variance is not constant, the habitual procedure to stabilize such variance is to transform the values of the variable by taking their logarithms—and sometimes, by taking the percentage change in the variable of interest.

With the plotting of the variable over time, one can easily determine whether the variable is stationary (in mean and variance). If the graph shows that the variable has a trend over time and/or shows fluctuations that increase over time or in certain periods of time, then that can be enough evidence to reject the assumption of a stationary variable. It would then be necessary to proceed with the suggested transformation of such a variable before including it in the transfer function or multivariate analyses. A much more careful way of finding whether the variable is stationary is by looking at the so-called *autocorrelation* and *partial autocorrelation functions*. This approach is beyond the scope of this manual; for more information consult the ARIMA models section in the list of references at the end of this book.

Once the transformed variables or time series are stationary, the Box–Jenkins methodology is used to identify the best ARIMA model by which the endogenous variable can be explained by its past values. The Box–Jenkins methodology focuses on precision in estimating the process that autogenerates the variation of the endogenous variable over time. This is also accomplished by examining the autocorrelation and partial autocorrelation functions and graphs. These functions can help in choosing between an autoregressive process or a moving average process. The notation is AR(p) for an autoregressive process of order p and MA(q) for the moving average process of order q. So that the AR(p) process that generates y_t is the following:

$$y_{it} = \phi_1 y_{it-1} + \phi_2 y_{it-2} + \cdots + \phi_p y_{it-p} \tag{7}$$

where y_{it} is a stationary time series and y_{it-s} are the past values of y_{it} at time $t - s$. The MA(q) process that generates y_{it} is specified as follows:

$$y_{it} = \varepsilon_{it} - \theta_1 \varepsilon_{it-1} - \theta_2 \varepsilon_{it-2} - \cdots - \theta_q \varepsilon_{it-q} \tag{8}$$

where y_{it} is a stationary time series and ε_{it} is a white noise error component, and ε_{it-r} are the past values of ε_i at time $t - r$.

ARIMA refers to the autoregressive process of order p (or AR(p)) and the moving average process of order q (or MA(q)) that jointly generate y_{it}. An ARIMA model (p, q) in general has the following functional form:

$$y_{it} = \delta + \phi_1 y_{it-1} + \phi_2 y_{it-2} + \cdots + \phi_p y_{it-p}$$
$$+ \varepsilon_{it} - \theta_1 \varepsilon_{it-1} - \theta_2 \varepsilon_{it-2} - \cdots - \theta_q \varepsilon_{it-q} \tag{9a}$$

where p and q denote, respectively, the orders of the autoregressive and the moving average components of the ARIMA model. Thus, the simpler ARIMA structure is the model ARIMA(1,1):

$$y_{it} = \delta + \phi y_{it-1} + \varepsilon_{it} - \theta \varepsilon_{it-1} \quad |\phi| < 1 \text{ and } |\theta| < 1 \tag{9b}$$

where the order of the autoregressive component is one as is the order of the moving average component. In this case, the ARIMA structure has four parameters to be estimated: the constant δ, the AR parameter ϕ, the MA parameter θ, and the variance of ε.

After the ARIMA model is estimated by the maximum likelihood procedure, a few available diagnostics can help in checking that the model fits the data satisfactorily by looking at both the significance of the coefficients in the ARIMA model and by looking at some overall goodness-of-fit statistics. In addition to these tests, one should carefully examine the random errors of the chosen ARIMA model and verify that these errors are "well-behaved" or "white noise," that is, normally distributed with a mean of zero and normal variance. The errors should not be correlated with each other over time either. Normally, a white noise process is denoted with ε_{it}.[14] To determine that the error term is white noise, again, one has to look at its autocorrelation and partial autocorrelation

---

[14] Remember that the white noise feature of the error is what guarantees that the OLS estimators are the linear estimators of minimum variance. When such errors are not white noise, then one of the basic assumptions of the classic model of regression has been violated and consequently the estimators are unreliable.

functions (and graphs). The Box–Ljung statistic is also a very useful and common diagnostic tool to test whether the errors of a given ARIMA model are well-behaved. It is a χ^2-test and the null hypothesis is that the error term is white noise. If there is any evidence that the error term does not behave like white noise, then the ARIMA model has not been specified properly and needs to be re-specified and estimated again using a different order. Typically, this is done by increasing the order of either its autoregressive component (AR) or of its moving average component (MA). Sometimes the proper specification of the model can be achieved only by increasing the order of both components.

Only after the best ARIMA model has been specified, estimated, and checked for its appropriate specification of the errors, can one begin controlling for the sources that generate current values of the endogenous variable from its past values. It is then appropriate to introduce the vector of explanatory variables X in the model to explain variation in the endogenous variable in the transfer function (which is simply a causal model).[15] If the coefficient associated to a given variable x_k is of great magnitude and significance, it then helps to explain some variation in the endogenous variable y_{it}, and one can conclude that variable x_k has a significant exogenous influence on the endogenous variable y_{it}.

The Box–Jenkins methodology highlights the importance of estimating the most parsimonious model, that is to say, the simplest ARIMA model with the smallest AR or MA orders *and* whose error term follows a white noise process. In the end, the final regression model has the following form:

$$y_{it} = \beta_0 + \sum_{k=1}^{K} \beta_k x_{kit} + \text{ARIMA}(p,q) \tag{10}$$

where $k = 1, 2, 3, \ldots, K$ is the number of independent variables of interest, $i = 1, 2, 3, \ldots, N$ is the number of individual cases, and $t = 1, 2, 3, \ldots, T$ is the number of time points.

As a researcher, one should decide the most parsimonious ARIMA process, by examining the autocorrelation and partial autocorrelation functions (and graphs) for the dependent variable under study. After the ARIMA process has been identified, the process is incorporated into the transfer function (as presented earlier in this section), and the entire model as formulated above is estimated by maximum likelihood. The estimated model is evaluated by looking at the significance level of the model coefficients. One must ensure that the square of the average error term of the final model is less than the square of the average error term of only the ARIMA process (i.e., the model that does not include any explanatory

---

[15] The structure of lags for the x_t effects is determined by the use of the cross-correlation function, which shows the correlation between the variation of the variable y_t (not generated or explained by its past values) and the variation of the explanatory variables (and their different time lags).

variables). A very common technique to evaluate the fit of the ARIMA model is the value of the Akaike Information Criterion (AIC) of the final model. This statistic is a goodness-of-fit indicator and is equal to minus two times the log-likelihood function plus two times the number of free parameters in the model, or: $-2 \log. L_0 + 2c$ where c is the number of parameters in the model. The lower the AIC, the better the fit of the model. Remember though, the most important requirement in the ARIMA model is that the errors of the final model are well-behaved with an average of zero, a constant variance, and no autocorrelation.

The estimation of these ARIMA models is easier with certain statistical packages than with others. For example, with SPSS, graphs can be easily implemented in order to determine whether any time series is stationary. SPSS can transform such variables and make them stationary. In the main *Pull*-down menu of commands in SPSS and with the following sequence of options: *Graphs/Time Series/Autocorrelations,* one can easily graph the autocorrelation function, and partial autocorrelation function for any variable in a matter of seconds. This is how the SPSS command box looks if one wants to create the autocorrelation function for variable WAGE:

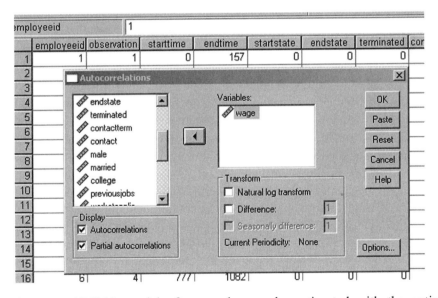

Also, any ARIMA model of any order can be estimated with the option *Analyze/ Time Series/Arima* (the user has many other choices under this option). To diagnose the appropriateness of the model, the SPSS program permits the creation of a variable that stores the error terms of each of the ARIMA models specified and estimated. This can easily aid in evaluating the autocorrelation functions later. In addition, one can calculate the Box–Ljung statistic to test whether residuals are white noise. In SPSS, this is the way the *Arima* box looks, where now WAGE, the dependent variables is modeled as a function of PREVIOUSJOBS (measuring previous months of employee experience) and the dummy variable MALE.

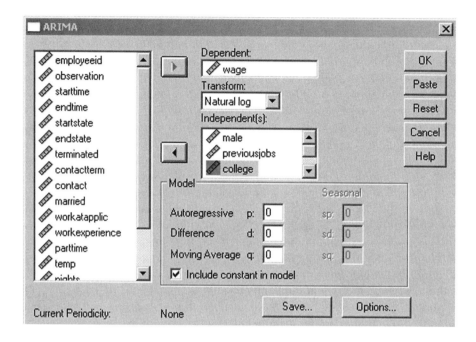

In sum, the Box–Jenkins technique permits estimating longitudinal models with great predictive power of some endogenous (dependent) variables in the case of time series. Nevertheless, there is not an easy way to apply such a technique to the analysis of cross-sectional cross-time data. In addition, the results of such models are notoriously complex to present and interpret in empirical studies. They have to be estimated and evaluated carefully (as explained above) and it is common for results to vary widely depending on the order of the ARIMA model included in the transfer function. Similar models that simply include lags of the endogenous variable y_{it} can sometimes be easier to estimate and interpret. An important limitation is that if there are not enough observations in the data, the ARIMA models might be difficult to estimate—this is why it is recommended that researchers analyze time series that have at least 50 time observations. This ARIMA method also requires the discrete time series data to be equally spaced over time and that there are no missing values in the time series. To learn more about ARIMA models beyond the scope of this book, I suggest a few advanced books in the list of references. To learn about the Stata commands for time series data, check the *Time Series Manual*. The first section of the manual covers univariate time series analysis; the second part covers multivariate time series.

6. Fixed Effects Model

The *constant coefficients models* reviewed in the previous section of this chapter assume that the coefficients do not differ by cases or panels in the sample. They also assume that such coefficients are time invariant—even though constant variance or the absence of serial correlation is not necessarily assumed. Although these models are simple enough and therefore appealing to use, the truth is that their listed model assumptions are quite restrictive empirically and theoretically speaking. Suppose a panel of data with a small temporal component but a high number of cases within each cross-section (i.e., the dominant component is the cross-section). In such a case, it could be of great interest to investigate if the coefficients of the regression model change for each of the time periods in the sample (though remain the same for all individuals or panels whether countries, organizations, or people). Alternatively, in the case of a longitudinal data set where the dominant component is time, one could be interested in investigating if the regression coefficients are different for each individual or case even if such coefficients are time-invariant.

The *fixed effects model* is the basic type of model based entirely on the presumption that the model coefficients (in particular, the constant or independent term in the regression model) may vary depending on the individual/panel or the time point under study. The fixed effects model thus easily allows examination of both the temporal and the cross-sectional variation in the dependent variable by including different constant terms in the equation model to be estimated. This is equivalent to treating the time and panel effect differences on the dependent variable as if they were deterministic.

In the case of panel data whose cross-sectional component is dominant (few social individuals observed at many points of time), the model can account for the existing variation in the sample by including $N - 1$ dichotomous variables d_i, whose corresponding coefficients are named α_i in the model—that is, one variable for each individual, except the individual or panel of reference, which is omitted from the main model. The variable d_1 thus takes the value of 1 for all observations of individual 1 and 0 for the rest of observations in the sample; variable d_2 takes the value of 1 for all observations available for individual 2 in the sample and 0 otherwise, and so forth. In the most general case, the variable d_i takes the value of 1 for the individual i's observations in the sample and 0 for the rest. The inclusion of such coefficients α_i is what allows the regression model to capture any variation in the constant β_0 across individuals or panels in the sample. Therefore, the effect of the main independent variables of interest is evaluated once individual differences in the constant term have been taken into consideration.

Similarly, the model can be specified so that it accounts for variation in the dependent variable entirely due to the passage of time. This can be easily accomplished with the inclusion of a set of dichotomous variables t_t with coefficients ϕ_t associated with each of the $T - 1$ dichotomous variables that are assumed to take the value of 1 at time t, and the value of 0 for the remaining time values. So, this model is helpful when the time component is dominant. An alternative model when the time component is the dominant one is to include time t as an additional continuous independent variable in the regression model to account for changes in y_{it} over time. This is very common especially when time t is a cardinal variable measuring duration.

The most generic case of a regression model is the following:

$$y_{it} = \beta_0 + \sum_{k=1}^{K} \beta_k x_{kit} + u_{it} \tag{1a}$$

where y_{it} is a linear function of K explanatory variables, $i = 1, 2, 3, \ldots, N$ social units, and $t = 1, 2, 3, \ldots, T$ time points. One could assume that the error term has the following structure in the case of a cross-sectional cross-time longitudinal data set:

$$u_{it} = \alpha_i + \phi_t + \varepsilon_{it} \tag{2}$$

where $\alpha_i = \sum_{i=1}^{N-1} \alpha_i d_i$ and $\phi_t = \sum_{t=1}^{T-1} \phi_t t_t$, so that with α_i, a series of $N - 1$ dichotomous variables is incorporated in the regression model for the purpose of controlling for the effect of each individual on the dependent variable. With ϕ_t, a series of $T - 1$ dichotomous variables is introduced in order to control for the effect of time on the dependent variable. Notice that the error u_{it} is no longer random; it now has an *individual fixed* component α_i that is invariable over time but varies across individuals. It also has a *temporal fixed* component ϕ_t that is invariable across individual cases or units, even though it does vary over time. Finally, u_{it} has a component ε_{it} that is completely random. In this sense, ε_{it} is the error term with the usual properties associated with a white noise process—i.e., the errors follow a normal distribution with a mean of zero, are homoskedastic, with constant variance, and are not correlated with the independent variables.

The final form of the fixed effects regression model can be specified as follows:

$$y_{it} = \beta_0 + \alpha_1 d_1 + \alpha_2 d_2 + \cdots + \alpha_N d_N + \phi_1 t_1 + \phi_2 t_2 + \cdots + \phi_T t_T$$

$$+ \sum_{k=1}^{K} \beta_k x_{kit} + \varepsilon_{it} \tag{1b}$$

or, in a much simpler version:

$$y_{it} = \beta_0 + \sum_{i=1}^{N-1} \alpha_i d_i + \sum_{t=1}^{T-1} \phi_t t_t + \sum_{k=1}^{K} \beta_k x_{kit} + \varepsilon_{it} \tag{1c}$$

The matrix formula for the same regression model is:

$$y_{it} = \alpha_i + \phi_t + \beta_i' X_{it} + \varepsilon_{it} \tag{1d}$$

where $i = 1, 2, 3, \ldots, N$ and $t = 1, 2, 3, \ldots, T$ and ε_{it} is "well-behaved."

Note that the structural differences among cases or panels are taken into consideration by including $N - 1$ additional independent terms (from α_1 to α_N, with the term α_i being the specific independent term for each individual in the sample except for the reference individual or panel). The time structural differences are taken into consideration by adding $T - 1$ independent terms (from ϕ_1 to ϕ_t, with a different term ϕ_t for every point in the time for which there is an observation except for the reference time, which can be chosen by the researcher). Models (1c) and (1d) are estimated by OLS, with $N + (T - 2)$ coefficients to be estimated in addition to the independent term β_0 and the k-slope coefficients or parameters of interest. β_0 is the independent term for the individual or panel whose term α has been omitted, at the point of the time whose term ϕ has also been omitted. The parameters α_i are the differences between the independent terms for each individual and β_0 (at any point in time), while the parameters ϕ_t are the differences between the independent terms for each time point and β_0 (for any given individual). This fixed effects model can be easily extended to also include interactions between individuals and temporal periods.

Table 2.3 reports the results of several of these fixed effects models compared with the classical OLS linear regression model using the OECD health care data. Remember that the data consist of 88 cases ($N = 22$ countries of the OECD, and $T = 4$). The main purpose of this regression analysis is to evaluate the effect of the health care expenses as well as the level of health cost coverage on population life expectancy for a given country. These regression models were calculated using SPSS this time (for illustration purposes). The first two columns of Table 2.3 display the OLS regression results (coefficients and standard errors). The other three double columns present three different fixed effects models. The first one is the fixed effects model with a set of time dummy variables only (i.e., the time fixed effects model). The second one is the fixed effects model with a set of country dummy variables only (i.e., the country fixed effects model). Finally, the third model is the fixed effects model with both time and country dummy variables (i.e., the country and time fixed effects model). To estimate these regression models in SPSS, I simply went to the *Analyze* pull-down menu, selected the *Regression/Linear* option, and entered the dependent variable LIFEEXP (in the Dependent box category), and the independent variables of interest plus the different dummy variables (in the Independent(s) box category). Right before hitting the *OK* button, the

Table 2.3
Regression coefficients for several fixed effects models compared with the OLS model coefficients

Independent Variables[a]	OLS Model (Model 1)		Fixed Effects Models					
			Time Fixed Effects		Country Fixed Effects		Time and Country Fixed Effects	
	B	St. Error	B	St. Error	B	St. Error	B	St. Error
Constant	67.179***	0.773	67.101***	0.736	64.186***	1.092	66.834***	1.111
GDP per capita (in thousands of $)	0.390**	0.126	0.289*	0.141	0.449**	0.1403	0.247	0.151
Health care expenditure per capita (in thousands of $)	−1.464	1.965	−0.619	1.799	−2.792	2.115	−2.454	1.640
Public coverage of health care expenses (% of population)	0.043***	0.009	0.036***	0.008	0.067***	0.0104	0.041**	0.010
Year 1970			2.002***	0.448			2.153***	0.355
Year 1980			1.098	0.689			2.044*	0.807
Year 1990			1.707	1.174			3.830*	1.492
Austria					−0.722	0.796	−0.984	0.618
Belgium					0.374	0.811	0.015	0.633
Canada					1.288	0.813	1.187	0.626
Denmark					1.141	0.795	1.083	0.612
Finland					0.064	0.807	−0.356	0.636
France					0.784	0.811	0.618	0.625
Germany					−0.104	0.806	−0.276	0.621
Greece					3.033***	0.842	1.439	0.883
Iceland					1.341	0.814	1.317*	0.627
Ireland					0.764	0.809	0.014	0.706

Italy	0.173	0.806	−0.063	0.626
Japan	1.594*	0.792	1.449*	0.612
Holland	3.400***	0.829	2.720***	0.665
New Zealand	−0.043	0.806	−0.257	0.630
Noruega[a]	2.346**	0.837	2.194***	0.649
Portugal	−0.608	0.862	−2.174*	0.869
Spain	2.203**	0.826	1.079	0.751
Sweden	2.350*	0.912	2.328**	0.702
Switzerland	1.044	0.808	1.240	0.648
United Kingdom	−0.080	0.791	−0.166	0.610
United States	3.268**	1.099	2.109*	0.889
Adjusted R^2	0.67	0.73	0.82	0.89
F-statistics:				
Compared with baseline model[b]	59.98***	40.01***	17.56***	28.10***
Degrees of freedom	(3 and 84)	(6 and 81)	(24 and 63)	(27 and 60)
Compared with OLS model (model 1)		8.78***	4.21***	9.92***
Degrees of freedom		(3 and 81)	(21 and 63)	(24 and 60)

$N = 88$ observations (countries per year).

Note: The levels of significance are: $*p < 0.05$; $**p < 0.01$; $***p < 0.001$ (two-tail tests).

[a] All independent variables of interest are delayed 5 years.

[b] The model labeled baseline is a model with no independent variables (only the constant term).

SPSS screen looked as follows—in the case of the first fixed effects model (i.e., the one with a set of time dummy variables only, namely YEAR1970, YEAR1980, and YEAR1990, given that YEAR1960 was taken as the reference category):

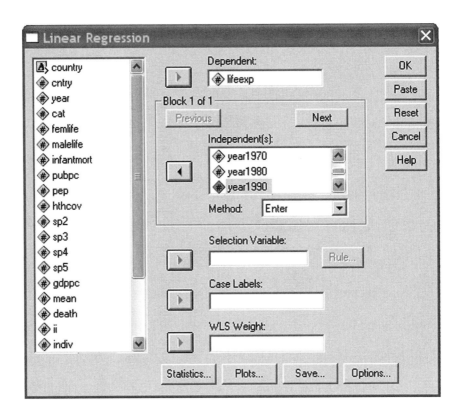

Similar results are obtained when these models are estimated using Stata. For the case of the model of country fixed coefficients model, the command line to use is the following (in ***bold italics***):

xtreg lifeexp gdppc pubpc hlthcov, i(cntry) fe

```
Fixed-effects (within) regression       Number of obs       =       88
Group variable (i): cntry               Number of groups    =       22

R-sq: within   =   0.8294               Obs per group: min  =        4
      between  =   0.2291                             avg  =      4.0
      overall  =   0.6613                             max  =        4

                                        F(3,63)             =   102.08
corr(u_i, Xb)  =  −0.1784               Prob > F            =   0.0000
```

```
----------------------------------------------------------------------
lifeexp |     Coef.   Std. Err.      t    P > |t|   [95% Conf. Interval]
---------+------------------------------------------------------------
  gdppc  |  .4498229   .1403768    3.20   0.002    .1693022   .7303435
  pubpc  | -2.791979   2.115246   -1.32   0.192   -7.01896   1.435003
 hlthcov |  .0667043   .0104423    6.39   0.000    .0458369   .0875716
  _cons  |  65.25959   .8357939   78.08   0.000    63.58939   66.92979
---------+------------------------------------------------------------
 sigma_u |  1.2469966
 sigma_e |  1.1185549
    rho  |   .5541371  (fraction of variance due to u_i)
----------------------------------------------------------------------
F test that all u_i=0:      F(21, 63) = 4.34      Prob > F = 0.0000
```

Notice that now the `xtreg` command is followed by the `fe` option. In the case of the time fixed coefficients model, the command line in Stata should look like this:

`xtreg lifeexp gdppc pubpc hlthcov, i(year) fe`

```
Fixed-effects (within) regression      Number of obs     =       88
Group variable (i): year               Number of groups  =        4

R-sq: within  = 0.2561                 Obs per group: min =       22
      between = 0.9058                                avg =     22.0
      overall = 0.6812                                max =       22

                                       F(3,81)           =     9.30
corr(u_i, Xb) = 0.4699                 Prob > F          =   0.0000
----------------------------------------------------------------------
lifeexp |     Coef.   Std. Err.      t    P > |t|   [95% Conf. Interval]
---------+------------------------------------------------------------
  gdppc  |  .2891002   .1409905    2.05   0.044    .0085734    .569627
  pubpc  | -.6194248   1.799747   -0.34   0.732   -4.200356   2.961507
 hlthcov |  .0356784   .0082166    4.34   0.000     .01933    .0520268
  _cons  |  68.30304   .9926022   68.81   0.000    66.32807     70.278
---------+------------------------------------------------------------
 sigma_u |  .88519159
 sigma_e |  1.3738868
    rho  |  .29334568  (fraction of variance due to u_i)
----------------------------------------------------------------------
F test that all u_i=0:      F(3, 81) = 7.06      Prob > F = 0.0003
```

Notice that identical coefficient results (except for the constant term, due to the difference in the estimation method) can be reached by estimating the regular linear regression model including a set of dummy variables accounting for COUNTRY (reported in Table 2.3, columns 5 and 6):

`xi: regress lifeexp gdppc pubpc hlthcov i.cntry`

```
i.cntry           _Icntry_1-22       (naturally coded; _Icntry_1 omitted)

   Source  |    SS       df     MS              Number of obs =      88
-----------+------------------------------       F( 24, 63)   =   17.56
    Model  | 527.237258   24  21.9682191        Prob > F      =  0.0000
 Residual  | 78.8234056   63  1.25116517        R-squared     =  0.8699
-----------+------------------------------       Adj R-squared =  0.8204
    Total  | 606.060664   87  6.96621453        Root MSE      =  1.1186
```

```
--------------------------------------------------------------------------
  lifeexp |    Coef.   Std. Err.     t    P > |t|   [95% Conf. Interval]
----------+---------------------------------------------------------------
    gdppc | .4498229   .1403768    3.20   0.002     .1693022    .7303435
    pubpc | -2.791979  2.115246   -1.32   0.192    -7.01896    1.435003
   hlthcov | .0667043  .0104423    6.39   0.000     .0458369    .0875716
 _Icntry_2 | -.7218542  .7962136  -0.91   0.368    -2.31296     .8692517
 _Icntry_3 | .3737209    .81066    0.46   0.646    -1.246254    1.993696
 _Icntry_4 | 1.288394   .8126798   1.59   0.118    -.3356173    2.912405
 _Icntry_5 | 1.140877   .795358    1.43   0.156    -.4485195    2.730273
 _Icntry_6 | .0636823   .8066294   0.08   0.937    -1.548238    1.675603
 _Icntry_7 | .7842305   .8108779   0.97   0.337    -.8361798    2.404641
 _Icntry_8 | -.1036344  .8058763  -0.13   0.898    -1.71405     1.506781
 _Icntry_9 | 3.032881   .8423683   3.60   0.001     1.349543     4.71622
_Icntry_10 | 1.341316   .8143882   1.65   0.105    -.2861087    2.968741
_Icntry_11 | .7644224   .8085019   0.95   0.348    -.8512398    2.380085
_Icntry_12 | .1731716   .8062372   0.21   0.831    -1.437965    1.784308
_Icntry_13 | 1.594349   .7924105   2.01   0.048     .0108428    3.177855
_Icntry_14 | 3.400081   .8291233   4.10   0.000     1.74321     5.056952
_Icntry_15 | -.0425529  .8060051  -0.05   0.958    -1.653226     1.56812
_Icntry_16 | 2.346083   .8370765   2.80   0.007     .6733194    4.018847
_Icntry_17 | -.6077486  .8617145  -0.71   0.483    -2.329748     1.11425
_Icntry_18 | 2.203484   .8261448   2.67   0.010     .5525652    3.854402
_Icntry_19 | 2.35021    .9118275   2.58   0.012     .528068     4.172352
_Icntry_20 | 1.044029   .8081763   1.29   0.201    -.5709826     2.65904
_Icntry_21 | -.0803817  .791196   -0.10   0.919    -1.661461    1.500697
_Icntry_22 | 3.26844    1.099262   2.97   0.004     1.07174     5.465141
     _cons | 64.18626   1.092505  58.75   0.000     62.00307    66.36946
--------------------------------------------------------------------------
```

where now, one can see the parameter estimates for 21 of the 22 countries in the sample. The xi command is extremely useful in Stata; it constructs those dummy variables on the fly. When I type i.cntry, the command tabulates CNTRY (or any other string or numeric variable) and creates indicator or dummy variables for each observed value of the variable, omitting the smallest value (in this case country with ID number 1). In this case it created _Icntry_2, _Icntry_3, _Icntry_4, ..., _Icntry_22. The two important parts of the xi command are the "xi:" right before any of the estimation commands such as regress; and the "i.varname," telling Stata which variable to decompose into a set of dummy variables. varname is defined as a categorical variable that identifies the different sub-groups desired to be compared (in the example above, I named such newvar variable CNTRY). Again, by default, i. omits the dummy variable with the smallest value for the variable. Nevertheless, one can change that and choose the most frequent value in the sample to be dropped by typing:

```
char _dta[omit] prevalent
```

or a particular value can be omitted by typing instead:

```
char _dta[omit] 5
```

where 5, in this case, is the country ID number to be omitted from the analysis—what has been called the reference group or case. The researcher can change the value for whichever value is desired (in the case of a string variable, one can type the text-value to be omitted from the estimation).

Similarly, one could run the regression model including only the set of time dummy variables in the regression model (reported in Table 2.3, columns 3 and 4) as follows:

xi: regress lifeexp gdppc pubpc hlthcov i.year

```
i.year          _Iyear_1-4           (naturally coded; _Iyear_1 omitted)

  Source |      SS       df     MS              Number of obs =      88
---------+------------------------------        F( 6,   81)   =   40.01
   Model | 453.167911    6  75.5279852          Prob > F      =  0.0000
Residual | 152.892753   81  1.88756485          R-squared     =  0.7477
---------+------------------------------        Adj R-squared =  0.7290
   Total | 606.060664   87  6.96621453          Root MSE      =  1.3739

-------------------------------------------------------------------------
 lifeexp |    Coef.    Std. Err.    t     P > |t|   [95% Conf. Interval]
---------+---------------------------------------------------------------
   gdppc |  .2891002   .1409905    2.05   0.044     .0085734    .569627
   pubpc | -.6194248  1.799747    -0.34   0.732    -4.200356   2.961507
 hlthcov |  .0356784   .0082166    4.34   0.000      .01933    .0520268
 _Iyear_2 | 2.001719   .4481336    4.47   0.000     1.110074   2.893364
 _Iyear_3 | 1.097984   .6897195    1.59   0.115    -.2743409   2.47031
 _Iyear_4 | 1.707489  1.174255     1.45   0.150    -.6289109   4.043889
   _cons | 67.10124    .7335567   91.47   0.000     65.64169   68.56079
-------------------------------------------------------------------------
```

Finally, the most complete model is the one that includes both country *and* time dummy variables in the regression model (reported in Table 2.3, columns 7 and 8) as follows:

xi: regress lifeexp gdppc pubpc hlthcov i.year i.cntry

```
i.year          _Iyear_1-4           (naturally coded; _Iyear_1 omitted)
i.cntry         _Icntry_1-22         (naturally coded; _Icntry_1 omitted)

  Source |      SS       df     MS              Number of obs =      88
---------+------------------------------        F( 27,   60)  =   28.10
   Model | 561.641996   27  20.8015554          Prob > F      =  0.0000
Residual | 44.4186679   60  .740311131          R-squared     =  0.9267
---------+------------------------------        Adj R-squared =  0.8937
   Total | 606.060664   87  6.96621453          Root MSE      =  .86041
```

```
---------------------------------------------------------------------
  lifeexp |    Coef.    Std. Err.     t    P > |t|   [95% Conf. Interval]
----------+----------------------------------------------------------
    gdppc |   .2467561    .151268    1.63   0.108   -.0558249    .5493371
    pubpc |  -2.453683   1.640111   -1.50   0.140   -5.734395    .8270277
   hlthcov |   .0414483   .0104859    3.95   0.000    .0204735    .0624231
  _Iyear_2 |   2.152749   .3551496    6.06   0.000    1.442344    2.863154
  _Iyear_3 |   2.043502    .806826    2.53   0.014     .42961    3.657395
  _Iyear_4 |    3.83013     1.4917    2.57   0.013    .846285    6.813975
 _Icntry_2 |  -.9835355   .6176205   -1.59   0.117    -2.21896    .2518895
 _Icntry_3 |   .0146352    .633093    0.02   0.982   -1.251739     1.28101
 _Icntry_4 |   1.187493   .6256136    1.90   0.062   -.0639201    2.438907
 _Icntry_5 |    1.08315   .6121744    1.77   0.082   -.1413806    2.307681
 _Icntry_6 |  -.3563115   .6359543   -0.56   0.577   -1.628409    .9157864
 _Icntry_7 |   .6179841   .6249813    0.99   0.327   -.6321645    1.868133
 _Icntry_8 |  -.2762857   .6210273   -0.44   0.658   -1.518525    .9659538
 _Icntry_9 |   1.439312   .8825299    1.63   0.108   -.3260105    3.204635
_Icntry_10 |   1.316573   .6266469    2.10   0.040    .0630926    2.570053
_Icntry_11 |   .0136753   .7059142    0.02   0.985   -1.398363    1.425714
_Icntry_12 |  -.0626389   .6262286   -0.10   0.921   -1.315283    1.190005
_Icntry_13 |   1.449058   .6119204    2.37   0.021    .2250349    2.673081
_Icntry_14 |   2.720496   .6653557    4.09   0.000    1.389586    4.051406
_Icntry_15 |  -.2568894   .6300072   -0.41   0.685   -1.517091    1.003313
_Icntry_16 |   2.194068   .6489771    3.38   0.001    .8959201    3.492215
_Icntry_17 |  -2.174204   .8685748   -2.50   0.015   -3.911612   -.4367957
_Icntry_18 |   1.079199   .7514642    1.44   0.156   -.4239533    2.582351
_Icntry_19 |   2.327501   .7017747    3.32   0.002    .9237425    3.731259
_Icntry_20 |   1.239875    .647774    1.91   0.060   -.0558659    2.535616
_Icntry_21 |  -.1657238    .610205   -0.27   0.787   -1.386316    1.054868
_Icntry_22 |   2.109461   .8885789    2.37   0.021    .3320388    3.886884
     _cons |   66.83412   1.111051   59.54   0.000    63.92816    68.37303
---------------------------------------------------------------------
```

As reflected by the higher adjusted R^2, the regression model that incorporates both time and country fixed effects is the one that provides greater information and therefore explains the highest variation in life expectancy. This country and time fixed effects model adds the individual effect that each country has as well as the effect that each time period has on the average population life expectancy. One can also look at each of these country-specific or time-specific effects. Thus, the dichotomous variable for Japan (in this case it is the coefficient associated to the `_Icntry_13` dummy variable automatically computed by Stata) indicates that life expectancy is significantly higher in Japan when compared with Australia during the 1960s (i.e., the reference country and year omitted from the model). The Japan effect is significant, all other things being equal or constant (in the case of the model, when both the public levels of health care expenditure as well as the population health coverage are constant).

Any of the fixed effects models seem to fit the data better than the OLS regression model, as reflected by the higher R^2 in comparison with the 0.67, which is the adjusted R^2 of the regular OLS regression model. The R^2 in case of the fixed effects model is the highest with both country and time controls, the R^2 value is then 0.92 (while the adjusted R^2 is 0.89). This country and time fixed coefficients model accounts for all possible structural or deterministic sources of variations in the dependent variable life expectancy. This complete model includes $21 + 3$ new parameters to be estimated (in addition to the coefficients associated with the four explanatory variables of interest and the constant term): One coefficient for each OECD country, except for Australia (the reference country), and one coefficient for each decade, except for 1960 (the decade of reference). So in this particular model, the constant β_0 is the average life expectancy for Australia in 1960.

In general, the *incremental F-statistic* can help test hypotheses regarding the significance of a few parameters in the model. They do this by comparing two nested models, one of which (the constraint model) is a subset of the other (the unconstrained or full model). To test whether the full model significantly improves upon a simpler model with h fewer parameters, one should compare the computed F-statistic value to a theoretical F-distribution with $df_1 = h$ and $df_2 = N - k$ (where N is the total number of observations in the sample). If the F-statistic is significant (that is, a low enough p-value), it indicates that the null hypothesis that the h-omitted independent variables have no effect (i.e., the h coefficients are equal to zero) can be rejected. This F-statistic is easy to estimate in Stata. To do so, first estimate the full model with all independent variables. Then run the `test` command followed by the name of the h variables whose combined null hypothesis of significance is to be tested.

For example, to test the null hypothesis that PUBPC=HLTHCOV=0, that is, a complex model including PUBPC and HLTHCOV versus a simpler model without them, type the following two lines of commands. The first line `quietly` estimates the full or more complex model.[16] One should then follow by typing the `test` command in order to use Stata to compute the incremental F-test, just as follows:

```
quietly xtreg lifeexp gdppc pubpc hlthcov, i(cntry) fe
test pubpc hlthcov
( 1)  pubpc   = 0
( 2)  hlthcov = 0
      F( 2,   63) =   25.80
        Prob > F = 0.0000
```

---

[16] The `quietly` command inserted right before the `xtreg` command or any other estimation command forces Stata to run the commands right after without having to display the results on the screen.

The probability of the F-statistic is below 0.0001, and consequently the null hypothesis that both the PUBPC and HLTHCOV coefficients are equal to zero is rejected. In other words, the full model significantly improves upon a simpler model without the PUBPC and HLTHCOV variables. Or put differently still, adding the PUBPC and HLTHCOV variables to the simpler model significantly improves the fit of the model.

In Table 2.3, I computed some incremental F-tests comparing several nested fixed effects models with the simple OLS model (that is, the regression model that only includes the independent variables GDPPC, PUBPC, and HLTHCOV but does not include any fixed effects). These F-tests are reported in the last two rows of the table titled "F-statistics with the OLS Model" and "Degrees of freedom." The F-statistics comparing the OLS regression model with each of the different fixed effects model (models that are nested) are significant at the 0.001 level, suggesting that the model equation controlling for fixed effects provides a better fit to the cross-sectional time-series data than the classical OLS regression model. This is usually true in the case of longitudinal data sets. For example, the F-test value comparing the time and country fixed effects model with the regular OLS model is 9.92 (with 24 and 60 degrees of freedom) and such an F-test is significant at the 0.001 level. One can compare these models because they are nested, meaning that the regular OLS is a constrained version of any of the fixed effects model. In other words, the fixed effects model includes the same variables as the OLS model plus all time and/or individual fixed effects.

Stata can fit more complicated fixed effects models. For example, with the `xtregar, fe` command, the user can fit a linear regression model when the error term is first order autoregressive like this one:

$$y_{it} = \beta_0 + \sum_{i=1}^{N-1} \alpha_i d_i + \sum_{t=1}^{T-1} \phi_t t_t + \sum_{k=1}^{K} \beta_k x_{kit} + u_{it} \qquad (3)$$

where $u_{it} = \rho_i u_{it-1} + \varepsilon_{it}$. For more information, look at the `xtregar` command in the *Stata Manual*.

7.   Random Effects Model

The *fixed effects model* (reviewed in the previous section) is a reasonable model when there is evidence that there are differences among the many individuals in the sample (and/or many points in time) and the constant term in the regression function. This assumption is especially useful when the number of units of analysis (or/and the time periods) is not very large. The example proposed here, where all OECD countries' health data is collected in four decades, demonstrates this. However, in other instances when the number of units of analysis is large, with many time observations, it is highly recommended estimating the *random effects models*. Unlike fixed effects models, in the case of *random effects models* the individual coefficients α_i and/or the time coefficients ϕ_t are no longer fixed effects associated with the independent term of the regression equation. Rather such effects are now allowed to vary in a random manner over time and across individuals. The random effects models therefore control for any individual influences on the dependent variable with some constant effect drawn from a distribution of individuals including all individuals in the sample. By doing this, the effects are assumed random, that is, such effects have been randomly selected from a population of individuals much larger than the sample of individuals under study.

There are several different types of random effects models. Some of these versions are not easy to implement and explain, especially in an introductory book like this one. However, the most popular random coefficients model is the so-called *error components model,* where the equation model is specified to account for several components of the error term. This model uses a time random component of the error, an individual random component, and finally a time-individual component that randomly depends on both time and individuals. All these components of the errors are included in the regression equation with the purpose of providing efficient and unbiased estimators of the regression coefficients for the main equation of interest. Again, going back to the most general case where the regression model can be specified as follows:

$$y_{it} = \beta_0 + \sum_{k=1}^{K} \beta_k x_{kit} + u_{it} \tag{1}$$

where y_{it} is a linear function of k explanatory variables, and the error term has the following structure:

$$u_{it} = \alpha_i + \phi_t + \varepsilon_{it} \tag{2}$$

where $i = 1, 2, 3, \ldots, N$ individuals or cases and $t = 1, 2, 3, \ldots, T$ time observations. As this last equation indicates, the error u_{it} has two random components now: (1) a *random individual component* that is invariable over time α_i (which characterizes each of the cases in the sample, and this is why this component is also referred to as the "between-groups" component) and (2) a *random temporal component* that is invariable across individuals ϕ_t (but that varies over time; also known as the "within-groups" component). u_{it} also has random component ε_{it} with a mean of zero and a constant standard deviation (i.e., white noise).

In contrast to the fixed coefficients model, the random effects model does not incorporate any deterministic components in the regression equation. Remember that a deterministic approach can be accomplished by incorporating in the regression model a series of $N - 1$ dummy variables (to control for the individual effects on the dependent variable) and also a series of $T - 1$ dummy variables (to control for time effects on the dependent variable). Each of these three components of the total error α_i, ϕ_t, ε_{it} now follow a normal distribution with a mean of zero. They are assumed to be uncorrelated with each other and also to be homoskedastic. Lastly, they are not correlated with any of the independent variables (or among themselves), or in formula notation, that:

$$E[\alpha_i] = E[\phi_t] = E[\varepsilon_{it}] = 0$$
$$\text{Var}[\alpha_i] = \sigma_\alpha^2; \text{Var}[\phi_t] = \sigma_\phi^2; \text{Var}[\varepsilon_{it}] = \sigma_\varepsilon^2$$
$$\text{Cov}[\alpha_i, \alpha_j] = 0; \text{Cov}[\phi_t, \phi_s] = 0; \text{Cov}[\varepsilon_{it}, \varepsilon_{js}] = 0$$
For all agents $i \neq j$, and for all points in time $t \neq s$
$$\text{Cov}[\varepsilon_{it}, \alpha_j] = 0 \text{ for all } i, j, \text{ and } t$$
$$\text{Cov}[\varepsilon_{it}, \phi_s] = 0 \text{ for all } i, t, \text{ and } s.$$

Consequently, the structure of the random errors u_{it} is the following:

$$\text{Var}[u_{it}] = \sigma_\alpha^2 + \sigma_\phi^2 + \sigma_\varepsilon^2$$

and the covariance for any two different individuals or cases i and j at time t is:

$$\text{Cov}[u_{it}, u_{jt}] = \sigma_\phi^2.$$

Similarly, the covariance for the same individual or case at two different points in time t and s can be formulated as follows:

$$\text{Cov}[u_{it}, u_{is}] = \sigma_\alpha^2.$$

When it comes to estimating any of the random effects models, it is again recommended that one uses the GLS estimation method (instead of the regular OLS). The GLS estimators are the efficient ones in this case.

The error components model presented above is *one* of the many random coefficient models preferred in practice, in which the specific effects for each individual α_i and/or for each period are treated like random variables. Many of these random coefficients models are more complex than the error components model. The most general case is the random coefficients model in which the slopes of the regression model (and not only the constant term of the equation) are different for each individual case or panel (this is $\beta_i \neq \beta_j$ for any i and j) and/or different at different points in time (this is $\beta_t \neq \beta_s$). In such general complex cases, these differences in slopes are considered different realizations of random variables taken from the same probability distribution. The most general model to consider here is the one proposed by Swamy in 1970. The *Swamy random coefficients model* is specified as follows—this time in matrix notation because the idea behind this technique is easier to grasp when presented this way:

$$y_{it} = [\beta_I + \alpha_i + \phi_t]'X_{it} + \varepsilon_{it} \tag{3}$$

where $i = 1, 2, 3, \ldots, N$ social units and $t = 1, 2, 3, \ldots, T$ points in time.

This method allows some random variation in the effects of each of the independent variables (including the constant term) on the dependent variable of interest. Such random variation can occur both across individuals and over time. The Swamy coefficients estimated by the generalized least squares method (GLS) are consequently a weighted average (or linear combination) of two estimators: The "between-groups" estimator and the "within-groups" estimator. The *between-groups estimators* are obtained when the components that vary across individuals (that is to say only α_i) are included in the model. These estimators are calculated by using solely the N observations in the estimation—one for each individual—calculated by the mean values of the dependent and independent variables for each of the individuals or cases:

$$(\bar{y}_i, \bar{X}_{ki})$$

so that the between-groups estimator β_{BG} is obtained by estimating the following equation by OLS:

$$\bar{y}_i = \beta'_{BG}(\bar{X}_{ki}) + \bar{\varepsilon}_i.$$

On the other hand, the *within-groups estimator* is estimated by using only the existing variation within each individual or panel, consequently ignoring any variation across groups, sections, or individuals. The within-groups estimator β_{WG} is obtained by estimating the following equation by OLS:

$$(y_{it} - \bar{y}_i) = \beta'_{WG}(X_{it} - \bar{X}_{ki}) + (\varepsilon_{it} - \bar{\varepsilon}_i).$$

However, the *Swamy random coefficients estimator* is the weighted average (or linear combination) of the between-groups and within-groups estimators. In particular, the vector of random coefficients β'_{RC} would be obtained when estimating the following regression model:

$$(y_{it} - \theta \bar{y}_i) = \beta'_{RC}(X_{it} - \theta \bar{X}_{ki}) + (\varepsilon_{it} - \theta \bar{\varepsilon}_i)$$

where θ is a function of the variance of α_i (that is, σ_α^2) and the variance of ε_{it} (or σ_ε^2) in the case that there is no time component in the variance.

This Swamy model is estimated by the GLS method. This GLS subroutine is already included in most statistical software packages that allow the analysis of longitudinal data. For example, in Stata, the command used to estimate the random coefficients models is `xtreg, re`. One can also run the command `xtgee`, because the estimators obtained with this command are asymptotically equivalent to those obtained by the command `xtreg, re`. To learn more about each of these commands in Stata, I recommend reading the entries for the commands `xtgee` and `xtreg` in the *Stata Reference Manual*.

To show how to estimate the random coefficients model using Stata, I will use the same OECD sample of countries that was used for the case of the fixed effects coefficients model. Below, one can see the Stata command line and the output when I incorporated the random component for each country to the `xtreg` command by using the `i(cntry)` and `re` subcommands after the comma:

```
xtreg lifeexp gdppc pubpc hlthcov, i(cntry) re
```

```
                                      Random-effects GLS regression
sd(u_id)            = 1.031156        Number of obs =      88
sd(e_id_t)          = 1.175465                     n =      22
sd(e_id_t + u_id)   = 1.563649                     T =       4

corr(u_id, X)       = 0  (assumed)    R-sq within   = 0.8073
                                      between       = 0.2683
                                      overall       = 0.6544

                                      chi2( 3)      = 266.02
              (theta = 0.5048)        Prob > chi2   = 0.0000
-----------------------------------------------------------------------
lifeexp |    Coef.    Std. Err.    z     P > |z|  [95% Conf. Interval]
--------+--------------------------------------------------------------
  gdppc |  .6250591   .1717148   3.640   0.000    .2885042    .961614
  pubpc | -4.169172   2.69197   -1.549   0.121   -9.445336   1.106993
hlthcov |  .0636947   .0088978   7.158   0.000    .0462554   .0811341
  _cons | 65.59608    .7939894  82.616   0.000   64.03989   67.15227
-----------------------------------------------------------------------
```

If I wanted to examine any time variation in the coefficients, I could have typed the following Stata command line (notice that I now include the variable time within the `i( )` subcommand in `xtreg`:

```
xtreg lifeexp gdpc pubpc hlthcov, i(time) re
```

In Table 2.4, I compare the coefficients obtained from the country random coefficients and fixed coefficients models, as previously estimated using Stata for the set of 88 observations (22 countries of the OECD). In the case of fixed coefficients models (reported in columns 2 and 3), I examined the *F*-statistic in order to assess the fit of the model. For the random coefficients models

Table 2.4
Regression coefficients for fixed effects and random effects models compared with the OLS model coefficients

Independent Variables[a]	OLS Model		Fixed Effects Model (Country Effects)[b]		Random Coefficients Model (Country Effects)	
	B	St. Error	B	St. Error	B	St. Error
Constant	67.179***	0.773	64.186***	1.092	65.596***	0.794
GDP per capita (in thousands of $)	0.390**	0.126	.449**	.1403	0.625***	0.172
Health care expenditure per capita (in thousands of $)	−1.464	1.965	−2.792	2.115	−4.169	2.692
Public coverage of health care expenses (% of population)	0.043***	0.009	.067***	.0104	0.064***	0.009
F-statistics						
Compared with baseline model[c]	59.98***		17.56***			
Degrees of freedom	(3 and 84)		(24 and 63)			
χ^2-statistics						
Compared with baseline model[c]					266.02***	
Degrees of freedom					3	

$N = 88$ observations (countries per year).

Note: The levels of significance are: *$p < 0.05$; **$p < 0.01$; ***$p < 0.001$ (two-tail tests).

[a] All independent variables of interest are delayed 5 years.

[b] For presentation purposes, the coefficients for the country fixed effects are omitted from the table (although they are included in the model estimation).

[c] The model labeled baseline is a model with no independent variables (only the constant term).

(in columns 5 and 6), the model χ^2-test is the goodness-of-fit model measure that needs to be examined. The random coefficients model is statistically significant at the 0.001 level (with a χ^2-test value of 266). Also the value of θ is estimated as 0.50. Substantively, this model predicts that the GDP per capita of a country has a positive and significant effect on a nation's population life expectancy. The same is found in the case of health care population coverage. It is worth noting here that the estimated coefficients are extremely similar regardless of the type of longitudinal regression equation estimated. This is not always the case, especially when observations are highly correlated across panels and/or over time.

In Stata, it is possible to test whether there are any systematic parameter differences between the fixed effects estimators and the random coefficients estimators by computing the so-called *Hausman specification test*. This is accomplished with the `xthaus` command in Stata as follows:

xthaus

```
Hausman specification test

                  -------Coefficients-------
             |        Fixed        Random
lifeexp |      Effects       Effects      Difference
--------+---------------------------------------------
   gdppc |      .624056       .625059      -.001003
   pubpc |    -3.858907     -4.169172      0.310265
 hlthcov |      .051435       .063694      -.012259

Test: Ho: difference in coefficients not systematic
            chi2( 3) = (b-B)'[S^(-1)](b-B), S = (S_fe - S_re)
                     =  7.45
           Prob>chi2 =  0.0588
```

In this OECD study in particular, it seems that the difference in the coefficients obtained by the method of the random coefficients and the method of the fixed coefficients is sufficiently small to accept the null hypothesis (or at least, not to reject the hypothesis) that the estimated coefficients by both methods are the same ($p > 0.05$). If the model is well specified and the variation of the term for each OECD country is not correlated with the other explanatory variables, then the coefficients obtained estimating fixed effects or random effects should not be statistically different.

In Stata, right after the estimation of the random coefficients model, the command `xttest0` can be used to test the hypothesis that $\alpha_i = 0$, that is, that the individual random component is zero. The χ^2-test obtained when I ran this command in my sample was 20.77 (with one degree of freedom); I can therefore reject the null hypothesis. It is therefore recommended introducing the random individual (or cross-sectional) component when estimating the model parameters of the main equation in this OECD data under study.

As in the case of fixed effects models, one can also use the `xtregar, re` command in order to fit a random effects model when the disturbance term is first-order autoregressive.

8.   Structural Equations Modeling

One of the main advantages of the Structural Equations Modeling (SEM) techniques is that the model coefficients can be specified to vary across individuals (or groups of individuals) and/or over time in a deterministic manner. Thus, these techniques distinguish how the effect of a given explanatory variable of interest on the dependent variables might differ across units or over time. In many instances, these structural equations help to determine the extent to which any given regression equation is applicable to two or more groups of individuals or at two or more time periods. In this case, one can use these models to estimate, evaluate, and compare both the functional form as well as the coefficients across each of the different estimated equations either by group and/or by time period. A second advantage is that these structural equation models allow the introduction (and estimation) of parameters about the correlation structure of the error terms of different equations.

For these two reasons, it is common to see these *structural equation models* used frequently in empirical studies in order to test whether the effect of certain independent variables on the dependent variables (i.e., specific coefficients) are the same across different estimated equations. This is especially so when analyzing longitudinal data sets with a dominant cross-sectional component. For example, one can use this method to examine whether any differences exist in the constant terms and one of the slopes associated with an independent variable of interest across each of the equations for each social unit. Alternatively, depending on the theoretical motivation of the study, one could estimate these structural equations in order to examine whether model coefficients change across time periods.

There is a simple way to accommodate for the possibility of any effect of a given independent variable to vary across panels or over time: It consists of including several interaction terms among explanatory variables of interest and dichotomous variables that identify certain individuals at different times. This can be done using only one equation model. *Interaction terms* are simply the multiplication of other independent variables, which can be included as additional explanatory variables. The most commonly used interaction terms are called *slope dummy terms* or variables, which are simply the product of a dummy (or indicator) variable times a continuous (or ordinal) variable. If d is a dummy variable and x is a continuous variable, one can easily create the new slope dummy variable dx (i.e., d times x) and include it with x in a regression model. This newly added term allows one to test for a difference in the slope or model of coefficient B_x on y between the two different categories as defined by d.

For example, suppose one is interested in estimating the following simple regression model where the dependent variable is WAGES and there is only one explanatory variable of interest, years of EDUCATION:

$$wages_{it} = \beta_0 + \beta_1 education_{it} + u_{it}$$

where β_1 measures the economic returns in terms of income for every additional year of education an individual has. In addition, assume that the data set under study contains information about women's and men's wages at two time periods: the 1980s and the 1990s. A possible regression model to estimate is this:

$$wages_{it} = \beta_0 + \alpha female + \phi_t year1990 + \beta_1 education + u_{it}$$

where the variable FEMALE is a dummy variable, which takes the value 1 if the individual interviewed is female, and 0 if the individual is male. The YEAR1990 variable takes the value 1 for any observation in the year 1990 and 0 for the year 1980. A hypothesis worth testing in this study is whether the effect of education on wage is different for men and women. If that is the hypothesis to be tested, the following model could be considered and estimated:

$$wages_{it} = \beta_0 + \alpha female + \phi_t year1990 + \beta_1 education + \beta_2 female\_educ + u_{it}$$

where a new independent variable has now been computed and included in the model. The FEMALE_EDUC equals variable FEMALE times EDUCATION. This FEMALE_EDUC is typically called the interaction term between the variable FEMALE and the explanatory variable EDUCATION. In Stata, computing the interaction term is easy; simply use the `generate` command below. If one wishes to interact FEMALE and EDUC (as termed in Stata), type:

```
generate female_educ = woman*education
```

where FEMALE_EDUC is the interaction or slope dummy variable just created in Stata, which can be included in any of the regression models discussed here. The coefficient β_2 explains the change in the slope β_1 for female respondents, accounting for how much the effect of education on wage differs when the individual is female. A t-test of H_0: $\beta_2 = 0$ helps to determine whether β_2, the coefficient on FEMALE_EDUC is significant; in other words, whether the difference in slopes between the FEMALE categories 0 and 1 is significant.

If both β_1 and β_2 are statistically significant, then the effect of an additional year of education on income is β_1 for males and $\beta_1 + \beta_2$ for females. This is true after controlling for time differences that might exist given the inserted dummy variable, YEAR1990, in the model. The coefficient α measures the existing differences of wage between women and men; in general, this coefficient equals the difference in Y-intercepts between the dummy variable categories 0 and 1. A t-test of H_0: $\alpha = 0$ helps to determine whether the two intercepts are significantly different.

The model could get more complicated. For example, one could test whether this female–education interaction effect decreases over time or not. To test such a hypothesis, one could include a three-way interaction term FEMALE times YEAR1990 times EDUCATION. This three-way interaction term would be computed in Stata as follows:

```
generate f_ed_90 = woman*education*year1990
```

where F_ED_90 is the newly created interaction variable as the product of variables FEMALE, YEAR1990, and EDUCATION. The magnitude and significance of the coefficient would give an idea of whether the gender difference diminishes or increases over time. As one can see, this is a simple way to test for differences in the effects of certain independent variables on the dependent variable across social groups, countries, or organizations as well as time differences in any longitudinal regression model.

There are situations in which one might wish to incorporate interaction terms for two or more groups of individuals as well as interactions for two or more points in time. In this case, I recommend using *structural equations modeling* or *simultaneous equations models*. This type of modeling, sometimes also called *path analysis*, helps to estimate causal models and evaluate the strength of variable relationships. In a system of simultaneous equations for the analysis of longitudinal data, one usually specifies a causal model and estimates such a model simultaneously for different groups of cases or panels—as long as there are a few panels, which is typically the case in longitudinal databases whose dominant component is temporal. The same model can be estimated at several time periods (when the dominant component of such longitudinal data is the cross-sectional one). So, in the general case of having a few q individuals (or groups of individuals or cases) in the sample (whose dominant component is temporal), the system of equations can be specified as follows:

$$y_{1t} = \beta_{10} + \sum_{k=1}^{K} \beta_k x_{k1t} + u_{1t}$$

$$y_{2t} = \beta_{20} + \sum_{k=1}^{K} \beta_k x_{k2t} + u_{2t}$$

$$y_{3t} = \beta_{30} + \sum_{k=1}^{K} \beta_k x_{k3t} + u_{3t}$$

$$\cdots\cdots\cdots\cdots$$

$$y_{qt} = \beta_{q0} + \sum_{k=1}^{K} \beta_k x_{kqt} + u_{qt}$$

where y_{qt} is the dependent variable for unit q over time; and each equation represents the linear function of y for each one of the individuals or panels in the sample. Such an equation may or may not contain the same number k of explanatory variables (it may even include different independent variables). In addition, the model needs to specify the error structure, including the variance and covariance of the errors of these different equations in the model to be estimated. A more complex structure of variances and covariances is the following:

$$\text{Var}[u_{1t}] \neq \text{Var}[u_{2t}] \neq \text{Var}[u_{3t}] \neq \cdots \neq \text{Var}[u_{qt}]$$
$$\text{Cov}[u_{it}, u_{jt}] = \sigma_{ij}^2 \neq 0 \text{ for each } i \neq j$$

where the variance and covariance of the error term are different for each equation. For example, in the case of time series for two individuals or groups of individuals 1 and 2, the system of equations consists of only two equations:

$$y_{1t} = \beta_{10} + \sum_{k=1}^{K} \beta_k x_{k1t} + u_{1t}$$

$$y_{2t} = \beta_{20} + \sum_{k=1}^{K} \beta_k x_{k2t} + u_{2t}$$

where y_{1t} (the dependent variable for individual 1 or groups of individuals 1 over time) and y_{2t} (for individual 2) are a linear function of the same explanatory variables k. For example, think about a study that estimates the same equation for the males and the females in the sample. The error term has the following structure (variance and covariance):

$$\text{Var}[u_{1t}] = \sigma_1^2 \neq \text{Var}[u_{2t}] = \sigma_2^2$$
$$\text{Cov}[u_{1t}, u_{2t}] = \sigma_{12}^2 \neq 0.$$

Such a causal model can also be estimated at different points in time (especially when the dominant component of the longitudinal database is the cross-sectional one and a few time observations are available for each individual or cross-section). In the example of two time periods, the system of equations should be the following:

$$y_{i1} = \beta_{01} + \sum_{k=1}^{K} \beta_k x_{ki1} + u_{i1}$$

$$y_{i2} = \beta_{02} + \sum_{k=1}^{K} \beta_k x_{ki2} + u_{i2}$$

where y_{i1} is the dependent variable at point of time 1 (or period 1 if grouping a few time units such as years together), and y_{i2} is the dependent variable at point of time or period 2. Such a dependent variable is a linear function of the same explanatory variables k, and the error covariance structure (including the two equations) is similar to the one presented above.

These models have been referred to as *unseemingly unrelated regression equations,* because the relations among each of the regression equations are not explicit. Rather such relationships come (unseemingly) from the correlation that might exist among the random errors for each of the different equations in the system. In other words, the equations are related through the error terms. An important part of the model is to be explicit in the specification of the correlation structure of the error terms among the different equations at the time of estimation. An alternative complication occurs when the different dependent variables are related to each other, not only through the correlation of their errors, but also because these dependent variables appear explicitly as independent variables in some other regression equations in the system of equations. These models are referred to as *non-recursive structural equations models* (for more information, see the list of references at the end of this book).

The system of equations is then estimated simultaneously (and jointly) because it is efficient to do so, especially when the correlations among the equation error terms are significant. Also, the results of this estimation are accurate even if the coefficients of the different equations are not related to each other. When there is no correlation between the error terms of the different regression equations, then the simultaneous estimation of the system of equations exactly coincides with the estimation of each of the equations one at a time (and independently) using OLS.

In sum, the two main goals behind these structural equation models is (1) to allow the model coefficients to vary across individuals or panels (or groups of individuals) and/or over time in a deterministic manner and (2) to introduce parameters about the correlation structure of the error terms of the different equations. Getting back to the practical example used in this chapter to illustrate some of the benefits of these structural equations, remember that the data set included information about the health care systems of all OECD countries in four decades. An example of a hypothesis one might investigate is whether the impact of health care expenditure (or the percentage of health care coverage) on population life expectancy is the same in every decade during the period of analysis. This is equivalent to testing whether such variable effects (of the two main independent variables of interest) are constant over time.

To compare the equation models for different groups or at different points of time (depending on which is the dominant component of the longitudinal data base), one normally uses the χ^2-statistic (more information about this test is given below). At the same time, when comparing these different equations, make sure to establish a hierarchy of models to be compared. Such a hierarchy of models could mimic the following one—where the models are ordered from the most to the least restrictive in terms of parameters to be estimated. So, for example:

Model 1. In the first model (the least restrictive one), the coefficients to be estimated are the same for all the different groups of interest (where observations are grouped either by groups of individuals or groups of time periods). In addition, the error terms of the different equations are assumed to be uncorrelated with each other.

Model 2. In the second model, the coefficients to be estimated are the same for all the different groups. But this time, the error terms of the different equations are allowed to be correlated with each other (according to a correlation structure that is specified in the model).

Model 3. In the third model, some of the coefficients to be estimated are hypothesized to be different for each of the groups of interest. In addition, the error terms of the different equations are correlated with each other (according to the same correlation structure specified in model 2).

Model 4. Finally, in the fourth and least restrictive model, all the coefficients to be estimated are still hypothesized to be different for each of the groups of interest. In addition, the error terms of the different equations are correlated with each other (according to the same correlation structure specified in model 2).

Obviously, this hierarchy of different models to be estimated can be modified depending on the main goal of the study. Any statistical package of simultaneous equations allows for testing (and comparing) these model possibilities by using the *incremental χ^2-statistic*. This test helps to determine if any null hypothesis can be rejected or not. Because the hypothesis to be tested includes a series of restrictions on a previously estimated model (restrictions about the parameters to be estimated), the models are termed *nested models*. In such a case, the two nested models can be compared by examining the value of the χ^2-statistic. The value of such a statistic is distributed like a χ^2-function in which the degrees of freedom are equal to the difference in the number of parameters to be estimated between the two nested models. In other words, such a number equals the number of parameter restrictions that are imposed in the equation model estimation. In the most general case, assume that model 1 has h parameters to be estimated while model 2 has k. Therefore, the χ^2-test has $k - h$ degrees of freedom.

As an example, in the case of the OECD database, one could compare the following set of nested models, which are ordered from the most restrictive to the least restrictive model:

Model A. The coefficients corresponding to the variables GDP, health care expenditure, and health care coverage are the same regardless of the time period when these are estimated (that is to say, these slopes are constant over time). In addition, the error terms of the four equations are uncorrelated with each other (even when assuming heteroskedasticity). In other words, the variance of the error terms for each equation is different. The model to consider in this case is the following one

$$lifeexp_{i1990} = \beta_{0,90} + \beta_1 gdppc_{i1990} + \beta_2 pubpc_{i1990} + \beta_3 hlthcov_{i1990} + u_{i1990}$$
$$lifeexp_{i1980} = \beta_{0,80} + \beta_1 gdppc_{i1980} + \beta_2 pubpc_{i1980} + \beta_3 hlthcov_{i1980} + u_{i1980}$$
$$lifeexp_{i1970} = \beta_{0,70} + \beta_1 gdppc_{i1970} + \beta_2 pubpc_{i1970} + \beta_3 hlthcov_{i1970} + u_{i1970}$$
$$lifeexp_{i1960} = \beta_{0,60} + \beta_1 gdppc_{i1960} + \beta_2 pubpc_{i1960} + \beta_3 hlthcov_{i1960} + u_{i1960}$$

where there are four equations (one for each decade) and the effects of the explanatory variables on life expectancy are assumed to be stable over time (i.e., the same for each and every decade). Notice, however, that decade differences are allowed in the constant terms. The variance and covariance structure is as follows:

$$Var[u_{i1960}] \neq Var[u_{i1970}] \neq Var[u_{i1980}] \neq Var[u_{i1990}]$$
$$Cov[u_{i1960}, u_{j1970}] = Cov[u_{i1970}, u_{j1980}] = Cov[u_{i1980}, u_{i1990}] = 0$$
$$Cov[u_{i1960}, u_{j1980}] = Cov[u_{i1970}, u_{j1990}] = 0$$
$$Cov[u_{i1960}, u_{j1990}] = 0.$$

Thus, the total number of parameters to estimate is 11: Three slope coefficients β_1, β_2, and β_3; four constant terms $\beta_{0,90}$, $\beta_{0,80}$, $\beta_{0,70}$, and $\beta_{0,60}$; and four different error variances. This is true given that covariances are constrained to be equal to zero.

Model B. In a second model, it is possible to assume that the equations are the same as in model A, but now the errors of the four equations are correlated among each other. Thus, in addition to differences in the variance of the errors of the different equations:

$$Var[u_{i1960}] \neq Var[u_{i1970}] \neq Var[u_{i1980}] \neq Var[u_{i1990}]$$

the following correlation structure is also assumed:

$$Cov[u_{i1960}, u_{j1970}] = Cov[u_{i1970}, u_{j1980}] = Cov[u_{i1980}, u_{i1990}] = \sigma_1^2 \neq 0$$
$$Cov[u_{i1960}, u_{j1980}] = Cov[u_{i1970}, u_{j1990}] = \sigma_2^2 \neq 0$$
$$Cov[u_{i1960}, u_{j1990}] = \sigma_3^2 \neq 0$$

where the correlation between any two equation error terms is assumed to be the same as long as there is the same time lag. This way, there are three additional parameters to be estimated in the model, parameters σ_1^2, σ_2^2, and σ_3^2. Notice that in this model, it is still assumed that the slopes are constant over time. The total number of parameters to be estimated is now 14. Also note that model A is nested in model B, so one can test whether model B better fits the data than model A by examining the value for an incremental χ^2-test with three degrees of freedom.

Model C. A third model could be one allowing the coefficients or slopes to be different for each of the regression equations. Thus, there are now nine additional parameters to estimate (in comparison with the previous model B) as long as the same correlation structure among the equation error terms, as in model B, is assumed. This last condition is necessary so that models B and C are nested models and therefore comparable. Hence, one has the ability to test whether model B is better than model C by using the incremental χ^2-statistic with nine degrees of freedom. Compared with model B, the magnitude of the slope coefficients can change every decade. So now, the system of equations is the following:

$$lifeexp_{i1990} = \beta_{0,90} + \beta_{1,90}gdppc_{i1990} + \beta_{2,90}pubpp_{i1990} + \beta_{3,90}hlthcov_{i1990} + u_{i1990}$$
$$lifeexp_{i1980} = \beta_{0,80} + \beta_{1,80}gdppc_{i1980} + \beta_{2,80}pubpc_{i1980} + \beta_{3,80}hlthcov_{i1980} + u_{i1980}$$
$$lifeexp_{i1970} = \beta_{0,70} + \beta_{1,70}gdppc_{i1970} + \beta_{2,70}pubpc_{i1970} + \beta_{3,70}hlthcov_{i1970} + u_{i1970}$$
$$lifeexp_{i1960} = \beta_{0,60} + \beta_{1,60}gdppc_{i1960} + \beta_{2,60}pubpc_{i1960} + \beta_{3,60}hlthcov_{i1960} + u_{i1960}$$

where the effects of the each of the explanatory variables on life expectancy now differ by decade, still allowing for model differences in the constant for every decade.

Graph 2.3(a) reports the results of model C when that model was estimated using the statistical program called AMOS. AMOS is a specific program used for the estimation of structural equations and needs to be purchased independently. It is a program module that is easily added to SPSS. After installing AMOS, start it up by clicking the *Start* icon on the Windows desktop, then *Programs*, and then select *AMOS*. In the latest versions of SPSS, one can also access AMOS from SPSS by going to the *Add-ons* pull-down menu and selecting the *AMOS* option (or by the *Statistics/Amos* pull-down menu). I believe that this is the easiest statistical package currently available to estimate structural equation models. It offers a useful graphical interface for the presentation and estimation of the model parameters. This program is quite powerful: One can

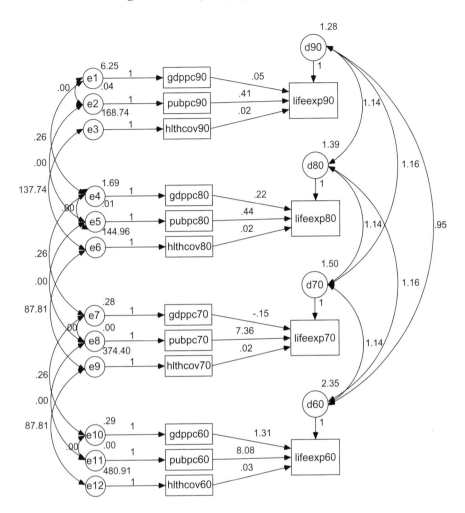

Chi-Square Test = 434.46; Degrees of Freedom = 100; Prob. = .00;

Graph 2.3(a)   Coefficients for model C of simultaneous equations using the OECD health longitudinal data.

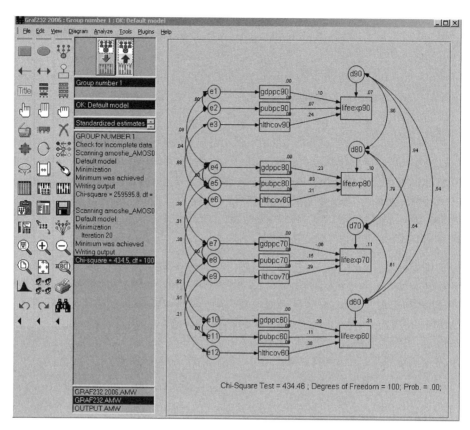

Graph 2.3(b)   Model C of simultaneous equations appearing in the AMOS 4.0 program window.

specify, estimate, assess, and present models by drawing an intuitive path diagram which shows the hypothesized relationships among variables. The program has a great drag-and-drop functionality. After AMOS has started, a window containing a large rectangle and several menu titles will appear (as shown in Graph 2.3(b)).

One can easily attach an Excel file by going to *File/Data Files* and browsing for it. AMOS 6 (as part of SPSS) automatically reads the current SPSS working file when AMOS is started directly from SPSS. For more information on how to use AMOS, consult the *Amos Programming Reference Guide*. This guide is typically installed in the Documentation subdirectory during the AMOS installation. In addition, the online AMOS guide can be purchased at the SPSS website (http://www.spss.com/amos/index.htm).[17] After installing AMOS, all

---

[17] Additional help may be accessible by visiting the website of the previous distributors of SPSS, SmallWaters Corporation (http://www.smallwaters.com).

files from the examples section of the AMOS Users' Guide are placed in the examples directory, below the AMOS program files. To read any of these files from Amos Graphics, choose the *File/Open* pull-down menu.

Again, Graph 2.3(a) reports the results of model C (as specified above) when using AMOS 6. The command file that I used to estimate the model of equations in my example is shown in Graph 2.3(c).

Since the χ^2-test with nine degrees of freedom (comparing model C and model B) was not significant, model B is preferred to model C. The χ^2-value of 11.2 with nine degrees of freedom is insignificant ($p = 0.26$). The χ^2, comparing models A and B, has the value of 63 (with three degrees of freedom) and is significant at the 0.001 level. So, adding the correlation structure (in which the terms of error of the four equations are correlated to each other) to model A seems to significantly improve the fit of the model; and consequently, model B is preferred to model A.

In Stata, one can also simultaneously estimate some of these seemingly unrelated equations by using the `suest` command. It is a bit more complicated than with AMOS (since it does not allow the estimation of these models graphically like AMOS), but it does allow the testing of cross-model hypothesis using the `test` command. Assume one wants to estimate two equations with the same dependent (LIFEEXP) and independent variables (GDPPC, PUBPC, and HLTHCOV). One equation will be estimated for the group of Southern European OECD countries and another equation for the rest of OECD countries. First, estimate both equations separately and store those model estimates using the `est store` command in Stata, as follows:

```
quietly xi: regress lifeexp gdppc pubpc hlthcov i.year i.cntry
if soeurope==1
est store south
quietly xi: regress lifeexp gdppc pubpc hlthcov i.year i.cntry
if soeurope==0
est store nonsouth
```

The first two command lines estimate and save the equation for Southern European countries (i.e., those sample observations for which SOEUROPE equals 1) in an equation called `south`. The same is accomplished in the next two command lines, this time saving the equation for the rest of the OECD countries (i.e., those sample observations for which SOEUROPE equals 0) in an equation called `nonsouth`.[18]

---

[18] Go to page 76 to learn how to use the `if` expression qualifier in Stata.

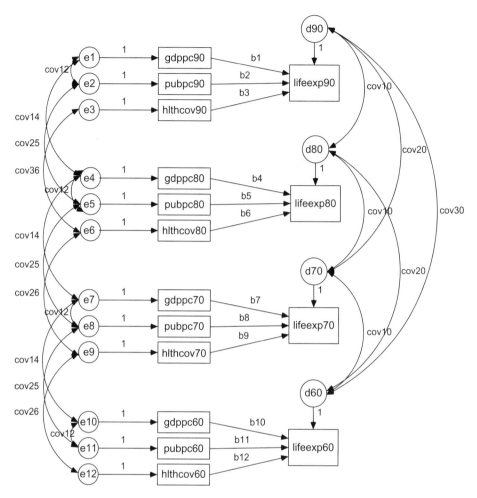

Graph 2.3(c) Example of command file in AMOS 4.0 used to estimate model C of simultaneous equations.

The second step is to use the `suest` command followed by the name of the two equations to be estimated simultaneously (a limitation here is the requirement that the regression models have to be able to produce equation-level scores):

```
suest south nonsouth
```

```
Simultaneous results for south, nonsouth
                                                    Number of obs = 88
------------------------------------------------------------------------------
             |               Robust
             |      Coef.   Std. Err.      z    P > |z|   [95% Conf.   Interval]
-------------+----------------------------------------------------------------
south_mean   |
      gdppc  |   .9975105   .1907618    5.23   0.000    .6236241    1.371397
      pubpc  |  -8.743048   2.003773   -4.36   0.000   -12.67037   -4.815724
     hlthcov |   .0567594   .0112089    5.06   0.000    .0347904    .0787284
     _Iyear_2|   2.376687   .3641517    6.53   0.000    1.662963    3.090411
     _Iyear_3|  -.2547233   .8725104   -0.29   0.770   -1.964812    1.455366
     _Iyear_4|  -.3203008   1.448739   -0.22   0.825   -3.159777    2.519175

[Here I am omitting the report of all country level dummy variables]

      _cons  |   62.35323   .7677052   81.22   0.000    60.84856    63.85791
-------------+----------------------------------------------------------------
south_lnvar  |
      _cons  |  -1.247344   .2109591   -5.91   0.000   -1.660816   -.8338714
-------------+----------------------------------------------------------------
nonsouth_m~n |
      gdppc  |   .2947431   .1766935    1.67   0.095   -.0515699    .6410561
      pubpc  |  -2.067065   1.814901   -1.14   0.255   -5.624205    1.490075
     hlthcov |   .0265014   .0111725    2.37   0.018    .0046037    .048399
     _Iyear_2|   1.890569   .2958192    6.39   0.000    1.310774    2.470364
     _Iyear_3|   1.562214   .5687668    2.75   0.006    .4474519    2.676977
     _Iyear_4|   2.679445   1.249335    2.14   0.032    .2307945    5.128096

[Here I am omitting the report of all country level dummy variables]
      _cons  |   67.60407   1.034165   65.37   0.000    65.57715    69.631
-------------+----------------------------------------------------------------
nonsouth_l~r |
      _cons  |  -.2344794   .2655324   -0.88   0.377   -.7549133    .2859546
------------------------------------------------------------------------------
```

In the final step, after both equations have been estimated simultaneously, one can start testing whether the impact of GDPPC on LIFEEXP is the same for both groups of countries (South Europe/non-South Europe), by using the `test` command as follows:

```
test [south_mean]gdppc = [nonsouth_mean]gdppc
(1) [south_mean]gdppc - [nonsouth_mean]gdppc = 0
        chi2( 1) = 7.30
     Prob > chi2 = 0.0069
```

Notice that the name of the model underscore mean is in square brackets to refer to the model, followed by the name of the variable whose coefficient is being tested. Thus, [south_mean]gdppc refers to the coefficient for GDPPC on LIFEEXP in the model equation called south. The null hypothesis tested is that such a coefficient equals the coefficient for GDPPC on LIFEEXP in equation nonsouth (or, in Stata terminology, [nonsouth_mean]gdppc). The χ^2-statistic (whose value is 7.3) is significant (at the 0.01 level) leading to the rejection of the null hypothesis that the coefficient for GDPPC is equal for Southern European and non-Southern European OECD countries. Thus, the difference in the effect of GDPPC between the South of Europe and the rest of OECD countries is significant (at the 0.01 level).

The same test could be performed in order to test any other hypothesis. For example, below I have tested whether the effects of PUBPC and HLTHCOV are the same regardless of which group of countries are being compared (in this example, Southern European OECD countries and the rest of the OECD countries). In both cases, I reject both hypotheses at the 0.10 significance level (see the tests below):

```
test [south_mean]pubpc = [nonsouth_mean]pubpc
(1) [south_mean]pubpc − [nonsouth_mean]pubpc = 0
       chi2( 1) = 6.10
     Prob > chi2 = 0.0135


test [south_mean]hlthcov = [nonsouth_mean]hlthcov

(1) [south_mean]hlthcov − [nonsouth_mean]hlthcov = 0
       chi2( 1) = 3.66
     Prob > chi2 = 0.0559
```

A common type of test to perform in this situation is the so-called *Chow Test*. Such a test helps to find out whether *all* equation coefficients vary between the two groups. To run this test, type:

```
test [south_mean = nonsouth_mean], cons

( 1) [south_mean]gdppc - [nonsouth_mean]gdppc = 0
( 2) [south_mean]pubpc - [nonsouth_mean]pubpc = 0
( 3) [south_mean]hlthcov - [nonsouth_mean]hlthcov = 0
( 4) [south_mean]_Iyear_2 - [nonsouth_mean]_Iyear_2 = 0
( 5) [south_mean]_Iyear_3 - [nonsouth_mean]_Iyear_3 = 0
( 6) [south_mean]_Iyear_4 - [nonsouth_mean]_Iyear_4 = 0
( 7) [south_mean]_Icntry_2 - [nonsouth_mean]_Icntry_2 = 0
( 8) [south_mean]_Icntry_3 - [nonsouth_mean]_Icntry_3 = 0
( 9) [south_mean]_Icntry_4 - [nonsouth_mean]_Icntry_4 = 0
(10) [south_mean]_Icntry_5 - [nonsouth_mean]_Icntry_5 = 0
(11) [south_mean]_Icntry_6 - [nonsouth_mean]_Icntry_6 = 0
(12) [south_mean]_Icntry_7 - [nonsouth_mean]_Icntry_7 = 0
(13) [south_mean]_Icntry_8 - [nonsouth_mean]_Icntry_8 = 0
```

```
(14)  [south_mean]_Icntry_9 - [nonsouth_mean]_Icntry_9 = 0
(15)  [south_mean]_Icntry_10 - [nonsouth_mean]_Icntry_10 = 0
(16)  [south_mean]_Icntry_11 - [nonsouth_mean]_Icntry_11 = 0
(17)  [south_mean]_Icntry_12 - [nonsouth_mean]_Icntry_12 = 0
(18)  [south_mean]_Icntry_13 - [nonsouth_mean]_Icntry_13 = 0
(19)  [south_mean]_Icntry_14 - [nonsouth_mean]_Icntry_14 = 0
(20)  [south_mean]_Icntry_15 - [nonsouth_mean]_Icntry_15 = 0
(21)  [south_mean]_Icntry_16 - [nonsouth_mean]_Icntry_16 = 0
(22)  [south_mean]_Icntry_17 - [nonsouth_mean]_Icntry_17 = 0
(23)  [south_mean]_Icntry_18 - [nonsouth_mean]_Icntry_18 = 0
(24)  [south_mean]_Icntry_19 - [nonsouth_mean]_Icntry_19 = 0
(25)  [south_mean]_Icntry_20 - [nonsouth_mean]_Icntry_20 = 0
(26)  [south_mean]_Icntry_21 - [nonsouth_mean]_Icntry_21 = 0
(27)  [south_mean]_Icntry_22 - [nonsouth_mean]_Icntry_22 = 0
(28)  [south_mean]_cons - [nonsouth_mean]_cons = 0
       Constraint 22 dropped
            chi2( 27) = 1750.11
         Prob > chi2 = 0.0000
```

Given the significance of the χ^2-test, I can reject the null hypothesis that all regression coefficients (including the constant term) are equal for the two groups of countries (Southern European OECD countries versus the rest of the OECD countries). Consequently, it makes a lot of sense to estimate one equation for each group simultaneously. For more information about seemingly unrelated estimation and the testing of cross-model hypotheses, check the *Stata Reference Manual* (especially look for additional information about the suest and test commands).

Additional statistical software programs such as LISREL, EQS, and SAS also estimate these types of simultaneous equations and perform a battery of tests in order to assess the significance of the model parameters. They also provide measures regarding the goodness-of-fit of each of these systems of equations and compare different nested models by computing the appropriate χ^2-tests. These multiple equation models have important theoretical and practical complications that go beyond the scope of this book. For more information about SEM, see the list of references at the end of the book (especially Bollen, [1989]).

9. There Is Still More

Despite all the different longitudinal techniques for the study of continuous dependent variables over time covered in this chapter, there is still more to be said at this point. Again, let me start with the most general regression equation to estimate given a cross-sectional cross-time data:

$$y_{it} = \beta_0 + \sum_{i=1}^{N} \alpha_i d_i + \sum_{t=1}^{T} \phi_t t_t + \sum_{k=1}^{K} \beta_k x_{kit} + u_{it} \tag{1}$$

where again there are $k = 1, 2, 3, \ldots, K$ independent variables of interest, $i = 1, 2, 3, \ldots, N$ individuals and $t = 1, 2, 3, \ldots, T$ observations over time, and also with the following error or disturbance structure:

$$\text{Var}[u_{it}] = \sigma_{it}^2$$

$$\text{Cov}[u_{it}, u_{js}] = \sigma_{ijts}^2$$

where the variance of the term errors is different for each individual or case and it is also allowed to vary over time. The covariance matrix is different from zero and differs by individuals i and j, and at times t and s.

This general specification is purely theoretical given that the parameters to estimate are way too many, making it impossible to estimate them using a real data set. Therefore a series of restrictions to the model parameters is needed so that the model is identifiable. Without such parameter restrictions, the proposed general model would be estimating more parameters than observations available in the sample. For this reason, the general regression equation does not have any direct empirical utility apart from illustrating the most general possible multiple regression model and see the many possibilities when taking such estimation into practice. The most obvious simplification of such a model is the classical linear regression model:

$$y_{it} = \beta_0 + \sum_{k=1}^{K} \beta_k x_{kit} + \varepsilon_{it} \tag{2}$$

where there is only one error term and where the slopes are constant and the same across individuals and over time. In addition, this model assumes that the error terms have a constant variance and that they are not correlated over time, nor they are correlated across individuals. If this was the correct model, the best linear unbiased estimators would be obtained by OLS. However, as explained in detail earlier in this chapter, this model is quite unrealistic in its assumptions when it comes to analyzing a longitudinal data set. One valid attempt at taking these considerations into account is to introduce the assumption that the disturbances are correlated and consequently estimate such a model by GLS where specifying the right variance and covariance error structure is key. This is referred to as the *constant coefficients model*.

Another valid variation of this model presented in this chapter is the one that incorporates *fixed effects* for each of the many individuals as follows:

$$y_{it} = \beta_0 + \sum_{i=1}^{N} \alpha_i d_i + \sum_{k=1}^{K} \beta_k x_{kit} + \varepsilon_{it} \tag{3}$$

This model can then be estimated by OLS, where α_i captures variations in y_{it} for each individual i. Still, this model assumes that the causal process operating might be different for each unit of analysis, but the same process applies over time. The estimation of this model can also be improved by incorporating a certain error structure to be estimated by GLS:

$$y_{it} = \beta_0 + \sum_{i=1}^{N} \alpha_i d_i + \sum_{k=1}^{K} \beta_k x_{kit} + u_{it} \qquad (4)$$

where now, some serial correlation can be specified (that is, for each individual or case, the error terms are correlated over time). Such serial correlation can be taken into consideration in the model, as I showed in this chapter, by adding an autoregressive process of first order, for example:

$$u_{it} = \rho_i u_{it-1} + \varepsilon_{it} \qquad (5)$$

where $i = 1, 2, 3, \ldots, N$ individuals and $t = 1, 2, 3, \ldots, T$ time observations, and the error ε_{it} is "well-behaved." These models are highly recommended when the dominant component of the longitudinal data is temporal. The alternative models can be estimated when the dominant component is the cross-section (for example, when a large number of individuals are interviewed in two time periods).

The fixed effects models may assume different constant terms for each individual case or panel that do not necessarily change over time (like α_i). In these models, it can also be assumed that such specific effects vary over time but not across units. The coefficients or slopes are, however, equal for all the cases in the sample at any point of time (this is $\beta_i = \beta_j$ for any i and j). There are certain models which allow the slopes to vary across sections and over time in a determinist way. These models allow an explicit examination of whether the effects of certain independent variables are different across individuals and/or over time. A simple way to incorporate this assumption in the previous dynamic models is by including some interactions between the explanatory variables of interest and the many dummy variables that identify certain individuals and/or certain time periods. For more complicated models with greater numbers of interactions, I have advised the use of a system of *structural equations* (in section 8). In such a system of equations for the analysis of longitudinal data, one can assume different regression models for each of the individuals or cases (if they are few) and/or each of the time periods (if these are few). Such equations are then simultaneously estimated.

The *random coefficients models* treat the specific effects for each individual unit α_i, and/or every time as random variables. This model is the simplest model within the wide range of other many possible random effects models. For example, there is one model in which the coefficients are allowed to be different for each individual so that $\beta_i \neq \beta_j$ for any i and j and/or for every point in time so that $\beta_s \neq \beta_t$. Such differences are considered like different draws of variables from the same probability distribution. From here, there exists a variety of mixed models with fixed and random coefficients. In these models, some of the coefficients are fixed and others are random with the same matrix of constant variance–covariance. However, the random effects models require a better specification of the error term correlation matrix than the fixed effects models require. These models are not always as easy to interpret as the fixed effects models, which are much more deterministic. For that reason, I recommend starting by estimating simpler models. The constant coefficients model is a good start, and one could later continue imposing different parameter restrictions on this model and perform tests to ensure that the model properly fits the longitudinal data.

This chapter presented several methods available for the analysis of longitudinal data when the dependent variable is continuous. Each technique begins with a series of assumptions, and one needs to be aware of the benefits and limitations behind each of the different model choices. The results of any model are appropriate only when the assumptions of such model are realistic and accurately reflect the process and data under study. Sometimes, imposing certain methodological techniques to the study of a social process or to the analysis of a given longitudinal data can be unsuitable and therefore unappealing. One must verify that the assumptions behind the chosen model correspond with the characteristics of the social process under study and that such assumptions make sense with the estimation of the dynamic causal model. In this chapter, I have shown what the main model assumptions are behind different models for change in a continuous dependent variable. The selection of model and estimation technique should depend not only on the nature of the data available for analysis, but also on the types of assumptions tolerated when formulating a statistical model. As always, start with simple models. Simplicity in the specification of longitudinal models is always preferable to more complicated models that are difficult to evaluate because the latter models (not covered in an introductory book like this one) are also more difficult to interpret and present to the reader of a study. Typically, these more complex models provide similar results to the more parsimonious and easily interpretable models.

This chapter covered the longitudinal models that compare alternative error covariance structures and provide strategies for choosing among them. I have described how to specify and estimate such models using the appropriate statistical software, as well as how to interpret their results. These models are typically the most common models used to examine change in a continuous variable.

To learn more about the statistical details behind the models presented in this chapter, I suggest looking at *Greene's Econometric Analysis* (1997, fifth edition). Chapter 12 "Serial Correlation" provides the theory behind the statistical models available to analyze time-series data. In Chapter 13 titled "Models for Panel Data," Greene surveys the most commonly used techniques for cross-section-at time-series data analyses in one single equation model. I have also discussed how to model change using SEM (also called "covariance structure analyses" or latent growth modeling in many other methodology books). To learn more about SEM, I recommend *Structural Equation with Latent Variables* by Bollen (1989).

There are additional models that can be estimated to analyze a longitudinal dependent variable measured as a continuous or ordinal variable. Many methodology books talk about these models as *multilevel models* for the analysis of change, in general, because they allow researchers to address within-case and between-case questions about change simultaneously. Many of these multilevel models specify two different sub-models. The first one describes how each case changes over time, and the second model describes how these changes differ across cases (for more information, see Bryk and Raudenbush, [1987]; Rogosa and Willett, [1985]; and most recently, Singer and Willett, [2003]; and Rabe-Hesketh and Skrondal, [2005]). Many more sophisticated models of change also assume that changes in the dependent variable are discontinuous or non-linear. In addition, there exist longitudinal negative binomial and Poisson models for analyzing change in cardinal variables (also known as "count variables") over time (not discussed in this introductory manual). All these models are appropriate extensions to the models already presented in this chapter.

The intent of this chapter was not to be comprehensive; consequently, I did not attempt to present a complete account of all possible longitudinal models for analyzing the change of a continuous variable over time. Instead, the goal has been to describe and illustrate the most important longitudinal models from beginning to end, walking through all the different steps necessary when specifying the model, fitting it to a longitudinal data set, and interpreting the results of the final model. This chapter provides a foundation on which to continue learning about advanced models to study change. For more information about these more advanced models for the analysis of quantitative variables, consult the list of references at the end of this book.

Chapter 3

Event History Analysis

The objective of this chapter is to introduce the various concepts and models available for studying change in variables of qualitative nature. This methodology is called *Event History Analysis* (henceforth, EHA), a term that refers to the group of techniques to study events. EHA allows researchers to examine the determinants or factors behind the occurrence of any type of social event over time and can consequently help to answer the questions that previously could not be answered using the classic linear regression model or the dynamic methods described in Chapter 2 (when the dependent variable is continuous). In the past decade, with the introduction of EHA commands in standard statistical software programs such as Stata and SPSS, EHA has become a set of popular techniques used by social scientists and professionals. This chapter is structured as follows: First, I introduce the unique language used in the discussion of events and the EHA methodology available to analyze such events. Next, I review several of the most commonly used EHA techniques, in detail, with some examples of commands.

EHA is used to study longitudinal data when the social process to study is the occurrence of an event. An "event" is a change from one state to another. Such an event is measured using a categorical dependent variable. EHA has also been called *survival analysis*, hazard analysis, and failure time analysis because biologists and epidemiologists were the first to use and develop this methodology in order to study the survival of organisms after certain treatments. EHA analyzes longitudinal data available for a sample of individual cases or units during a period of time when a series of events may occur. The focus of this event study is to understand the main determinants behind the occurrence and timing of such events. The variables measuring information about individual cases in a sample can take different values over time for each one of the cases. In particular, there is at least one variable that indicates whether an event takes place—or never takes place—during the period of time under analysis (sometimes multiple events can happen during said period).

More formally, in this EHA methodology, "event" is referring to the change in the value of a categorical (or discrete) random variable $y(t)$ (also denoted as

y_t or y_{it} because the values of y change over time and across cases) that can occur at any time during the interval of time under study. This change is also termed "transition." Remember that categorical variables generally take integer values (or whole numbers), so each value represents a certain property, category, or group characteristic. In the case of EHA, the discrete variable of interest can take a few mutually exclusive values. The set of values that the $y(t)$ (or y_t) variable can take is called the *state space*. For example, a discrete variable can be marital status, whose space of possible states may include being single, married, separated, divorced, or widowed. Other examples of discrete variables at the individual-level type of analysis may include the labor market situation of individuals (whether one is employed, self-employed, or unemployed), the purchasing of an apartment (as opposed to renting), or the change of city location. The units or cases of analysis do not necessarily have to be people. Cases can include territories, companies, organizations, relations between different organizations or individuals, documents, etc. At the country level, for example, think about a country's legalization of abortion, joining of an international organization, or signing of a multinational treaty. Although this statistical procedure is applied most frequently to analyze nominal variables, it can also be applied when analyzing ordinal or cardinal variables as long as the number of possible values is not too large. When the range of possible values within the state space is large and such values have some ordinal features, it is advised that one considers using the longitudinal models available for the study of continuous variables, as reviewed in Chapter 2.

    Graph 3.1 illustrates the hypothetical life history of one case pertaining to the employment status of an individual with case ID number 1, a female in the sample of interest. One can see how her employment situation changes over time. The X-axis represents the passage of time. In this case, the individual's age is used to measure time; the Y-axis represents her labor status at a particular age, $y(t)$. This individual illustrated on the graph became part of the sample at the age of 16, at time t_0. She was a student until the age of T_1, when she decided to leave school and look for a job. She was unemployed between the ages of T_1 and T_2. At time T_2 she decided to go back to school until she found a job at the age of T_3. At T_4 she became unemployed once again. After a period of $T_5 - T_4$ years of unemployment, this individual finds another job; she is employed until shortly after the age of 65 (T_6), when she retires and stops being part of the labor force. This person is not part of the labor force from age 65 until she dies at the age of 75. In this hypothetical example, six different employment events (or transitions) occur in the individual's professional life: (a) from student to unemployed, (b) from unemployed to student, (c) from student to employed, (d) from employed to unemployed, (e) from unemployed to employed, and (f) from employed to retired. If one were to collect employment data like this for a sample of individuals in the

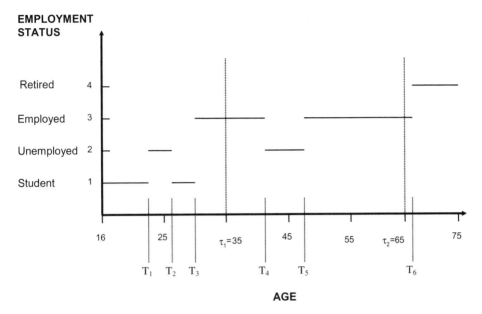

Graph 3.1 Hypothetical life history including all changes in the employment status of one individual.

United States, there would be a set of employment events to study over time. Obviously, the type and timing of such events would vary by individual case. For example, certain individuals will experience more employment changes than others will with distinct combinations and sequences of events as well as different timings in the occurrence of these employment transitions over their lives. One of the many possible studies of such data could be to understand the factors contributing to individual unemployment. One could then imagine designing a study that tries to model how individual-level characteristics (such as gender, race, age, experience, years of school, etc.) affect the likelihood of becoming unemployed over time. A different, but related, study could consist of analyzing the duration of events; that is, an analysis of the duration of employment. EHA analyzes rates at which events occur (i.e., whether events occur or not) as well as duration (i.e., timing of events, or how slowly or quickly events occur).

1. Event History Data

A crucial element in analyzing events over time is the format of the longitudinal data set to be used to study such events. This format includes the type and time pattern of both the dependent and independent variables to be analyzed. Some longitudinal data are more complete than others when it comes to the coding of dependent and independent variables over time while some are more aggregated than others over time. For obvious reasons, incomplete data are more difficult to analyze than complete or disaggregated data. For the group of key techniques for the estimation of EHA that I will introduce in this chapter, the data set should contain the following two fundamental variables as well as a set of time variables, all three together identifying the most important aspects of the events to be examined over time. The first fundamental variable uniquely identifies each of the individuals or cases under study (this is the case/unit/individual/panel or even cross-section ID number). The second fundamental variable is the categorical variable measuring the different state (or stage) realizations for each of the individuals in the sample over time; this variable measures whether a certain individual case is at state k or state j at a given point of time. This is what one could call the dependent variable of interest in the case of EHA. Finally, the set of time variables records the times when such state changes occur for each individual case in the sample. Typically, each record in the sample measures the start and end time, during which a set of different variables (including the dependent variable) have the same value. Together, these variables provide the necessary information about the event and its timing in order to perform longitudinal modeling. This type of data is also referred to as *event history data*. These variables can provide information about the different independent variables of interest (which may or may not vary over time).

To illustrate the importance of these variables in a given individual record or observation, look at the simple longitudinal data set stored in Excel on page 139. This basic longitudinal data set is the one that was used to plot Graph 3.1, where age was on the X-axis and employment status on the Y-axis for case ID number 1, a female in the sample. Case ID uniquely identifies the individuals in the sample; there is one ID number for each individual case. In this case there are three different individuals: individual 1 (with ID number 1), individual 2, and individual 3. "Start Age" measures the timing when the employment situation of each individual changes. Individual 2, for example, was a student until the age of 23 when he became employed; he was employed until the age of 65 when he retired. Notice that within each record, the value of the different variables is constant. In addition, the multiple records for a given individual case in the sample will vary depending on how often there are changes in any of the values in the variables' coding information about each of the cases in the sample. The data set contains information about each individual case in the sample with multiple records (also called "spells"), each of which records changes in different measurements or variables of interest over time. This data format has been referred to as *spell* or *episode data*.

	A	B	C	D	E	F
1	Case ID	Start Age	Final Age	Employment Status	Employment Description	Gender
2	1	16	23	1	Student	Female
3	1	24	26	2	Unemployed	Female
4	1	27	30	1	Student	Female
5	1	31	42	3	Employed	Female
6	1	43	47	2	Unemployed	Female
7	1	48	66	3	Employed	Female
8	1	67	75	4	Retired	Female
9	2	16	22	1	Student	Male
10	2	23	64	3	Employed	Male
11	2	65	72	4	Retired	Male
12	3	16	89	3	Employed	Male

As one can easily infer, this type of longitudinal data for the study of events can be expensive to compile, given that it not only requires measuring the events (and changes of states) but also their timing over the observation period of study. Such a data set entails, first, a clear definition of the temporal process with a categorical variable that changes over time; and second, an observation period when the change may happen for a sample of individuals or cases with the time variables recording the timing of such changes or events. Thus, the most complete longitudinal data set has detailed historical data containing records or observations at the most frequent points of time possible for a sample of individual cases. Of course, it is common for such historical data to be incomplete, with some time aggregation and even some data gaps. The *event count data,* for example, is a different form of longitudinal historical data that informs the researcher about the number of events that have happened to the individuals in a sample during a time interval when the data were collected. A characteristic of this type of data is that it does *not* detail the exact timing when these events happen or in what order they occur; rather, these data present the number of events that have happened during a given time interval. Thus, in real life one could encounter multiple types of longitudinal data sets recording events.[19]

At first sight, one may think of formulating and estimating different statistical models depending on the type of longitudinal data collected. However, the social process to be studied is the same in all cases, the only difference being the format in which the data are available to model such a social temporal process. Nevertheless, the best approach is to develop and use the EHA statistical models to study the same social process independent from which data type is available. The estimation and understanding of such methods of analysis become much easier when the longitudinal data of events is as complete as possible. This is because such data provide complete information about the event process during the period of analysis, unlike other forms in which data can be collected with some information deficiency. In substantive terms, complete data of events allow researchers to perform the most comprehensive examination of theories and testing of hypothesis possible regarding the process of change over time.

Look, again, at the previous Excel table containing information about a set of labor employment situations throughout three individuals' lives. Such a table provides an example of the most basic type of complete data set for the study of employment events. The life employment history of individual 1 is presented graphically in Graph 3.1. A data set like this one has the so-called *episode format*. This episode format is the most common (and ideal) format used to store data containing EHA. Such data include one record or *episode* for each individual case during a time interval; during that time interval there is no change in any of the variables measuring different features and characteristics of a given individual or unit, including the dependent variable of interest. In this case, the dependent variable measures the four possible labor states of any individual over time (broadly speaking): student, unemployed, employed, or retired. Each row in the file records when there is some change in any of the variables measured for each individual, and most importantly, the qualitative variable of interest measuring the event. In this particular example, the variable measures the labor situation of the person. Understandably, the number of records differs by individual, with more records corresponding to more changes in the dependent and independent variables for any given individual case in the sample. This way, the data set is recording time episodes when the values of all variables are time

---

[19] There are many other types of data sets not that relevant to the introductory purpose of this volume. For example, the *current status data* record whether an event has happened by a certain moment. An example of such data is whether individuals have obtained a job by the age of 20. Some other data sets might only detail the length of time since an individual case has been in a certain state. For example, looking at the employment status of a person at the time of the survey interview, one could measure how long that person has been in the current employment status (without information about previous employment status).

constant. In general, some of these variables, such as employment status, educa-tion level, or salary, will naturally change over time. Some others might typically be time invariant, such as gender, ethnicity, birth date, and location. In these databases, the observations are not independent from each other simply because there might be multiple observations or records for a given individual case.

Going back to the previous illustration, the data file contains, for example, one single record for those individuals who worked from age 16 until they died (such as the individual with ID number 3 on the table). Such data file contains three records for those people who were students, became employed, and worked for a set of companies until they retired (like the individual with ID number 2, where there is one record for each labor situation). In the case of the person with ID number 1, there are seven records, one for each different labor situation at dif-ferent time points. Record 3 on the table shows that individual 1 became a stu-dent again at the age of 27 (until she was 30 years old). Notice that in this data set in particular, gender is time-invariant. Most of the available statistical soft-ware packages that perform some form of survival analysis or EHA work great with data stored in this episode format.

I can now begin to introduce some important terms and formulas in the study of events over time using mathematical notation. An *event history* describes the values that a given qualitative variable $y(t)$ (i.e., categorical or discrete variable) may take during some observation time that goes from T_1 to T_2. In basic statis-tical terminology:

$$w[\tau_1, \tau_2] \equiv \{y(u); \tau_1 \leq u \leq \tau_2\}$$

where I follow the convention of using capital letters to represent a random vari-able and lowercase to represent one of the many possible realizations of such a random variable. The set of values that $y(t)$ or y_t can take is again referred to as the *state space* of y. The number of values that y can take over time gives the size of the state space and has been denoted with ψ. In this chapter, I will focus on the analysis of variables of a qualitative nature that have a countable and limited number of possible values. By countable I mean that each value in the state space is associated with a whole number. For simplicity, I will start with the analysis of one type of event, which is a variable that takes the value of 1 when the state or stage has changed, 0 otherwise (i.e., there is no state change). This is the case of dummy or indicator dependent variables. However, in the most general case, y can take any whole number between 1 and ψ. In the small sam-ple used up to now, four categories describe the labor employment situation of any given individual in the sample, with 1 being equal to the student state, 2 the unemployed state, 3 the employed state, and 4 the retirement state. Suppose one is interested in understanding unemployment, in this case, individual 1 is unemployed between T_1 and T_2 (since y equals 2 during that period) and between T_4 and T_5 (since y equals again 2). However, this individual is employed between T_3 and T_4 and between T_5 and T_6 (since y is 3 then). The employment situation of the person at any other moment of time can be found both in the graph and in the episode data file.

The *event* is used to represent changes in the value of a variable of interest over time (i.e., change of state or situation). For example, in the Graph 3.1, the first event takes place when individual 1 changes from employment situation 1 to situation 2 at time T_1, that is to say, when individual 1 transitions from student to unemployed. The number of events that can happen to any person in a given time interval $(s, t]$ is denoted as a random variable $N(s, t]$. When noting a time interval, the parenthesis indicates that the value is excluded from the interval, whereas the bracket is used to indicate that the value is included in the interval. Thus, $(s, t]$ denotes all the time points between times s and t, including time t but excluding time s from the time interval. In the figure, $n(16, 75]$ is equal to 6 employment events. The random variable T_n represents the point in time when event n or the nth event occurs. In the case of the graph, the six observed events occur in times T_1, T_2, T_3, T_4, T_5, and T_6. The beginning of the observation period is t_0 and is considered arbitrary. The random variable y_n measures the state right after the nth event has occurred and is equal to $y(t_n + \Delta t)$, where Δt is positive and infinite. In Graph 3.1, $y(4) = 2$, indicating that individual 1 becomes unemployed during the period of time that goes from T_4 to T_5. The initial state at the beginning of the observation period is $y(t_0)$. Again, an *episode* or *spell* occurs when there is a change in any of the variables in the data set, including the variable measuring the events; so the length of the episode will be shorter or longer depending on when that change happens in at least one of the variables. Episode n refers to the period of time between the interval $n - 1$ and n. Finally, a key concept in the analysis of events is the *risk set* or *group at risk*. This is defined as the group of individual cases at risk of experiencing an event at a given moment of time.

Graph 3.1 illustrates the complete labor employment situation of a given individual in a sample. Having such complete information is quite unusual. Assume that there is only information about the employment situation of individual 1 during a time interval that goes from age 35 to 65 as represented in Graph 3.2 (this is, $\tau_1 = 35$ and $\tau_2 = 65$). In such a case, not all duration periods are complete. For example, the duration for episodes 4 and 6 are incomplete since information is not available now about the employment status for the individual before the age of 35 and after the age of 65. $y(t)$ varies before and after the time interval considered but those values were not observed (in other words, they exist but they are unknown to the researcher), the longitudinal data are *censored*. The censoring is therefore a particular limitation of the data set under analysis but not a

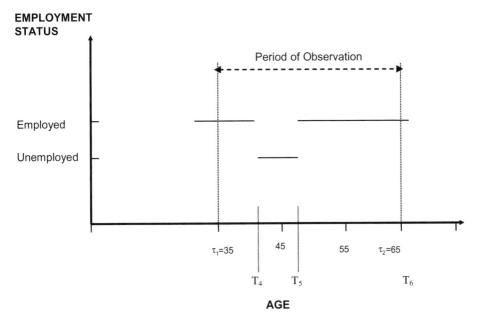

Graph 3.2 Hypothetical life history including changes in the employment status of one individual from age 35 to 65: Only employed and unemployed categories.

feature of the population under study. When there is no data about $y(t)$ before moment τ_1 (before 35 years, in the example of Graph 3.2), the data are said to be *left-censored*. When such information is not known after τ_2 (after the 65 years in the example), the data are *right-censored*. Most longitudinal data sets are right-censored—since at some point the time period of observation needs to be closed and one might not be able to know what happens to each individual case in the sample after that time.[20] To keep it simple, in this chapter I will assume that the analyzed data is only right-censored, that is to say that $\tau_1 = t_0$.

Graphs 3.1 and 3.2 provide a simple way to describe the history of events for one particular individual or case in a sample. Another way to present information about these events is to record the following information for each individual case: (1) the status at the beginning of the observation period, $y_0 \equiv y(t_0)$; (2) the number of events that happen during this observation period, $n \equiv n(t_0, \tau_2]$;

---

[20] Frequently the data can also be left-censored. This data censoring creates special problems when it comes to the analysis of data. In the case of right-censored data, this problem can be solved easily using the techniques in this chapter; and statistical software programs typically take care of this issue (for more information, see Tuma and Hannan [1978]). The problem becomes more difficult to solve when analyzing left-censored longitudinal data; this case is beyond the introductory scope of this book.

(3) the moment in time when each of the n events occur, $\{t_1, t_2, t_3, \ldots, t_n\}$, or alternatively, the delay time for each of the events, $\{u_1, \ldots, u_n\}$; and finally (4) the new state after the event has happened represented by $\{y_1, \ldots, y_n\}$. Such information allows for inferring the exact status for each individual at each moment of time during the period of observation. In general, this can be expressed with the following formula:

$$w[t_0, \tau_2] = \{t_0, y_0; t_1, y_1; \ldots; t_n, y_n\}.$$

A complete history about the status of any given individual, the times at which each individual changes status, is frequently referred to as the *sample-path*. The sample-path therefore permits the following of the behavior of any individual case over time (from t_0 to τ_2). The complete path for individual 1 in the sample is the following (assuming again that her entire life employment history is observed):

$$W_1[16, 75] = \{16, 1; 24, 2; 27, 1; 31, 3; 43, 2; 48, 3; 67, 4\}.$$

In any given sample, there exist differences in the possible sequence of events across individuals. The EHA provides the tools necessary to examine the evolution of such states and come up with some specific predictive model (beyond the simple observation of each of the individuals in the sample). Still, and before undertaking any of the advanced EHA techniques to model the occurrence of events, I highly recommend examining some of these individual trajectories of possible states for a group of individuals in the sample. This is what I did in Graphs 3.1 and 3.2 in this chapter. This can provide a good understanding of the sample and the social event process under study.

2. A Practical Example

To illustrate some of the methodological techniques available to study dependent variables of qualitative nature, I use a real sample of employees in a service organization in the United States (for more information about the research setting and the sample under analysis, see Castilla [2005]). One of the basic research questions in the study was to investigate the role of employee social networks in the hiring and post-hiring of employees. It is quite common and widespread for organizations to recruit new employees by implementing employee referral programs where a current organizational employee refers someone for a job opening. In this study, I examine the performance implications of hiring new workers via employee referrals, using referrals as indicators of pre-existing social connections. The study also provides a further understanding of how worker interdependence impacts their performance. I structure my argument as follows. First, I test the central prediction of the "better match" theory in economics. The proposition here is that if referrers help to select better-matched employees, one would expect that, after controlling for observable human capital characteristics, workers hired via employee referrals should be more productive than non-referrals at hire. Second, I test a more sociological proposition that presumes that the interaction between the referral and referrer at the workplace enriches the match between the new hire and the job.[21]

Using unique employees' hiring and performance data in a call center in the United States, I show some basic estimated models to evaluate how the employee-level characteristics influence employee turnover. In understanding employee turnover, I pay special attention to the effect of having been hired through the referral program and to whether the person who referred the hired employee is in the call center or not. More specifically, I evaluate the impact of referrers on the risk of an employee to turnover, once I control for other important employee characteristics of the individual such as age, marital status, years of education, or wage, among other controls.

The data was collected for over 300 call center employees of a large service organization in the United States. The job I study is the Phone Customer Service Representative, whose duties consist of answering customers' telephone inquiries about credit card accounts. The sample includes 325 workers who were hired between January 1994 and December 1996. The period of analysis includes a total of 1104 days. The event of study is whether the employee is terminated or not. At the end of the observed period, 165 hires had been terminated, a little above 50% of the total hires during the time between 1994 and 1996. The set of independent or explanatory variables include those that record information coded from the employees' job applications. Each job applicant must fill out an application asking for age, level of education, marital status, work experience, computer experience, and language skills. Another important vector included in the analysis refers to job-specific characteristics, such as wage, part-time status, and night shift. Finally, I include the two central variables of interest following the proposed research question. The first variable, CONTACT, is a dichotomous variable, which takes the value of 1 if the employee was hired into the company via the company's referral program, and 0 if the applicant was hired through other recruitment channels such as job fairs, newspapers, or internet. A second independent variable of interest, CONTACTT, is a measure of whether the

---

[21] For more information about the theory behind these two testable hypotheses, I recommend reading my published article, Castilla [2005]. The full cite is: Castilla, E.J., 2005, Social networks and employee performance in a call center, *American Journal of Sociology* 110 (5), 1243–1283. Copyright © 2005 by The University of Chicago Press.

contact that referred the employee for the job has already left the organization (that is, the employee contact or referrer has been terminated). Consequently this variable takes the value of 1 when the employee organizational contact (or referrer) is terminated; 0 if the employee contact (or referrer) still works for the company.

The main definitions and descriptive statistics for the variables that I use in the examples throughout this chapter appear in Tables 3.1 and 3.2. Table 3.1 includes the descriptive statistics for the dependent variable and those independent variables that vary over time such as age, part-time status, and whether the contact has been terminated. Table 3.2 includes all independent variables that do not vary over time in this setting, such as gender, marital status, salary, number of previous jobs, and whether the employee was referred for the job or not (all of these variables are time invariant because they were coded at time of application). Typically, it is useful to use two different tables, one for the time-varying case characteristics and another for the time-invariant characteristics or variables of the individual cases in the sample. It is also very common for the time-varying characteristics to be summarized in a descriptive table at few points in time (mainly means and frequencies; see Chapter 1 for more about this). This is like taking a few snapshots in order to provide summary statistics for those variables that change over time to get a sense of variance over time for the sample under study. In the two descriptive tables, it can be seen that on average, an employee at the call center is 28 years old. Around 22% are male, 42% are married, and 12.7% have a college degree. Moreover, the initial hourly wage is $6 per hour, with a standard deviation of $2.7.

The variable of interest (or dependent variable) in this study will be the risk of being terminated. TERMINAT is a dichotomous variable that takes the value of 1 when the employee is terminated, 0 otherwise. By looking at Table 3.1, the reader can learn that about 50% of the employees at the call center were terminated during the period of study. For some of the dynamic analysis I will perform in this chapter, I have created two additional new variables. The variable STARTSTA or "start state" measures whether or not the employee is working for the organization at the *beginning* of the employee episode. The variable END-STATE or "end state" measures the same this time at the *end* of the employee episode. Any of these two variables equal 1 when the employee is terminated; 0 otherwise (i.e., the employee is working for the organization). It is also important to include the time variables STARTTIM or "start time," which measures the tenure of the employee at the beginning of each episode or spell (measured by number of days since the employee was hired, given that the day has been chosen as the time unit). ENDTIME or "end time" refers to the employee tenure at the end of the episode. In this study, I happen not to be interested in historical time, meaning the day or month of the year, but rather I am interested in the time

Table 3.1

Descriptive statistics for the dependent variable and the time-varying covariates in the employee data set

Variables	Percent of All Employees	For All Observations in Time[a]		Description
		Mean	Standard Deviation	
Age		28.47	8.71	Age in years
Night shift		0.16	0.37	If employee works the night shift = 1; 0 otherwise
Part-time status		0.08	0.26	If employee works part-time (less than 40 hours per week) = 1; 0 otherwise
Referrer is terminated		0.08	0.28	If the employee referrer is terminated = 1; 0 if referrer is still around in the organization
Terminated	50.76%			If the employee is terminated during the period of analysis (three years) = 1; 0 otherwise

$N = 325$ employees/751 observations

[a]The mean and standard deviation for all these variables are calculated using all 751 employee episodes/records, even when the total number of employees in the sample is 325, with many episodes at different point times for each employee.

Table 3.2
Basic descriptive statistics for the time-constant covariates in the employee data set

Variables	Percent of All Employees	Mean	Standard Deviation	Description
Male	22.70%			If employee is male = 1; 0 if female
Married	42.28%			If employee is married = 1; 0 otherwise (at time of hire)
College degree	12.65%			If employee completed her college degree = 1; 0 otherwise (at time of hire)
Hourly wage		6.09	2.69	Starting wage per hour (in dollars)
Number of previous jobs		3.01	1.23	Number of previous jobs before applying for the current position
Referral	52.31%			If the employee was referred for the job position at time of application = 1; 0 otherwise
$N = 325$ employees				

since the employee started working for the company. So, in the first record or episode for any employee in the company, STARTTIM is 0; after that, START-TIM can take any value from 1 to 1104 (which is the maximum number of days any employee could be observed given the time observation design of this study).[22]

The variable MALE is a dichotomous variable and takes the value of 1 if the employee is male, 0 if the employee is female. AGE is a continuous variable with an average of almost 29 years (as measured at time of application). There are two additional dichotomous variables: MARRIED or "marital status" (to signal whether the employee is married) and COLLEGE or "college education" (to signal if the employee finished some college education before starting working for this service organization). Two important variables are used to control for the basic job features: PARTTIME job, if the employee works less than 40 hours of work a week and NIGHTS, referring to "night shift," if the employee works a night shift. All four variables are dummy or indicator variables with 1 equals "yes" and 0 equals "no". WAGE refers to the hourly salary; and PREVIOUS is the "number of previous jobs" at the time of job application.

Table 3.3 presents 8 employees and 28 episodes (or observations) randomly selected from the sample of employees in the organization (of all 325 workers, with 751 episodes—since there are, on average, 2.3 episodes for each employee hire). In this case, each episode in the data set shows the values of all variables for a given employee during the time interval that goes from STARTTIM (or "start time") to ENDTIME (or "end time"). Again, the episode is the period of time between any changes in any of the variables in the data set, including the period of time between any two successive events. For example, individual with ID number 1 is a female employee; she is 24 years old at the time she is hired, does not have a college degree, and was referred by a current employee in the organization. After working at this company for 158 days, her personal connection is terminated. Individual 1 is terminated in her 255th day of tenure in the organization. Another example in the sample is employee with ID number 18 who also has a personal connection at the time of applying for the job. In her 16th day of tenure, her hourly wage is increased by 8%; again her wage is increased after being employed in the organization for almost a month (on day 25 to be exact). After 42 days in the company, she earns a wage that is 33% higher than her starting salary. On her 86th day, she is terminated. These are all examples of how to read and code the employee data information presented in episodes.

---

[22] In many other examples of studies, one may be interested in historical time. In such cases, the variables pertaining to time now measure the specific historical date in day/month of the year and are included in the EHA models.

Table 3.3
Part of a longitudinal database sorted by employee ID and observation number (ascending order)

Employee ID	Obser- vation	Start Time (in days)	End Time	Start State (1 = Terminated)	End State (1 = Terminated)	Personal Contact (1 = Yes)	Personal Contact Terminated	Male (1 = Yes)	Age (in years)	Married (1 = Yes)	College Degree (1 = Yes)	Hourly Wage (in $)	Number of Previous Jobs	Part-time Job (1 = Yes)	Night Shift (1 = Yes)
1	1	0	157	0	0	1	0	0	24	0	0	8.50	1	0	0
1	2	158	162	0	0	1	1	0	24	0	0	9.00	1	0	0
1	3	163	255	0	1	1	1	0	24	0	0	9.00	1	0	0
4	1	0	856	0	1	0	0	0	25	0	0	8.40	4	0	0
12	1	0	212	0	0	0	0	1	27	0	0	6.00	4	0	0
12	2	212	905	0	0	0	0	1	27	0	0	7.00	4	0	0
12	3	906	952	0	0	0	0	1	27	0	0	8.00	4	0	0
12	4	953	1054	0	1	1	0	1	27	0	0	8.00	4	0	0
18	1	0	15	0	0	1	0	0	18	0	0	6.00	3	0	1
18	2	16	24	0	0	1	0	0	18	0	0	6.50	3	0	1
18	3	25	41	0	0	1	0	0	18	0	0	7.00	3	0	1
18	4	42	86	0	1	1	0	0	18	0	0	8.00	3	0	1
22	1	0	107	0	0	0	0	1	24	0	0	9.40	3	0	0
22	2	108	237	0	1	0	0	1	24	0	0	9.40	3	0	0
33	1	0	56	0	0	1	0	0	22	0	0	6.00	3	0	0
33	2	57	128	0	0	1	0	0	22	0	0	7.00	3	0	0
33	3	129	163	0	0	1	0	0	22	0	0	7.50	3	0	0
33	4	164	783	0	0	1	1	0	22	0	0	8.00	3	0	0
33	5	784	1060	0	0	1	1	0	22	0	0	8.00	3	0	0
35	1	0	267	0	0	1	0	0	20	1	0	8.00	1	0	1
35	2	268	730	0	0	1	1	0	20	1	0	8.20	1	0	1
35	3	731	785	0	0	1	1	0	20	1	0	8.00	1	0	1
35	4	786	837	0	0	1	1	0	20	1	0	8.50	1	0	0
39	1	0	16	0	0	1	0	0	31	1	0	9.00	4	1	0
39	2	17	49	0	0	1	0	0	31	1	0	7.90	4	1	1
39	3	50	121	0	0	1	0	0	31	1	0	7.90	4	1	1
39	4	122	231	0	0	1	0	0	31	1	0	7.90	4	0	1
39	5	232	764	0	1	1	0	0	31	1	0	7.90	4	0	0

Note that in the table extraction from the entire data set included in Table 3.3, employees' data with ID number 33 and 35 are right-censored, meaning that it is unknown how long they remained at the company after the period of observation. What is known, however, is that these workers did not leave the company during the time window of analysis. In contrast, the other six employees (with ID numbers equal to 1, 4, 12, 18, 22, and 39) were terminated at some point during the period of study.

3. Basic Methodology for the Analysis of Events

Social researchers are almost never interested in whether a particular individual is hired or is terminated at any point in time. Instead, they are interested in analyzing the probability that a given employee gets hired or is terminated. Generally, the purpose of any research project is to be able to generalize from a sample about the behavior of any population of interest. In the case of categorical dependent variables, researchers are typically interested in estimating the probability that a given individual or case reaches a given state or situation. *State probability* is therefore the probability that an individual is at a given state e at time t; or in mathematical notation:

$$P_e(t) \equiv \Pr[y(t) = e]$$

where e takes any value within the space of ψ possible states, stages, or categories. Since the state space ψ is limited and includes a set of mutually exclusive states, then the sum of all state probabilities is equal to 1 at any given point of time, as indicated in the formula below:

$$\sum_{e=1}^{\psi} P_e(t) = 1, \quad \text{for each point of time } t.$$

In the case of a longitudinal data set, the number of cases with $y(t) = e$ (a given state e) divided by the size of the sample provides an unbiased estimate of $p_e(t)$ or state probability. The state probability, however, does not provide any information about the process of change in the dependent variable *per se*. When modeling the factors affecting such state probability, it is quite common to introduce a set of independent or explanatory variables as predictors of such state probability. The equation model can be written as follows, where the state probability is now a function of a vector of independent variables X_k (where k denotes the number of independent variables included in the vector X_k):

$$P_e(t \mid x) \equiv \Pr[y(t) = e \mid X_k(t)].$$

The state probability depends on the time t when such probability is being measured, thus allowing for that probability to change over time. Such probability also depends on a set of characteristics or factors, as measured by some independent variables $X_k(t)$ (also denoted as X_{kt} or X_{kit} because their values may or may not change over time and across individuals or cases) which can either increase or decrease such state probability. The model formulation of $P_e(t \mid x)$ is not difficult at all as long as one can obtain and analyze complete information about the values of $X_k(t)$ at each of the points of time under study.

If the data were cross-sectional (i.e., with no time component and therefore collected at one point of time t), the analysis would be easy. Both the *logit* and *probit regression models* are the appropriate (and commonly used) methods to use in the empirical analyses of such cross-sectional data sets. The general logit regression model with a polytomous dependent variable takes the following functional form:

$$L_e = \log \frac{P_{ei}}{P_{1i}} = \alpha_e + \sum_{k=1}^{K} \beta_{ek} x_{ki} + u_i \tag{1a}$$

or, in matrix notation:

$$\log \frac{P_{ei}}{P_{1i}} = \alpha_e + \beta_{ek}' X_{ki} + u_i \tag{1b}$$

where $e = 2, 3, 4, \ldots, E$, depending on the final state being examined, with $e \neq 1$; e can be equal to 1 in either the initial state or the chosen state of reference (for some specific theoretical reasons); x_{ki} refers to any of the k independent variables with their respective coefficients β_{ek}.[23] The subscript e in the vector β_{ek} is referring to the set of model parameters estimated when modeling the probability of being in a given state e. Finally, p_e is the state probability of state e. This following model implies that the probability of being in state e is the following:

$$P_{ei} = \frac{\exp[\alpha_e + \beta_{ek}' X_{ki}]}{\sum_{e=2}^{E} \exp[\alpha_e + \beta_{ek}' X_{ki}]}, \quad \text{with } e \neq 1.$$

In the case of a dichotomous dependent variable y measuring two possible states 1 and 0, with 1 being the event or state of interest, then p_1 denotes to probability of y being equal to 1, and the probability of y not being equal to 1 is $1 - p_1$.

---

[23] Again, I use β_k to denote the column vector containing the model coefficients, $\beta_0, \beta_1, \beta_2, \ldots, \beta_k$. X_{ki} or simply X_i is the column vector with k rows (one for each independent variable) for any given individual i; that is, the transpose of the ith $1 \times k$ row of the matrix X_{ik} (of dimension $N \times K$).

The odds of getting y equal to 1 are therefore:

$$\frac{p_1}{1 - p_1}.$$

By taking the natural logarithm of the odds, the so-called logit or L—which ranges from $-\infty$ (when $p_1 = 0$) to ∞ (when $p_1 = 1$)—is obtained:

$$L = \log \frac{p_1}{1 - p_1}.$$

So now, the logit regression refers to the following model:

$$L = \log \frac{p_1}{1 - p_1} = \alpha + \sum_{k=1}^{K} \beta_k x_{ki} + u_i \qquad (2a)$$

or, in matrix notation:

$$L = \alpha + \beta_k' X_{ki} + u_i. \qquad (2b)$$

Given that the logit (L) is a linear function of x_k variables, the probability p is a non-linear S-shaped function. The following figure plots the predicted probability of y as a logit function of one independent variable x_1; one can observe that the predicted probabilities never exceed the minimum of 0 and the maximum of 1. Consequently, this logit model provides a better fit to the modeling of probability than that of the classic OLS regression model.

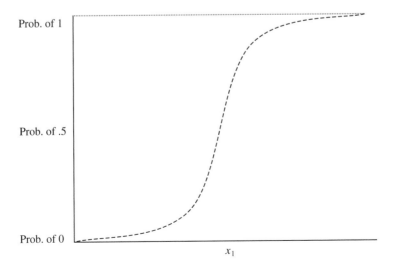

Parameters in the above logit model are a bit more complicated to understand than OLS coefficients. α can be used to compute the probability that y equals 1, which is:

$$p = \frac{1}{1 + \exp(-\alpha)}.$$

One can generalize the formula given above and use it to get the probability that y equals 1 for any particular individual or case scenario, with a certain combination of values for the several independent variables x_k:

$$p = \frac{1}{1 + \exp[-(\alpha + \beta_k' X_{ki})]}.$$

But a researcher is always more interested in interpreting any coefficient as $\exp(\beta_k)$, which tells that each-unit increase in x_k multiplies the odds favoring y being equal to 1 by $\exp(\beta_k)$. Another popular way of reporting these coefficient results is by computing:

$$100\% \times (\exp(\beta_k) - 1).$$

This expression indicates that the odds or probability of y being equal to 1 changes by $100\% \times (\exp(\beta_k) - 1)$ percent with each one-unit increase in x_k.

Logit models are estimated by maximum likelihood (ML) instead of ordinary least squares. The *likelihood function* measures the probability of getting the observed sample as function of the model coefficients. It is the product of the individual contributions of the cases in the sample. Statistical software programs, through an iterative process, help estimate the coefficients that generate the highest possible values for the logarithm of the likelihood function (also called *log-likelihood*; for more information, see reading recommendations for logit and probit models in the list of references of this book). I will bypass the details of such an estimation and focus instead on hypothesis testing using logit models. Two of these tests are crucial to understand. The first one is the Wald test or *z-test* that is used to test the statistical significance of each model coefficient. The *z*-test strategy resembles that of the OLS *t*-test: To test the null hypothesis that an individual population coefficient equals zero:

$$H_0: \beta_k = 0$$

one would have to examine the *z-statistic*:

$$z = b_k / s_k$$

where b_k is the sample regression coefficient (i.e., the estimate of β_k using the data sample under study) and s_k is the estimated standard error; both coefficients and standard errors are now calculated using the ML estimation routine. The alternative to the above null hypothesis is that the population coefficient is not zero, or:

$$H_1: \beta_k \neq 0.$$

The z-test formula above indicates how far the sample regression coefficient b_k is from the hypothesized population parameter in estimated standard errors. If H_0 is true, then $\beta_k = 0$, and this z-statistic is asymptotically normally distributed. In general, any coefficient for which the obtained z-test p-value is less than 0.05 is said to be *statistically significant*.

The confidence intervals for logit models follow the same logit as in OLS. So the confidence interval goes from $(b_k - z\text{-value} \times s_k)$ to $(b_k + z\text{-value} \times s_k)$. Or, as follows:

$$b_k - z\text{-value} \times s_k \leq \beta_k \geq b_k + z\text{-value} \times s_k.$$

To test the significance of a set of independent variables, the logit regression also employs a strategy similar to the F-statistic in OLS. The null hypothesis that g independent variables have no effect on the probability of y being equal to 1, or:

$$H_0: \beta_1 = \beta_2 = \cdots = \beta_g = 0$$

is tested with the so-called *log-likelihood ratio statistic* or χ^2-test, comparing the log-likelihood of a full model (with $k + g$ parameters) with the log-likelihood of the reduced model (i.e., a model with k parameters or g fewer variables). Such a ratio statistic follows a theoretical χ^2-distribution with g degrees of freedom. A low enough p-value (for example, $p < 0.01$) indicates that the null hypothesis that all g logit regression coefficients are equal to zero can be rejected (at the 0.01 level). This means that *at least one* of such coefficients is different from zero— but not necessarily that each and every one is different from zero. If the test rejects the null hypothesis, the full model fits the data better than the reduced one.

Notice that up to now, I have not mentioned the word *time*. This is because time t does not appear explicitly in the right-hand side of the model equation. It is fine, methodologically speaking, to ignore time when the social process under study is stable and consequently does not change over time (and the data set is cross-sectional). In such cases, the probability that a given individual is in a certain state e is therefore constant over time as well. However, such an assumption is rather unrealistic, and such probability may indeed change over time. When the social process under study is not constant, most researchers include time directly in the functional form or equation as one of the elements of $X_k(t)$. This is indeed what the so-called *panel logit regression models* accomplish for the analysis of longitudinal data. The easiest option is to include time t as an additional independent variable in the functional form to account for any changes in the state probability over time; just as follows:

$$\log \frac{p_{ki}}{p_{1i}} = \alpha_e + \sum_{k=1}^{K} \beta_{ek} x_{ki} + t + u_i \tag{3}$$

where time t is a cardinal variable included in the regression-like model controlling for the passing of time. Another common methodological possibility is to include a series of time dichotomous or dummy variables in the regression model predicting the state probability. So that, when a given time-dummy variable takes the value of 1, one is referring to a given period of time t (0 for the rest of time periods in the sample):

$$\log \frac{p_{ki}}{p_{1i}} = \alpha_e + \sum_{k=1}^{K} \beta_{ek} x_{ki} + \phi_t + u_i \tag{4}$$

where $\phi_t = \sum_{t=1}^{T-1} \phi_t t_t$ and t_t are the different time-dummy variables in the model (one for each point of time t).

This last specification is helpful in predicting probability differences in the outcome over time. One could even incorporate interaction terms between these time-dummy variables and any independent variables of interest. These interactions help to assess whether the effect of certain independent variables on the state probability varies over time.

The panel logit regression approaches are relatively easy to implement using any statistical program that performs logit and probit analyses, and they provide quick and quite satisfactory results. But one needs to understand that despite their simplicity, these techniques were designed for the study of static social processes where observations are not correlated with each other over time. These models, when estimated using event history data, violate many of the basic assumptions of the logit or probit regression models. The most serious violation occurs when the error terms are correlated across observations in the sample (depending on the cases and the time periods). Furthermore, these logit and probit models have important shortcomings even when trying to incorporate time in the prediction of state probabilities. First, such models ignore that the probability of being in a certain state e at time t is rarely independent of the probability of being in state f during the same time period or the probability of being in state e during the previous time period $t-1$. These models also ignore any process of inertia that might exist in the social process under study, that is, the tendency for social cases to remain in a state f after they have been in such state e for a while.

Finally, these models do not correct the problem that some omitted variables might be highly correlated with time. For all these reasons, one should be cautious when using these static probability models for analyzing events over time. In the case of longitudinal data sets, one needs to evaluate whether the main assumptions behind the logit and probit models are violated and/or make sense.

In many other instances, one may be interested in analyzing another kind of probability, the so-called *transition probability*, defined as:

$$P_{fe}(u, t) \equiv \Pr[y(t) = e \mid y(u) = f].$$

The transition probability is the probability that an individual or case will transit from state f at time u to state e at moment t. When examining transition probabilities, one of the three limitations associated with the logit analysis mentioned above is overcome because the model is now controlling for the previous state at moment u. For example, a researcher could be interested in analyzing the probability that an individual is married by the age of 40 given that he was not married at the age of 30. This probability could be expressed as $P_{01}(30, 40)$. Similarly, the probability that an individual is unemployed at the age of 35, given that she was employed at the age of 25, could be denoted as $P_{10}(25, 35)$. The subscripts are 0 and 1 in the first example and 1 and 0 in the second example because married (employed) previously was codified as 1 and unmarried (or unemployed) was codified as 0. In both examples, one is interested in explicitly studying the transition from a given state f at time u to another different state e at a later time t.

As with the case of modeling state probabilities, one can then evaluate the influence of a series of explanatory variables on the transition probability using logit and probit regression models or, even better, *multinomial logit regression* models when the space of possible states includes more than two different states. Since these classic static approaches were meant for the analyses of cross-sectional data sets, they still do not account for the passage of time. However, one can introduce the variable time as an additional independent variable (or set of time-dummy variables) in such logit or probit models to correct for the process of time (as I did earlier in this section).

I now refer back to the example data set to study employee turnover and illustrate the results obtained estimating some of the logit models presented above. Several logit models (including time variables as independent variables) could be estimated in order to understand which factors contribute to employee turnover at this service organization. Table 3.4 reports the number of employees terminated each semester for every year under study (for a total of 3 years). Of the total population of 325 employees, 162 are still in the company at the end of the time window; these are right-hand-censored cases. The estimated logit

Table 3.4
Time distribution of all employee terminations during the first three years of tenure on the job

Year	Semester	Terminated Employees (Number)	Total Number of Employees (at Risk of Termination)	Estimated Risk of Termination
1	1	53	325	0.163
1	2	51	272	0.188
2	1	32	221	0.145
2	2	14	189	0.074
3	1	10	175	0.057
3	2	5	165	0.030
4	(Very beginning)		160	
Total		165		

$N = 325$ employees.

regression model regresses the probability of a given employee to be terminated on several explanatory variables (i.e., the dependent variable is TERMINAT). Eight of these explanatory variables are control variables (such as gender, age, or education). The other two explanatory variables help to understand the impact of personal connections on employee termination, that is, CONTACT (if the employee was referred at the time of application, this dummy variable equals 1, 0 otherwise) and CONTACTT (if the referrer has been terminated, this dummy variable equals 1, 0 otherwise).

In this example, the risk set includes all employees at risk of being terminated. Thus, in the first semester, all 325 hired employees are at risk of being terminated. Fifty-three of those employees were terminated during that first semester, with 272 remaining in the risk set of the following semester. In the second semester of that first year, 51 employees were terminated, leaving 221 in the call center, and so forth. Obviously, at the end of each semester, the risk set decreases in the number of employees at risk of being terminated from the previous semester. One could easily compute the same event table yearly, monthly, and even daily depending on the research agenda. However, keep in mind that the purpose of tables like Table 3.4 is to start providing some basic summary description of the events occurring in the sample to the readers of the study or report. For the case of employee termination, I show some basic statistics like the ones provided in Table 3.4 before starting any more advanced quantitative analysis.

Table 3.4 clearly indicates that the number of employees at risk of termination decreases from 325 in first semester of the first year to 165 in second semester of the third year. The total number of events to analyze is 165 employee termi-nations. In any event history data analysis, there are three key numbers to report to the readers. The first one is the total number of individual cases at risk at the beginning of the study (in this example, 325 employee hires in a three-year period); the second is the total number of individual records (or spells) in the sample to analyze (or 751 in the example). Finally, the table reports the total number of events occurring in the time period under study (or a total of 165 in the three-year period under analysis).

In addition to state probability, an important concept in EHA is the *estimated risk* of termination. This is the probability that any particular employee in the organization is terminated at a given moment of time, given that such an employee is at risk of being terminated. The risk of employee termination could be estimated for each time period. To estimate the risk of any particular event at a point of time *t* for a sample of individual cases, divide the total number of events at a given time *t* by the total number of individual cases at risk (of such event) at time *t*. For example, the risk of termination during the first semester of the second year of work at the bank is 0.145 (that is, 32/221, or 14%). Estimations of the risk in other semesters are provided in the last column of Table 3.4. As one can see, such a termination risk diminishes over time. It should also be noted that even when the risk set gradually decreases in size, the risk of termination can still increase over certain time periods. For example, in this par-ticular employee sample the estimated termination risk during the second semes-ter of the first year is greater than the risk during the first semester of that same year, even though the absolute number of terminations is practically identical in both semesters.

Table 3.5 reports the estimated coefficients for two different logit regression models estimated using the sample of employees. I used the Stata software pro-gram in order to estimate the logit regression model that fits the employee termi-nation data. I use the command `logit`, immediately followed by the name of the dependent variable, followed by the list of independent variables (or their com-puter variable names). Model A is the classic logit regression model when esti-mated for a longitudinal data set—*as if* the analyzed data set is a cross-sectional data set, and consequently this model ignores any temporal component of such database. To estimate model A, I typed "`logit terminat male age married college wage nights parttime previous contact contactt`" in the Stata Command box once the employee data set was read by Stata. A summary of the main logit regression results is reported in Table 3.5.

Two columns are relevant in order to understand employee terminations. The first column of Table 3.5, headed by "*B*" or "Coef.," gives the estimated model parameters of interest; the second column gives their standard errors. The coefficients are not standardized, meaning that their magnitude or size depends

Table 3.5
Logit regression predicting employee termination

Independent Variables	Model A		Model B	
	B	Standard Error	B	Standard Error
Constant	−0.310	0.507	−1.143	0.586
Male	0.349	0.236	0.363	0.239
Age	−0.012	0.012	−0.010	0.012
Married	−0.354*	0.202	−0.346*	0.204
College degree	−0.270	0.309	−0.342	0.314
Hourly wage	−0.037	0.040	−0.026	0.041
Night shift	−0.725**	0.313	−0.739**	0.316
Part-time	0.035	0.065	0.030	0.065
Previous jobs	0.015	0.092	0.022	0.093
Personal contact	−0.603***	0.207	−0.694***	0.210
Contact is terminated	0.814***	0.334	0.994***	0.344
First year (of tenure)			0.972***	0.314
Second year			0.563**	0.334
χ^2 statistic:				
Compared with baseline model[a]		25.076***		36.931***
Degrees of freedom		10		12
Compared with model A				11.855***
Degrees of freedom				2

$N = 325$ employees.
Note: All significant levels are the following: *$p < 0.05$; **$p < 0.01$; ***$p < 0.005$
(one-tailed tests).
[a] The model labeled baseline is a model with no independent variables (only the constant).

entirely on the unit of measurement used for each of the independent variables. The two main independent variables of interest in this study have a significant effect on the probability of employee termination; both coefficients are significant at the 0.005 level. Employees who were recruited through the referral program (i.e., those who had a personal contact in the organization at the time of applying for a job) have a lower probability of termination, other things being equal. However, when the personal contact leaves the organization, the likelihood of termination increases: To the extent that those employee referrals whose employee contact (or referrer) left the organization are even more likely to be terminated than those non-referral applicants.

Again, the following formula can help to transform the estimated model coefficients into meaningful probability changes (in percent). Such percent changes inform about increases or decreases in the probability of any event happening. In general:

$$100\% \times [\exp(\beta) - 1].$$

In the example, for a referral employee, the probability of termination decreases by 45% when compared to non-referral employees ($100\% \times [\exp(-0.603) - 1]$). When the referrer (or personal contact) leaves the organization, the probability of termination of the referred employee then increases by 24% compared with an employee who was not referred to the job by a current organizational member ($100\% \times [\exp(0.814 - 0.603) - 1]$).

Model B allows for changes in such probability of termination over time. This is accomplished by explicitly incorporating two dummy or dichotomous variables in the logit regression analysis, one for the first year and another for the second year (the third year-dummy variable is omitted because it is the reference year). This new logit regression controlling for time shows that the termination probability does vary significantly from year to year. Both year coefficients for the two dichotomous variables are positive and statistically significant, and therefore they are statistically different from zero. There is a temporal trend in the employee termination probability.

Understandably, the introduction of the year-dummy variables greatly improves the fit of the model. The χ^2-statistic comparing models A and B allows for testing whether adding the temporal variables improves the goodness of the model. Hypotheses tests for logit regression (using χ^2-tests) employ a similar model comparison strategy to the F-tests in OLS regression models. This incremental χ^2-test can only be applied when the two models to compare are nested, meaning that model B includes the same variables that model A includes plus some additional independent variables. In this case I added two year-dummy variables to model A in order to get model B. The null hypotheses or H_0 is that the year variables have no effect on the probability of termination:

$$H_0: d_{\text{First year}} = d_{\text{Second year}} = 0$$

where I use d to denote a year-dummy variable (the reference category is the third year under study). As explained earlier, the χ^2-statistic is calculated by the method of ML. The value of the χ^2-statistic comparing models A and B is 11.86. The degrees of freedom are two, since this is the number of restrictions that distinguishes model A from model B: So model A is model B when the first-year and second-year coefficients are restricted to zero. The value of the χ^2-statistic is over the critical value for the 0.005 significance level. Therefore, it is important to control for any possible temporal variations in the probability of termination. This incremental χ^2-test is the standard way of comparing two models by testing whether adding a sub-group of explanatory variables into the model significantly increases the fit of the model in the case of models estimated by ML. This incremental χ^2 is also helpful for comparing models using the different EHA techniques discussed in the rest of this chapter.

In this particular case, also note that including the year-dummy effects increases the magnitude of the two sociological variables of interest (i.e., the two variables evaluating the effect of personal networks on the process of employee termination). I chose to include year-dummy variables; I could have chosen to incorporate month or semester-dummy variables instead. The specification, estimation, and interpretation of the model would not change much if that were the way I had chosen to control for time differences in the employee termination probability.

4.   Dynamic Methodology for the Analysis of Events

So far, I have reviewed how to use some available regression techniques for the case of categorical variables such as logit or probit regression models to analyze the risk of events. These regression methods were designed for the analysis of a cross-sectional data set though. An entire dynamic methodology exists, which is specifically appropriate for the analysis of longitudinal data where the main dependent variable is of a qualitative nature. These methods, broadly known as EHA, are described in detail in the rest of this chapter. The EHA includes a set of techniques to explain how variation in a categorical dependent variable y_{it} is influenced by a vector of independent or explanatory variables X_{kit} when a cross-sectional cross-time data set is available for the study. Unlike the standard logit and probit analyses, these EHA techniques include the time variable explicitly in the functional form of the model at the time of estimation.

When analyzing a longitudinal data set with some events, there are two main EHA approaches or sets of EHA techniques. The first category of EHA techniques refers to the methods used to describe and summarize the basic time component of the data and the event as the unit under analysis. These techniques are called *non-parametric methods* or *exploratory analysis*. By non-parametric techniques, I refer to the group of methodological techniques that describe in detail the temporal dimension of the event under study, including both the frequency and the pattern whereupon events happen. These standard techniques include life tables, graphs of the survivor function, graphs of the hazard rate, and many other key EHA functions. These techniques are valuable because they help researchers to learn about the risk of the event as it changes over time. Understanding this risk is crucial for its later parametric specification and estimation using multi-variate models. These standard techniques are also extremely useful for comparing hazard rates and risk over time across a few groups of individuals or cases in the sample. In the example used in this chapter, one could easily have plotted the termination risk for two groups of employees, the referrals versus the non-referrals. And, go so far as to compare those referrals whose referrer left the organization to those whose referrer is still around. The idea behind this first set of techniques is to "explore" the timing and pattern of events in the sample by estimating and plotting some of the EHA non-parametric functions over time for both the entire sample and for groups of interest.

Such non-parametric analyses can be, by themselves, the final objective of the research project. However, they are often the first (and, I would say, necessary) steps before getting into the second EHA approach, which includes a set of techniques for the further investigation of events over time. These models are called *parametric methods* or *multivariate* (even confirmatory) *analysis* because they explore how additional factors or variables (other than time) account for differences in the risk of any event happening in the population under study in some multivariate fashion. More specifically, these parametric methods model the risk of an event as a function of a set of explanatory variables over time. The general EHA model is typically formulated as follows:

$$g(t) = f(t, X)$$

where $g(t)$ can be any of the EHA functions including the hazard rate, the survivor function, or the density function—I will describe these in the next section of this chapter. Notice that each EHA function is specified as a function of time t and a vector of independent or explanatory variables X, including variables that may or may not change over time. The most frequently modeled expression of $g(t)$ is the *hazard rate* or $r(t)$. This hazard rate is the most popular EHA function because it represents the risk of experiencing a categorical change or event in a given moment of time. Most importantly, this rate is probably the most intuitive measure one can use for analyzing events over time as a function of explanatory variables.

In this chapter, I help the reader model the hazard rate—even though the same specification and estimation techniques can easily be applied to the modeling of other EHA functions of interest without much difficulty. So, if one is interested in specifying the hazard rate as a function of time, the formula is now:

$$r(t) = f(t, X_k)$$

where $r(t)$ is the hazard rate or risk of an event e happening at time t. The central idea behind this model is that the risk of an event happening or hazard rate is a function of time and a vector of *k* independent variables. Given that this is a type of dynamic methodology, the formula needs to account for time explicitly in the right-hand side of the equation. In addition, this hazard rate depends on a vector of explanatory variables, so the causal logic behind the change in such a rate is:

$$\Delta X(t_{t-1}) \rightarrow \Delta r(t) \quad t_{t-1} < t.$$

Changes in some of the explanatory variables in the past change the risk of an event happening or the rate of transition into a given state in the future since these concepts describe the propensity of individual cases under study to change. With this definition, notice that it can never be the other way around. It is crucial to seriously consider the temporal order when analyzing any social process. The hazard rate or rate of transition $r(t)$ at time t depends on past conditions or factors (at t_{t-1}) or current conditions or factors (at time t), but this rate should never depend on the future (that is, after t). How far in time past values affect the current rate depends entirely on theory or substantive reasoning as well as on the data availability for a study. In addition, some of the variables X included in the model can be individual constant variables (that is, variables that do not change over time). The causal relationship could now be represented as follows:

$$X \rightarrow \Delta r(t).$$

In this case, the goal of the researcher is to identify differences in the hazard across individual time-fixed attributes or characteristics.

There are many other possibilities when it comes to specifying this functional form $f(t, X)$. It is classic (and well-established) for the analysis of events to specify the hazard rate as an exponential function of the independent variables multiplied by a function of time, just as follows:

$$r(t) = \exp(\beta' X_{it}) \, q(t). \tag{1}$$

The exponential model, which will be discussed later in this chapter, is chosen because the estimated or predicted risk cannot be negative. For obvious reasons, a negative risk value does not make much sense. The widespread models identified in the above formula are also called *proportional models*. There are several types of *proportional models* depending on how the dependency of the hazard rate on time represented as $q(t)$ is modeled, including the Gompertz, Weibull, and logistic time models. The *Cox model* is a dynamic regression technique that is *partially* parametric; this is so because the time functional form represented by $q(t)$ is not directly specified nor estimated in the EHA model. Finally, the *non-proportional hazard rate models* and the *piecewise exponential models* also allow to further examine any possible time variations in the effects of certain explanatory variables on the risk of an event happening over time. In such cases, the $q(t)$ or function of time is specified as $q(t, Z_{it})$, where Z_{it} now includes any vector of explanatory variables—which can be identical or different from X_{it}—that influences the time component of the hazard rate. I further discuss all these multivariate EHA models in greater detail in this chapter.

5. Fundamentals of EHA: The Hazard Rate

There is still some additional terminology relevant to understanding the EHA methodology. In the analysis of events, the space of possible states an individual unit can occupy at any given time is discrete and can change over time. Thus, the dependent variable y_{it} is a discrete variable whose values change over time as the result of changes in some explanatory variables. Consequently, the most basic form of summarizing change in the sample is by estimating the probability of change in y_{it}. More specifically, one could specify the change that takes place when individuals transition from an origin state f into a final state e at some point in time.

I demonstrated earlier that a useful concept that permits the description of the process of change at a given point of time is the *probability of an event* (or state probability)—where an event is defined explicitly as the change in the value of a discrete random variable y from f to e. To define such an event probability concept, consider T as a random variable that represents the time duration from the start time (or origin) t_0 until the change in the dependent variable y_{it} takes place (that is to say, when the event defined as transitioning from an initial state f to an end state e takes place). T is also the moment of time when an event takes place. Thus, the probability of an event can be properly defined as follows:

$$\Pr(t + \Delta t > T \geq t \mid T \geq t).$$

This formula represents the probability that the event or transition from one state to another occurs during the time period that goes from t until $t + \Delta t$, given that the event did not occur before time t. This describes the temporal aspect of the process of change under study. The definition includes all moments in time when such a change can possibly occur, and it therefore expresses the idea that change is continuous. This definition entirely depends on any past information about the process of change, that is, any information about what happened up until the moment of time t. The previous formula can be used to describe the process even before it has been completed for all individual cases in the population under analysis.

As individual cases are examined during a continuous time interval, the previous probability expression could take any time interval Δt of interest. As this time interval becomes smaller and consequently approaches the value of 0, the concept of probability of change in the dependent variable tends to disappear because the probability that a change occurs in such a short time interval is 0. So as Δt approaches 0, so does the probability of an event. One can, however, consider the ratio of the probability of an event and the time interval or duration Δt (as a measurement of future changes in the dependent variable in a given time interval). Thus, the following is the mathematical definition:

$$\frac{\Pr(t + \Delta t > T \geq t \mid T \geq t)}{\Delta t}.$$

If one takes the limit of this quantity, one of the pivotal key concepts in EHA can be formulated, the *hazard rate* $r(t)$, also known as the *failure rate*, or simply as the *risk*. As Δt approaches 0 in the limit, the hazard rate is defined as follows:

$$r(t) = \lim_{\Delta t \to 0} \frac{\Pr(t + \Delta t > T \geq t \mid T \geq t)}{\Delta t}.$$

Notice that the hazard rate is the instantaneous probability or risk of an event occurring in a given moment of time t, conditional on the fact that the event has not occurred before time t. This concept is useful in describing the evolution of a process of change over time: The rate or hazard $r(t)$ is interpreted as a measurement of the instantaneous probability that an event occurs in a moment of time t, always with the condition that this event or change of state has not happened before t. I should emphasize that this rate of change is calculated for each group of cases still at risk of an event in a moment of time t—that is, the set of individual cases that can experience the event because they have not experienced it prior to time t. For this reason, this concept can be estimated empirically by simply calculating the proportion of individual cases that have experienced the event during an interval of time, divided by the duration of the time interval. The assumption is that the estimated hazard rate is an approximation or estimation of the rate for all individuals in the population under study.

When the event is defined as the transition or change from state f to another state e, the hazard rate concept is also called the *transition rate*. One typically uses the notation, $r_{fe}(t)$ to indicate the rate at which individuals change or transition from state f to e at time t, assuming that the transition has not taken place before time t. When there are *competing risks*, meaning that there are multiple states into which any given individual could transition (from a given initial state f), this formulation is much more attractive than simply $r(t)$ because it clearly shows what the start and end states are. There are other statistical concepts of interest here. One of these concepts can be written as $m_e(t)$, representing the probability that any individual case changes to state e (regardless of the initial state) at time t. Another useful concept measures the intensity of the instantaneous rate of transition to e. Again, in the case of competing risks, this rate notation is much preferred to the hazard rate. In any of the previous scenarios, researchers are interested in looking into factors or variables that either increase or decrease the hazard of an event happening.

The hazard rate is the key concept in EHA in contemporary empirical analyses, but there are other frequently used and equally useful statistics, such as the *cumulative distribution function*, the *survivor function*, and the *density function*. Each of these three EHA functions is discussed just ahead because they are helpful in describing the same temporal pattern of events in different meaningful ways. Consider any point of time t within the period of observation for an individual case or unit of analysis i. Again, the time when the first event occurs is a random variable T in the case of a single non-repeatable event, such as, for example, getting a first job or having a first child.[24] One fundamental concept describing the timing of this event is the *cumulative distribution function*, or:

$$F(t) \equiv \Pr(T \leq t) = \int_0^t f(u)\, du.$$

The $F(t)$ function represents the probability that the event happens before time t. The function $f(t)$ (lowercase f) is the *density function* (of $F(t)$) and is used to denote the probability that the event occurs at time t. $F(t)$ can be calculated by adding together all the events that have occurred until moment t (including the number of events at moment t) divided by the number of individual cases in the sample under study at the outset. The function $f(t)$ is calculated by dividing the number of events that happen at a given moment of time t, by the total number of individuals in the sample at the outset. For example, suppose that 100 employees begin working for one organization in the year 2000: 20 employees leave in their first year, 10 the second year, 2 in the third, and 1 in the fourth year. Then $F(3)$ is $(20 + 10 + 2)/100$ or 0.32 (or 32%) whereas $f(3)$ is $2/100$ or 2%.

---

[24] I simply start with presenting the EHA models for the analyses of non-repeatable events. The techniques learned in this chapter can easily be extended (and understood) when analyzing repeatable events over time.

Similarly, the probability distribution of T is defined as a complement function of $F(t)$. This function is usually called the *survivor function* (also called *survival function*):

$$S(t) \equiv 1 - F(t) \equiv \Pr(T > t).$$

$S(t)$ is the probability of survival up to time t, that is, the probability of "surviving" (or not experiencing) the event or change until time t. In other words, $S(t)$ is the probability that the event has not happened for a given individual before time t. Understandably, at the beginning of the period under analysis t_0, the probability of survival is 1: $S(t_0)$ is equal to 1. Theoretically, the probability of survival after an infinite amount of time has passed is 0: $S(\infty)$ is equal to 0, meaning that in the very long run no individual case survives the event. However, very frequently $S(\infty)$ is greater than 0, which means that the probability that an event does not occur is different from 0, with many cases surviving the event forever. In the example given above about the 100 new employees, $S(3)$ is equal to $(1-0.32)$ or 0.68; in other words, 68% of all employees are still working in the organization three years after 2000. Again, the survivor function is 1 at the beginning of the time period under analysis and gradually diminishes over time. That also means that the number of "survivors" or non-participants in the social change under study diminishes as events occur over time.

The *cumulative distribution function* and the *survivor function* are mathematical complements. One can easily be obtained from the other by subtracting from 1. The survivor function is often preferred mainly because it provides a more intuitive description of the process of change. Imagine a population of individuals beginning all at the state of origin f at the same moment t_0 (when equal to 0). As time goes by and various events occur, many of the individual cases change state, leaving the initial state f. This process can be described and captured by the survivor function, so that if N is the total number of individual cases in the population at time 0, then $N \times S(q)$ is the number of cases that have not yet experienced the event or change under study at time q. Therefore, $N \times S(q)$ is the number of individuals remaining at the initial state f at time q (i.e., "surviving"). Remember that the individual cases who have survived the event constitute the *risk set*, or the group at risk (of experiencing the event).

Since T is a continuous random variable, its distribution can also be described using the *density function*, which in turn is related to the cumulative distribution function (or the survivor function) as follows:

$$f(t) \equiv \frac{dF(t)}{dt} \equiv \frac{dS(t)}{dt}.$$

The density function is the *unconditional* probability that an event occurs at time t. The relative frequency distribution of events that occur at time t (histogram of frequencies over time) provides an empirical approximation of the probability density function over time. The meaning of the density function is similar to that of the hazard rate, only that the density function is an unconditional probability. Remember that the hazard rate measures the probability that an event occurs at time t, conditional on such an event not having occurred before t. This last condition is what distinguishes the hazard rate from the probability density function.

The other common concept in EHA studies is the integral of the hazard rate (also known as the *integrated hazard rate*), $R(t)$ (that is represented by capital R), also called the *cumulative hazard rate*. It can be viewed as the sum over time of the time-specific hazard rates. It is also expressed as minus the logarithm of the survivor function, or:

$$R(t) \equiv \int_0^t r(u)\,du = -\ln(1 - F(t)) = -\ln(S(t)).$$

This last expression allows the survivor function to be expressed as:

$$S(t) = \Pr(T \geq t) = \exp\left[-\int_0^t r(u)\,du \right].$$

The hazard rate $r(t)$ is the unconditional probability that the event occurs at time t, $f(t)$ or density function, divided by the probability of surviving the event at time t, $S(t)$ or survivor function.[25] Empirically, this hazard rate can be computed by dividing the number of events that have occurred in the sample up until moment t by the number of individual survivors that have not experienced the event at time t. In the employee sample above, the estimated hazard rate at year three or $r(3)$ is 0.029 (or 2/68). In general, the formula including all three relevant EHA concepts is the following:

$$r(t) \equiv \frac{f(t)}{S(t)}.$$

---

[25] While the unconditional probability that the event occurs at time t, $f(t)$ or density function of T, can be formally represented by the formula:

$$f(t) = \lim_{t \to 0} \frac{\Pr(t + \Delta t > T \geq t)}{\Delta t} = r(t) \exp\left[-\int_0^t r(u)\,du \right].$$

This simple formula demonstrates that knowing two of the three major EHA functions, $r(t)$, $S(t)$, or $f(t)$, enables one to obtain the third function.[26] In fact, all three functions provide different useful ways of describing the same distribution of events over time.

6.   Exploratory Analysis of Events

As of now, I have covered the most important EHA concepts such as the survivor function, integrated hazard rate, and hazard rate. All these concepts are extremely useful in understanding the statistical theory behind EHA. Moreover, they are key when performing more advanced dynamic analyses of event history data. By "exploratory analysis" or "descriptive analysis," one typically refers to non-parametric type of analyses, i.e., the analyses that incorporate only a few explanatory variables when describing the pattern and frequency of events in the sample under study. These methods are mainly used to describe the temporal component of the process under investigation. Here, it is insightful to examine the graphical representation of the survivor function and the hazard rate. These descriptive analyses are also convenient when comparing the process of change among different sub-groups—whichever explanatory variable is used to divide any population sample into meaningful groups of interest. For example, one could test whether the shape of the survivor function when studying hiring is the same for female and male employees in one organization; one could plot the different survivor functions for employees in different occupational groups; or one could plot the hazard rate of new companies going bankrupt by industry and region. All of these distinct analyses are exploratory in nature and provide some useful insights about patterns of events in a sample.

---

[26] The density function can be written as follows:

$$f(t) = \lim_{\Delta t \to 0} \frac{F(t + \Delta t) - F(t)}{\Delta t} = \lim_{\Delta t \to 0} \frac{\Pr(t + \Delta t > T \geq t)}{\Delta t}.$$

Comparing this last expression to the hazard rate, one can observe that the hazard rate is the conditional density function (conditional on the event not having occurred before time t). Formally, its expression is:

$$\Pr(t + \Delta t > T \geq t \mid T \geq T) = \frac{\Pr(t + \Delta t > T \geq t)}{\Pr(T \geq t)}.$$

Applying the definition of the hazard rate, one can see the relationship among all three EHA functions:

$$r(t) = \lim_{\Delta t \to 0} \frac{\Pr(t + \Delta t > T \geq t \mid T \geq t)}{\Delta t} = \lim_{\Delta t \to 0} \frac{\Pr(t + \Delta t > T \geq t)}{\Delta t} \frac{1}{\Pr(T \geq t)} = \frac{f(t)}{S(t)}.$$

Thus, one can derive that the hazard rate function is the conditional density function $f(t)$ divided by the survivor function $S(t)$.

Next, I provide some information about the main techniques available in order to perform the exploratory analysis of events. I start with life tables as a common statistical tool for the description of events in a given sample over time. Then I introduce the Kaplan–Meier and Nelson–Aalen estimators of some of the basic EHA functions. At the end of this section, I present some practical ways of testing the equality of a few of these EHA functions across groups or categories of interest as a means of starting to evaluate and test theoretical hypothesis of interest.

6.1 Life Tables

One of the simplest methods to describe the occurrence of events is the *life tables method*. This method consists of estimating the values of any of the main EHA functions during any set of fixed time intervals. The basic idea behind the life table is to subdivide the period of observation into smaller time intervals or pieces and to use the events happening in those time intervals to estimate the EHA functions. For each interval, all individual cases in the risk set are used to calculate the probability of the event occurring in that interval. The probabilities estimated from each of the intervals are then used to estimate the overall probability of the event occurring at different time points. The best way to understand the rationale behind these life tables is to look at an example.

In Stata, the `ltable` command allows the user to estimate, display, and graph life tables. Here is the Stata command line followed by its output:

```
ltable endtime terminat, survival intervals(45) tvid(employee)
hazard
```

Interval		Beg. Total	Deaths	Lost	Survival	Std. Error	[95% Conf.	Int.]
0	45	325	10	0	0.9692	0.0096	0.9436	0.9833
45	90	315	15	0	0.9231	0.0148	0.8883	0.9474
90	135	300	13	0	0.8831	0.0178	0.8429	0.9135
135	180	287	12	0	0.8462	0.0200	0.8021	0.8811
180	225	275	14	0	0.8031	0.0221	0.7555	0.8423
225	270	261	19	0	0.7446	0.0242	0.6935	0.7885
270	315	242	10	0	0.7138	0.0251	0.6614	0.7597
315	360	232	8	7	0.6889	0.0257	0.6354	0.7361
360	405	217	10	6	0.6567	0.0264	0.6021	0.7057
405	450	201	11	8	0.6200	0.0272	0.5643	0.6707
450	495	182	5	18	0.6021	0.0275	0.5458	0.6537
495	540	159	6	6	0.5789	0.0281	0.5219	0.6317
540	585	147	3	8	0.5668	0.0283	0.5093	0.6202
585	630	136	4	7	0.5497	0.0287	0.4915	0.6040
630	675	125	1	14	0.5450	0.0289	0.4867	0.5996
675	720	110	5	5	0.5197	0.0297	0.4599	0.5760
720	765	100	6	4	0.4878	0.0306	0.4267	0.5461
765	810	90	1	13	0.4820	0.0308	0.4205	0.5407
810	855	76	3	6	0.4622	0.0315	0.3994	0.5226

Interval		Beg. Total	Deaths	Lost	Survival	Std. Error	[95% Conf. Int.]	
855	900	67	2	15	0.4467	0.0323	0.3825	0.5087
900	945	50	2	7	0.4274	0.0337	0.3608	0.4923
945	990	41	2	15	0.4019	0.0362	0.3308	0.4719
990	1035	24	0	11	0.4019	0.0362	0.3308	0.4719
1035	1080	13	1	6	0.3617	0.0501	0.2649	0.4591
1080	1125	6	0	6	0.3617	0.0501	0.2649	0.4591

Interval		Beg. Total	Cum. Failure	Std. Error	Hazard	Std. Error	[95% Conf. Int.]	
0	45	325	0.0308	0.0096	0.0007	0.0002	0.0003	0.0011
45	90	315	0.0769	0.0148	0.0011	0.0003	0.0005	0.0016
90	135	300	0.1169	0.0178	0.0010	0.0003	0.0004	0.0015
135	180	287	0.1538	0.0200	0.0009	0.0003	0.0004	0.0015
180	225	275	0.1969	0.0221	0.0012	0.0003	0.0006	0.0018
225	270	261	0.2554	0.0242	0.0017	0.0004	0.0009	0.0024
270	315	242	0.2862	0.0251	0.0009	0.0003	0.0004	0.0015
315	360	232	0.3111	0.0257	0.0008	0.0003	0.0002	0.0013
360	405	217	0.3433	0.0264	0.0011	0.0003	0.0004	0.0017
405	450	201	0.3800	0.0272	0.0013	0.0004	0.0005	0.0020
450	495	182	0.3979	0.0275	0.0007	0.0003	0.0001	0.0012
495	540	159	0.4211	0.0281	0.0009	0.0004	0.0002	0.0016
540	585	147	0.4332	0.0283	0.0005	0.0003	0.0000	0.0010
585	630	136	0.4503	0.0287	0.0007	0.0003	0.0000	0.0013
630	675	125	0.4550	0.0289	0.0002	0.0002	0.0000	0.0006
675	720	110	0.4803	0.0297	0.0011	0.0005	0.0001	0.0020
720	765	100	0.5122	0.0306	0.0014	0.0006	0.0003	0.0025
765	810	90	0.5180	0.0308	0.0003	0.0003	0.0000	0.0008
810	855	76	0.5378	0.0315	0.0009	0.0005	0.0000	0.0020
855	900	67	0.5533	0.0323	0.0008	0.0005	0.0000	0.0018
900	945	50	0.5726	0.0337	0.0010	0.0007	0.0000	0.0023
945	990	41	0.5981	0.0362	0.0014	0.0010	0.0000	0.0033
990	1035	24	0.5981	0.0362	0.0000	.	.	.
1035	1080	13	0.6383	0.0501	0.0023	0.0023	0.0000	0.0069
1080	1125	6	0.6383	0.0501	0.0000	.	.	.

The `ltable` command is followed by the time variable ENDTIME in this example, which is measuring the duration of the episode. The second variable to list is the variable TERMINAT, which indicates whether the employee is terminated (value of 1) or not (value of 0). The `tvid( )` option helps to identify observations by individual case (in the case of multiple-observations per individual). In this case, the variable EMPLOYEE refers to the employee ID number. The `hazard` option is also used to ask for the hazard values to be displayed on the life table. Specifying `failure` as a command option displays the cumulative failure table.

Another popular software program, TDA (which stands for Transition Data Analysis), can be used for the analysis of event history data. TDA is an executable DOS-like program specifically designed to analyze longitudinal data. It is relatively simple to use (especially for those already familiar with DOS programs) and allows for the estimation of these basic life tables and other parametric EHA models which are discussed later in this chapter. The main reason for introducing this TDA program in this book is because TDA facilitates estimating some more advanced EHA models than Stata or SPSS. The table is estimated using the command `tp = 0 (45) 1140` (and outputting the table into a word document using the command `ltb`). Below, I show the TDA input file used to read the same employee database (using the `dfile=` command), including the commands that define the variables to be created by TDA (with the `v1 (label) = expression` command).

```
# Test    Different model with covariates

dfile = employee.txt;       data file

v1  (employeeid)=c1; This variable uniquely identifies each employee
v2  (observ)    =c2; This variable identifies the observation order
v3  (startt)    =c3; Start time for each record
v4  (endt)      =c4; End time for each record
v5  (starts)    =c5; Start stage
v6  (finals)    =c6; Final stage (Terminated = 1; Not terminated = 0))
v7  (contact)   =c7; Referral employee = 1; Non-referral = 0
v8  (contactt)  =c8; If employee contact or referrer is terminated = 1
v9  (male)      =c9; 1 if employee is male; 0 if employee is female
v10 (age)       =c10; Employee age in years
v11 (married)   =c11; 1 if employee is married (0 otherwise)
v12 (college)   =c12; 1 if employee has a college degree (0 otherwise)
v13 (wage)      =c13; Hourly wage in dollars
v14 (prevjob)   =c14; Number of previous jobs
v15 (partt)     =c15; 1 if employee works part-time (0 otherwise)
v16 (night)     =c16; 1 if employee works night-shift (0 otherwise)

# commands to define a multi-episode data

id  = v1  ; define case-id
sn  = v2  ; define episode or observation number

org = v5  ; define origin state
des = v6  ; define destination state
ts  = v3  ; define starting time
tf  = v4  ; define ending time

# add commands for life table estimation

tp = 0 (45) 1140;  time intervals 0, 45, 90 up to 1140
ltb = lifetab1.ltb; life table is written to lifetable1.ltb
```

Note that the "#" sign is used to include comments in the input file. The semicolon helps to separate each of the commands.[27] The file also includes the following six commands to define the multi-record data set (in ***bold-italics***):

```
id  =   v1  ; define case-id
sn  =   v2  ; define episode or observation number
```

---

[27] Each command in a command file must be terminated by a semi-colon. Any text after the semi-colon of a command line is ignored. Empty lines and lines where the first character is a # or a * are also interpreted as comments and consequently ignored.

```
org = v5  ; define origin state
des = v6  ; define destination state
ts  = v3  ; define starting time
tf  = v4  ; define ending time
```

All episodes that have the same value of the `id`-defined variable (in the TDA example directly above, this variable is v1) refer to the same individual case in the data set. The subsequent episodes or records in the data set are distinguished by their serial number given in the `sn` command line. The serial numbers must be positive integers and they should be contiguous. For more information about TDA, consult *Techniques of Event History Modeling* by Blossfeld and Rohwer (1995). The appendices in *Techniques of Event History Modeling* are very helpful and provide basic to-the-point information needed to run TDA.

One can also estimate these life tables using SPSS. Simply go to the *Analyze* pull-down menu and choose *Survival/Life Tables*. Then enter the time variable followed by the specification of the time intervals to be examined. Next, select the "Status variable," which is the variable coding the different possible state values. Click on the *Define Event* box in order to specify the value of the status variable that indicates that the event has occurred. Like in Stata, one can also select a categorical variable in order to compute different tables for each category of such variable. The Life Tables SPSS box should look like this:

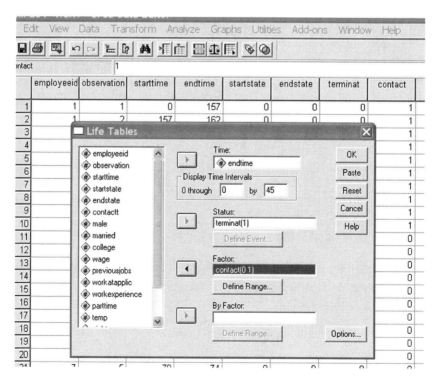

To explore the kind of event history analyses that can be performed in SPSS, the Survival Module needs to be installed first. Then, go to the *Analyze* pull-down menu and choose *Survival* with its current four options: Life Tables, Kaplan–Meier Estimators, Cox Model, and the Cox Model with Time-Varying Covariates. I discuss the last two parametric types of regression models in the next section of this chapter. I will not be using SPSS to estimate parametric survival models though. The main reason is that currently SPSS cannot be easily used to estimate many of the advanced multivariate EHA regression models I will cover in this book using cross-sectional cross-time data sets. The latest version of SPSS can only estimate the Cox Model. Furthermore, the SPSS parametric procedures assume the independence of observations (in other words, one observation for each individual case).

The *life tables method* requires a defined series of fixed intervals in a period of time determined either arbitrarily or based on some rationale. Obviously the longer the time intervals, the less descriptive the table would be of the social process under study; but likewise the shorter the time intervals, the more difficult to read and grasp the main evolution of the values for the different EHA functions over time. Within those time intervals then (in the example earlier in this section these intervals are 90 days long), I estimate the values of the survivor function, density function, and hazard rate. In the last TDA output example of 325 employees, during the first 90 days, there were 53 terminations. The hazard rate of termination during that first 90-day interval is estimated as 0.0867.

Before estimating some useful EHA techniques (described in the next section), there are other useful Stata commands to describe some event history data. In Stata, the `stset` command declares the data to be event history data (or "survival-time" data using SPSS terminology). This command informs Stata of the key variables and their future roles in the EHA analysis. In the typical case of multiple records per individual case or subject in the sample, follow the `stset` command with the variable measuring the duration of the episode (also known as the time variable). The `id( )` option is then used to specify the individual ID variable so that equal numbers refer to the same individual cases or subjects in the sample. The `failure( )` option specifies the condition indicating the occurrence of the event. Getting back to the example employee data set, I use the next command line (in ***bold italics***) to get it to be analyzed in Stata using EHA:

stset endtime, id(employee) failure(terminat==1)

ENDTIME is the employee tenure in the organization at the end of the spell or record, and EMPLOYEE is the employee ID number uniquely identifying each employee in the sample. The dependent variable here is TERMINAT, that is, the variable measuring whether the employee is terminated or not. It takes the value of 1 if any given employee is terminated (0 if the employee remains working in the organization). Notice that in the parentheses immediately following the `failure` option, I specified the condition that when true, informs Stata that the event or "failure" has occurred for a given individual case in the sample—i.e., `terminated==1` or the variable TERMINAT equals 1.[28] Note that Stata requires two "=" signs in the mathematical expression.

This example command line is one of the simplest ways of informing Stata about the longitudinal nature of the events in a data set. Other options and sub-commands within the `stset` command might be relevant depending on the structure of your event data. To learn more about these additional options, consult the online help menu about the command `stset` or read the *Survival Analysis and Epidemiological Tables Reference Manual.*[29]

Right after running the `stset` command, check the output that Stata produces for consistency. This is the best way to ensure that Stata has read the data set properly and that the variables used to declare the different aspects of the longitudinal data make sense. This is the output I obtained after running the `stset` command using the employee data set:

```
stset endtime, id(employee) failure(terminat==1)

            id:  employee
 failure event:  terminat == 1
obs. time interval:  (endtime[_n-1], endtime]
 exit on or before:  failure
------------------------------------------------
   751  total obs.
     2  obs. begin on or after (first) failure
------------------------------------------------
   749  obs. remaining, representing
   325  subjects
   165  failures in single failure-per-subject data
171381  total analysis time at risk, at risk from
                                 t =          0
            earliest observed entry t =          0
                  last observed exit t =       1104
```

The Stata output is quite informative. The first line identifies the ID variable used followed by the line indicating the condition when the event happens. One should also check that the reading of the data includes the correct number of

---

[28] If the `failure( )` option is omitted, Stata assumes all episodes end in an event. If `failure(terminat)` is specified (without any logical expression), Stata assumes that there is an event when TERMINAT is greater than 0; if TERMINAT is equal to 0 or missing, Stata assumes no event.

[29] Two options can be particularly helpful when running the `stset` command: `origin( )` and `enter( )`. The `origin(time 0)` is the default option (if not specified as part of the options in the `stset` command); such an option defines the time when each individual case becomes at risk and the default time is equal to zero. Information prior to `origin( )` is ignored. The `enter( )` option allows to specify when individuals in the sample became at risk.

total observations (or spells), and the correct number of subjects or individual cases within all those observations (obviously, this number will be identical to the number of observations when there is one observation for each individual case in the sample). The output also reports the number of "failures" or events in the sample. In this case, the output acknowledges that I am looking at single-failure-per-subject data, that is, once any individual case experiences the event, this individual is taken out of the risk set. It is possible to allow the individual to go back to the risk set in the case of multiple-events-per-subject when repetitions of events for the same individual can occur. If this is the case in a data set, use the exit option within the stset command to inform Stata of the multiple-failures-per-subject nature of the data. If this option is not specified, the default is exit(failure) meaning that the individual case is removed from the risk set, even if the individual has subsequent records or spells in the data and even if some of those subsequent records indicate other events. When specifying the exit(time .) option, the Stata program keeps all individual records after the event happens for any given individual. In other words, individuals remain in the risk set even after they have experienced the event or "failure" (as referred to in Stata). This option is of great help when modeling repeatable events, such as the risk of moving to a new city or the risk of changing jobs.

Now that Stata has read the employee termination data, I can demonstrate how to examine and summarize this data using some key commands. Two commands that should immediately be run after setting the data set for analyses are the stdes and the stsum commands. The first is used to describe the data set that sits in the memory of the Stata program and is a helpful command to ensure that the data and its description make sense. Here is how the output appears in the case of my data set:

stdes

```
     failure _d:  terminat == 1
analysis time _t:  endtime
             id:  employee
```

			per subject		
Category	total	mean	min	median	max
no. of subjects	325				
no. of records	749	2.304615	1	1	11
(first) entry time		0	0	0	0
(final) exit time		527.3262	3	480	1104
subjects with gap	0				
time on gap if gap	0	.	.	.	.
time at risk	171381	527.3262	3	480	1104
failures	165	.5015385	0	1	1

The output for the `stdes` command also documents the number of individual cases or subjects in the sample, 325. It also gives the mean number of records for individual case, 2.3, with a minimum of one record and a maximum number of 11 records in this particular data example. The median number of records per individual is 1. The first entry time and final exit time are also useful variables to read in this case: It can be seen that all employees were included in the risk set at time 0 (or with no tenure in the organization), which makes sense. Some individuals exited the organization almost immediately after having been hired; the minimum observed tenure was 3 days and the mean tenure was 527 days. The last output line reveals the number of events in the sample, a total of 165 terminations or "failures" (again using Stata jargon).

The `stsum` presents a different, yet useful, set of summary statistics including time at risk, incidence rate, number of subjects, and the 25th, 50th, and 75th percentiles of survival time (which are obtained from the Kaplan–Meier product-limit estimate of the survivor function discussed later in this chapter). The 25th percentile is obtained at the minimum value of time t such that $S(t)$ is lower than or equal to 0.75, the 50th percentile is obtained at time t so that $S(t)$ is lower than or equal to 0.50, and so forth. The `by( )` option is employed to request separate summaries for several groups (of cases) as defined by whichever categorical variable one is interested in using. For example, I chose to examine the basic statistics for the referral and non-referral hires in the sample. In that case, I typed the following command line in Stata:

stsum, by(contact)

```
        failure _d:  terminat == 1
  analysis time _t:  endtime
               id:  employee
```

contact	time at risk	incidence rate	no. of subjects	|---- Survival time ----|		
				25%	50%	75%
0	80404	.0010696	155	256	707	.
1	**90977**	**.0008464**	**170**	**267**	**914**	**.**
total	171381	.0009511	325	266	745	.

The bold line of the output, corresponding to CONTACT is equal to one, states that among all the referral employees, the termination incidence is 0.0008 (in the "Incidence Rate" column), and the median survival time was 914 days (in the "Survival Time 50%" column).[30] In contrast, one can already see that the

---

[30] Time at risk is the sum of the times at risk for all cases in the sample. The incidence rate is the ratio of the number of events and the time at risk; it therefore measures the mean number of events per one time unit.

termination incidence is much higher for non-referrals than referrals (0.001 versus 0.0008) and the median survival time is much lower (707 versus 914). This seems to be a good indication that there might be significant differences in the termination pattern between referral and non-referral employees in the sample.

6.2. The Kaplan–Meier Estimators

Life tables can be calculated using Stata, SPSS, or TDA. One of the disadvantages of this methodology, however, is that one must specify the intervals in a fixed manner. In addition, this method provides reliable estimates of the different EHA values *only* when the number of events in each one of the time intervals considered is sufficiently large. In the case of small data sets, these estimates might not be reliable (especially in those time intervals where very few events occurred).[31]

Mainly for these reasons, more attention has been devoted to the method of the *Kaplan–Meier estimator* (or product-limit estimator) of the survivor and the hazard rate functions. When using this methodology, one does not need to define the time intervals; the Kaplan–Meier method is based on value calculations of the functions for the group at risk in each time point when at least one event occurs. Consequently, this method makes an optimal use of the events contained in the empirical data. To illustrate some of these concepts, think of a sample of 100 students who begin studying for their college degrees in a given year. Suppose that the students are observed during a period of four years (that is, 48 months) and that at the end of the entire period of observation, 45 students have already quit college. To study the social process of dropping out of college, one could sort the 45 student cases in the temporal order in which they quit college. The first drop takes place in month t_1. Then the time intervals between events are given in the following manner:

$$[0, t_1), [t_1, t_2), \ldots, [t_m, t_{m+1})$$

with m being the point of time when the last student dropped out of school. The number m does not have to be equal to 45 (the total number of students who quit) since several events are likely to occur at the same time; for example, three students could drop out of school after one year in college. The probability of not dropping out of college at any moment t_e of time can be calculated using the following formula:

$$p_{te} = 1 - \frac{N_e}{R_e}$$

---

[31] For a more detailed discussion of life tables, read pages 51–66 in Blossfeld and Rohwer [1995].

where N_e represents the number of events that have occurred in time t_e, and R_e represents the group at risk or number of students at risk of dropping out of school before time t_e (before t_1, R is 100). Defining the survivor function only makes sense during the time period until time m, when the latest event occurs. Below, I show how the values of the survivor function at four specific time intervals can be computed; I assume that the time intervals are measured each month:

Time	Interval	R_e	Events	p_{te}	Estimation of $S(t)$
$t < t_1$					1.00
1	$[0, t_1)$	100	1	$1 - (1/100) = 0.99$	$1 \times 0.99 = 0.99$
2	$[t_1, t_2)$	99	1	$1 - (1/99) = 0.9899$	$1 \times 0.99 \times 0.9899 = 0.98$
3	$[t_2, t_3)$	98	4	$1 - (4/98) = 0.9592$	$1 \times 0.99 \times 0.9899 \times 0.9592 = 0.94$
4	$[t_3, t_4)$	94	2	$1 - (2/94) = 0.9787$	$1 \times 0.99 \times 0.9899 \times 0.9592 \times 0.9787 = 0.92$

In this example, I observe how during the first month ($e = 1$) a student drops out of college. During the second month, another student drops out (i.e., two events). During the third month ($e = 3$), four students drop from school, and at the fourth month, two more students drop. The last column shows how one could estimate the value of the survivor function based on this event information. In general, one of the main goals consists of obtaining the different values of the survivor function given the available data sample. The Kaplan–Meier estimation of the function of survival $S(t)$ is unbiased and asymptotically consistent, and it is defined as 1, for $t < t_1$ (this is before any event occurs), and at later times, the Kaplan–Meier version of the survivor function or $S_{KM}(t)$ is estimated as follows:

$$S_{KM}(t) = \prod_{t_k < t} \left[1 - \frac{N_k}{R_k} \right], \quad \text{for } t_{k-1} < t < t_k$$

where the subscript k makes reference to the kth event that occurs in the sample, when events have been previously ordered over time (in ascending order). t_k is the time when the kth event occurs, N_k is the number of events happening at time t_k, and R_k is the number of cases at risk before t_k.

$S_{KM}(t)$ is a step function that connects the different empirical point estimates. However, most of the statistical programs connect those point estimates for you in order to obtain a picture closer to the underlying smooth survivor function. The utility of the Kaplan–Meier estimation is even greater because it estimates not only the values of the survivor function but also its variance. The most common estimator of the variance of $S(t)$ (also the Kaplan–Meier estimate) is the following:

$$\text{Var}[S_{\text{KM}}(t)] = [S_{\text{KM}}(t)]^2 \sum_{t_k < t} \frac{N_k}{R_k(R_k - N_k)}, \quad t_{k-1} < t < t_k.$$

So, knowing the point estimates and the variance of these point estimates helps in deriving the two-sided 95% confidence interval of the survivor function at any time t_e. This interval is useful for understanding the trend and its variation over time. This 95% confidence interval can be calculated using the following formula:

$$S_{\text{KM}}(t_e) \pm 1.96 \sqrt{\text{Var}[S_{\text{KM}}(t_e)]}.$$

In Stata, after declaring the data to be event history using the `stset` command (presented in the "Life Tables" section), the `sts` command allows the user to easily generate, graph, and test the survivor function using the Kaplan–Meier technique. Below, I present the graph of the Kaplan–Meier Survivor function using Stata. The first command line, `sts graph`, helps to calculate and graph the survivor function for all employees in the organization; whereas the second command line, with the inclusion of the `by( )` option, estimates two different survivor functions, one for referral hires and another one for non-referral hires:

`sts graph`

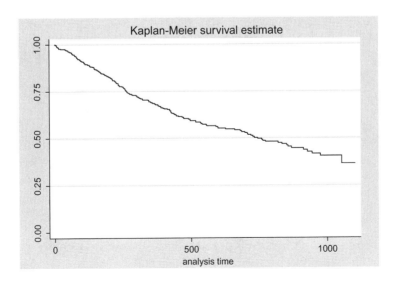

```
sts graph, by(contact)
```

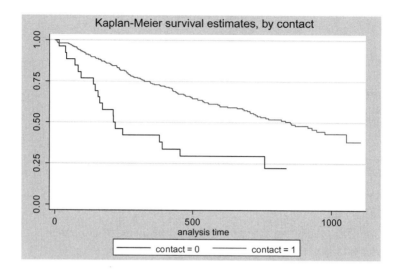

Note how the survivor function for all employees (first plot) roughly reaches 0.6 at the 500th month, when all employees are considered. Past 900 days, the survival curve becomes less smooth. There are fewer employees who have been with the company for that long, so there is less information available (in the estimation) and thus the curve is blocky. The by( ) option in the `sts graph` command gives a visual representation of the effect of any covariate. The second plot I estimated provides the survival curves for each of the values of the variable CONTACT, in this case 0 or 1. By looking at the estimates of the values of the survivor function by referrals (CONTACT equals 1) versus non-referrals (CONTACT equals 0), the survivor function is lower for referrals in comparison with non-referrals. After 250 days of tenure, for example, less than 50% of the non-referrals remain in the organization. This number is around 75% for referral hires. The survivor function for referrals never goes below (nor reaches) the 35% value.

One can easily run these graphs in Stata and paste them directly in the main report or study. One could even generate a new variable in Stata by saving these survivor function estimated values. For example, the Kaplan–Meier survivor function estimates can be saved in a variable called SURVKM by typing the following command line:

```
sts gen survkm = s
```

where s is the name of the variable in which Stata automatically stores the estimates of the Kaplan–Meier survivor function at different points in time. This new variable can be used to list or plot some of the values of the function with time or to compute basic descriptive statistics such as the mean, median, maximum, and minimum values of the survivor function. For example, one could summarize the values of the survivor function by CONTACT and obtain the following Stata output:

```
sts gen survkm = s
sort contact
by contact: sum survkm

-> contact = 0

Variable      Obs      Mean    Std. Dev.      Min     Max
survkm        313   .7009494    .1809131    .3650842    1

-> contact = 1

Variable      Obs      Mean    Std. Dev.      Min     Max
survkm        436   .7132691    .1779985    .3650842    1
```

While the minimum and maximum values of the survivor function are similar for referrals and non-referrals (0.36 and 1, respectively), the mean value of the survivor function for referrals seems to be a bit higher than for non-referrals (and has a lower standard deviation).

Graph 3.3 was constructed in TDA using the values of the survivor function estimated by the Kaplan–Meier method using the same employee data. Observe

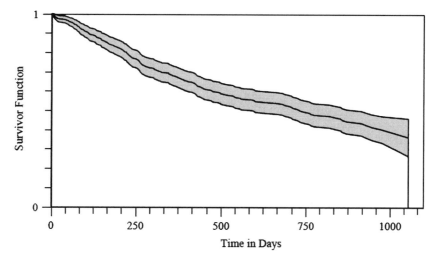

Graph 3.3 Survivor function (Kaplan–Meier estimate) for all employees in one organization.

that the survivor curve decreases relatively fast until the 500th day, when the period of observation ends (three years). After three years of work in the company, more than half of the employees had been terminated.

6.3.   Comparing Groups over Time

The main advantage of using exploratory analysis like the life tables or the survivor function estimates is that the point estimates of the survivor function for different sub-groups of interest can be very simply calculated (and plotted). Graph 3.4, for example, compares the values of the survivor functions for referrals and non-referrals (calculated using TDA this time; in the previous section the same survivor function was estimated using Stata). In this example, the confidence intervals do not overlap except at the very beginning of the job employment contract; one can consequently be confident that the distribution of T differs for referrals and non-referrals, with referrals being less likely to be terminated. This result is not surprising given the literature on social networks that argues that referrals are more likely to be better matched to their jobs than non-referrals, ultimately accounting for significant differences in their termination rates (see Castilla [2005] or Fernandez, Castilla, and Moore [2000]).

Notice that the curves for referrals and non-referrals are quite similar in shape, with the curve for referrals further to the right than the one for non-referrals. In many other examples, it could be possible that curves cross over time. In

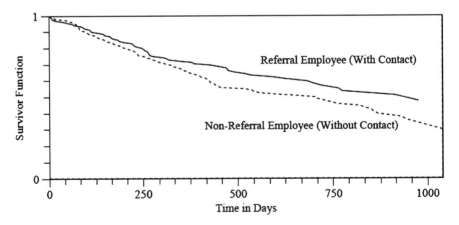

Graph 3.4   Survivor functions for two groups of employees in one organization: Referrals and non-referrals.

addition, the differences between survival curves are not always as clear as in this example. Therefore, an easy way to compare the survivor functions across different groups is by calculating the confidence intervals for the survivor function for each of the groups. One can then verify visually whether such functions overlap over time.

A series of statistics have been proposed to test whether the survival curves for g groups (at least two groups) are significantly different from each other. Most of these statistics are based on the null hypothesis that the values of the survivor function are the same for all g groups of individual cases in the sample. These statistics can be computed with TDA, Stata, or SPSS (among others). TDA computes four different statistics (with the name of the scholar who first proposed each test in parentheses): Log-Rank, Wilcoxon (Breslow), Wilcoxon (Tarone–Ware), and Wilcoxon (Prentice). In addition to these four tests, Stata also computes the Generalized Fleming–Harrington test for equality of survivor functions. All these statistics follow a χ^2-distribution with $g - 1$ degrees of freedom. The null hypothesis H_0 is that the survivor functions are all equal across groups. In the example, the number of groups g, is 2, referrals versus non-referrals. All these tests are global tests in the sense that they do not test the equality of the survivor functions at any specific point in time, instead they compare the overall survival functions. Thus, each of the tests differs only in how they weigh each of these individual comparisons over time.

After looking at Graph 3.4, I illustrate how to obtain some of these χ^2-tests using Stata in order to test the equality of the survivor functions across groups. Again, I use the `sts` command which has already been discussed. To compare the survivor functions for referrals and non-referrals, I include the CONTACT variable in the `sts` command, just as follows:

```
sts test contact
sts test contact, wilcoxon
sts test contact, tware
sts test contact, peto
```

Each of the four command lines provides a different χ^2-statistic. So the output for the first `sts` command would look like this:

```
sts test contact

      failure _d:  terminat == 1
analysis time _t:  endtime
              id:  employee
```

```
Log-rank test for equality of survivor functions

           |    Events           Events
contact    |   observed         expected
-----------+-----------------------------
0          |       86              76.43
1          |       77              86.57
-----------+-----------------------------
Total      |      165             163.00

                  chi2(1)  =      16.70
                  Pr>chi2  =      0.0000
```

The `sts test` command provides the Log-Rank test. The value for the Wilcoxon χ^2 is 19.32; the Taroen–Ware test is 18.36; and finally the Peto–Peto test is 18.87. All these tests are χ^2-tests with one degree of freedom (two groups minus one). The values of these statistics suggest that the differences in the survivor functions between referrals and non-referrals are statistically significant (in all cases at the 0.001 level). Thus, the null hypothesis of equality of survivor functions is rejected. Below, I present part of the output that was obtained in TDA (only the part of the output that refers to these four statistics). This time I use the TDA command `csf`:

```
Episodes
SN  Org  Des  Group    Label            Events   Censored
---------------------------------------------------------
1    0    1     1       Non-referral       86       227
1    0    1     2       Referral           79       359
Sum                                       165       586

SN   Org   Des   Test Statistic               Statistic   DF   Significance
---------------------------------------------------------------------------
1     0     1    Log-Rank  (Savage)            5.4634      1      0.9806
1     0     1    Wilcoxon  (Breslow)           4.2220      1      0.9602
1     0     1    Wilcoxon  (Tarone-Ware)       4.7931      1      0.9713
```

where the estimated probability is now equal to the significance value minus 1. In the case of the Log-Rank test statistics, the probability is approximately 0.02. The four statistics suggest that the null hypothesis—i.e., that the survivor functions for referrals and non-referrals are the same—should be rejected at the 0.05 significance level. Given that the difference between the two survivor curves is clear-cut, there does not seem to be much difference in the values of the four statistics as calculated by TDA. However, each statistic has a region of different sensitivity and therefore pays more or less attention to differences in the survivor curves at different moments in time. For example, the Wilcoxon test pays greater attention to differences at the beginning of the time period under observation, whereas the Log-Rank statistic pays greater attention to differences at the end of the period of observation.

Examining the values of the survivor function on a plot is equally useful when exploring differences across groups in the timing of such events. This is especially so given that these values are calculated at the time when certain events occur. However, this methodology is not appropriate when examining time differences in the likelihood of an event happening over time, given that the event has not happened yet. Moreover, all survivor functions tend to look quite the same, starting at one and beginning to decline towards zero over time (as it could be seen earlier in this chapter when I graphed the survivor function using the Kaplan–Meier technique using the sts graph command in Stata). This is why at this exploratory level of analysis, one should examine the time pattern of the other important EHA functions. One way to examine time variation in the likelihood of an event happening is to plot the values of the integrated hazard rate $R(t)$ over time. The $R(t)$ function is the integral of the hazard rate function, which also equals the minus logarithm of the survivor function. The *Nelson–Aalen estimator* has been proposed in this case because this estimator has good asymptotic properties and is unbiased:

$$R_{NA}(t) = -\log S_{NA}(t) = \sum_{t_k < t} \frac{N_k}{R_k}, \quad t_{k-1} < t < t_k.$$

This estimator differs only slightly from the minus logarithm of the Kaplan–Meier estimator of $R(t)$—except when the sample is too small or on the other extreme, when the sample is large. Just as with the Kaplan–Meier estimator of the survivor function, the Nelson–Aalen estimator of the integrated hazard rate is a step function that connects empirical point estimates. However, current statistical software programs easily connect these points to give the presumed underlying smooth version of the function. One can also compute the confidence intervals for the integrated hazard rate using its estimation of the variance as below:

$$\text{Var}_{NA}(t)] = \sum_{t_k < t} \left[\frac{N_k}{R_k} \right]^2, \quad t_{k-1} < t < t_k.$$

Stata makes it simple to start generating, graphing, and testing not only the survivor but also the cumulative hazard and the hazard rate functions. Below, I present important sts commands that allow to perform useful exploratory analyses of these EHA functions.[32] I also used the by( ) option to provide a visual representation of the plot of the hazard curves for each value of the CONTACT variable. To graph the Nelson–Aalen cumulative hazard function, first for all employees and then for the two groups of employees of interest (referrals

---

[32] For more information about all the different possibilities available with the sts command, check the *Stata Reference Manual*.

versus non-referrals), one could type the two following command lines (immediately followed by the Stata graphs of the cumulative hazard rate functions):

```
sts graph, na
```

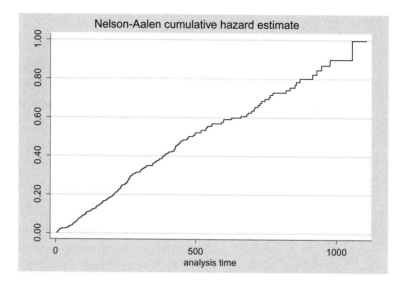

```
sts graph, na by(contact)
```

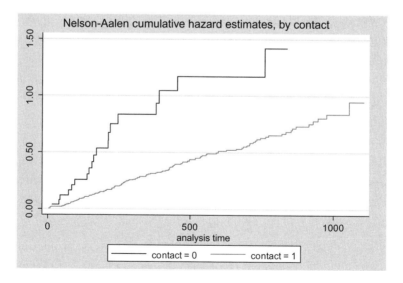

The previous two graphs display the values of the Nelson–Aalen estimates of
the integrated hazard rate of termination for all employees first, and then for
employees with and without employee contact or referral in the organization. The

vertical axis shows the cumulative hazard equal to the negative log of the survival probability. Beyond 900 days, the cumulative hazard curve, like the survival curve, becomes less smooth for the same reason (because fewer events occurred during that time period). If the slope of the function were constant (or flat), that would mean that the hazard rate of termination does not vary over time. As one would expect, the slope clearly varies over time—indicating that the termination hazard rate varies over time. It is also clear that the hazard function is quite different for referrals versus non-referrals.

The plot of the integrated hazard rate does not give us any idea, though, of the manner of variation in the likelihood of the event happening over time. This is why I recommend estimating the hazard rate directly. Empirical plots of the point estimates of the hazard rate over time can be confusing as they tend to give a series of spikes. Advanced statistical programs therefore plot the smoothed version of the hazard rate function over time. Several methods of estimation of the hazard rate function are available, including the Kaplan–Meier estimator. The Nelson–Aalen estimation is calculated using the following formula:

$$r_{\mathrm{NA}}(t) = \frac{1}{\Delta t_k} \frac{N_k}{R_k}, \quad t_{k-1} < t < t_k.$$

In Stata, graphing the estimated (smoothed) hazard functions is simple—type the two following command lines (immediately followed by the Stata graphs of the hazard rate functions):

```
sts graph, hazard
```

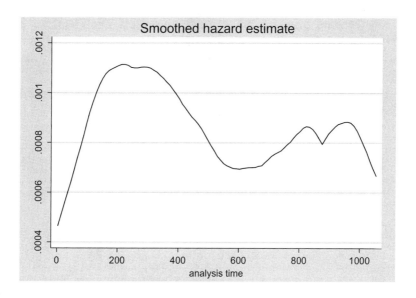

`sts graph, hazard by(contact)`

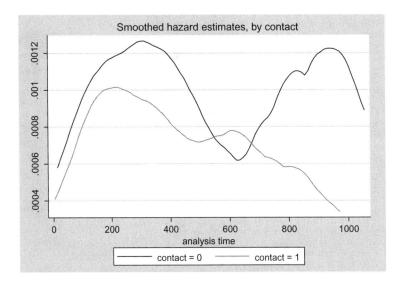

Smoothed hazard estimates, by contact

These two smoothed hazard functions estimated from the employee data on terminations for referrals versus non-referrals show that the time manner in which termination events happen is remarkably similar for both groups of employees up until month 15: It increases over the tenure of employees and later decreases during the first year and a half. Even when this pattern is similar for the two groups of hires, the hazard rate is remarkably higher for non-referrals than referrals. Notice, however, that around month 30, the hazard rate is much higher for referrals than for non-referral employees. Because the increase after day 600 is based on a few observations (as the number of employees remaining in the company has decreased), the hazard estimates might not be meaningful then.

I have illustrated these exploratory techniques when examining the first event happening in the sample. These techniques can be extended in order to study subsequent events or repeated events.[33] They can also be applied when there are several possible outcomes of interests or *competing risks*. For example, in the case of termination, one could compare voluntary versus involuntary termination. In the case of competing events, the conditional probability $m_k(t)$ can be estimated the same way as the previous probabilities. The estimation of the hazard

---

[33] For example in Stata, as I mentioned earlier in this section, use the `exit` option in the `stset` command to inform Stata of the multiple-events-per-case nature of the data. If this option is not specified, the default is `exit(failure)` meaning that the individual case is removed from the risk set, even if the individual has subsequent records or spells in the data and even if some of those subsequent records indicate other events.

rate $r_k(t)$ is similar to the estimation of the hazard rate, except that now N_k refers only to events where individuals transition to one particular state of interest (in this case state k).[34]

7. Explanatory or Multivariate Analysis of Events

Although a good first step in the analyses of events, the use of the non-parametric or exploratory EHA methods previously reviewed has its own limitations. Researchers are often interested in comparing the three EHA functions across different social groups of interest. As the number of sub-groups (as defined by more than one categorical variable) increases, the more complex it becomes to compare functions across groups. This method may also be inefficient because one might be estimating too many functions to compare the many groups. As the number of cases in each one of these groups becomes smaller, the estimation of these functions gets more difficult, and the results of such estimations may be highly unreliable because of the low statistical power. Additionally, the systematic comparison and interpretation of a number of these EHA functions across groups may be slow and extremely cumbersome. When one wants to compare the shape of these functions for different values of certain continuous variables (for example, wage, age, or education), such variables must previously be converted into ordinal variables (with a group of categories) in order to be able to compare the survivor functions across a reasonable number of groups. For example, in understanding how turnover relates to years of education, think of dividing a sample into primary school, secondary school, some college, college, and graduate work and plot different survivor functions. However, this transformation of a continuous variable such as years of education into an ordinal variable with a few categories ignores the effect of the exact number of years of education for each individual. To overcome these disadvantages associated with the use of exploratory methods, one can perform explanatory or multivariate methods to study the occurrence of events in a sample.

In this section, I present the *parametric EHA models*, also known as *multivariate models*, for analyzing events. The *multivariate analysis* provides an extremely useful way to analyze a process of qualitative change. In many cases, this analysis has also been denominated as *confirmatory analysis* because it is typically used for confirming the study of hypotheses. The ultimate goal of this multivariate analysis is to specify and estimate some regression-like models that examine the effect of a vector of explanatory variables or X_{it} on the chances or risk of occurrence of the event or social process under study. These techniques are popular and have been used extensively to investigate events over time in many

---

[34] This can be easily done in Stata by running the command `stset, failure (varname==k)` several times, each time specifying a particular state of interest k.

empirical papers in the past few decades. Indeed, they have become powerful tools to analyze events. One should *not* conclude, though, that the descriptive analyses of events (as presented earlier in this chapter) are unnecessary and should consequently be skipped. On the contrary, these analyses constitute a fundamental first step for understanding the basic temporal pattern of events (even the basic comparison across groups) *before* any multivariate analysis of events is undertaken. These descriptive techniques allow the observation of the pattern of temporal variation of the different EHA functions such as the survivor or hazard rate functions. Thus, later I further explore how this basic temporal form is affected by several explanatory variables of interest.

Like the exploratory analyses, the multivariate analysis of events is developed around the basic concepts defined in the EHA, typically the survivor function or the hazard rate. The majority of these multivariate models of analysis of events have been defined using the hazard rate or the instantaneous probability of an event happening during the period of observation. This is simply because the hazard rate is easier to interpret among all of the different EHA functions. Remember that this hazard rate was represented as $r(t)$ or $r_{jk}(t)$: It measures the probability that any given individual in the sample transitions from a state j to a state k.

7.1.   *Proportional Hazard Rate Models*

The *proportional hazard rate models* begin by specifying some equation where the dependent variable is the hazard rate, as formulated below:

$$r(t, x) = r_{tx} = \theta(X_{it})\, q(t) \tag{1}$$

where again X_{it} is the vector of explanatory or independent variables that are hypothesized to increase or decrease the risk or chance of the event happening. These explanatory variables x_{it} have i as a subscript to indicate that they vary across the individual cases in the population under study and t is used to indicate that the values of these explanatory variables may also vary over time. Equation (1) shows how the risk of the event has a temporal component, i.e., the hazard rate changes over time, following the functional form defined by the function $q(t)$. In addition, this equation indicates that the hazard rate depends on a set of explanatory variables as determined by the function $\theta(X_{it})$. This model has been denominated as the "proportional hazard rate" model because the function of the vector of explanatory variables $\theta(X_{it})$ multiplies the value that the hazard rate takes over time based on the function $q(t)$. So, the hazard for two different values of x_k, for example, is a function of variable x_k only and not of the function of time $q(t)$. A non-proportional specification of the hazard model can also be proposed. The most common form of specification of the non-proportional model consists of allowing the time function $q(t)$ to also be a function of some independent variables z, as follows:

$$r(t, x) = \theta(X_{it}) \, q(t, Z_{it}). \tag{2}$$

I will come back to these non-proportional models later in the chapter. But for now, let me describe the proportional version of the multivariate hazard rate model as specified by equation (1). When specifying the proportional hazard rate model, two functions have to be first defined and estimated. The first function is the so-called *function of explanatory variables* $\theta(X_{it})$; the second is the *function of time* $q(t)$, which has also been referred to as the *baseline hazard rate* or *time distribution*. Ideally, how one defines these different functional forms depends entirely on the dynamic theories interested in testing. Nevertheless, too often these functions are defined on the basis of convenience and ease in the adjustment and estimation of the model.[35] It is important, though, to always verify that the functional form $q(t)$ used in the specification of equation (1) is valid for the social process under study. Again, one can start checking this by running exploratory EHA techniques such as life tables and, especially, the plots of the EHA estimators over time.

Initially, one should proceed with the specification of the first key function in equation (1), the *function of explanatory variables* or $\theta(X_{it})$. By definition, the hazard rate (or instantaneous probability of an event happening) must be non-negative at any moment of time for any of the values that either t or x_{it} take. Clearly, it does not make much sense to predict that the hazard of an event occurring is negative. For this reason, the function of explanatory variables $\theta(X_{it})$ should be non-negative, no matter what the values of x_k are. Accordingly, this explains why the following specification of a linear multiple regression model is not used (even though it could be used as long as the parameters β are restricted so that $\theta(X_{it}) > 0$ at any moment):

$$\theta(X_{it}) = \beta_0 + \beta_1 x_{1it} + \beta_2 x_{2it} + \beta_3 x_{3it} + \cdots + \beta_k x_{kit} \tag{3}$$

or, in matrix notation:

$$\theta(X_{it}) = \beta' X_{it}$$

where $i = 1, 2, 3, \ldots, N$ and $t = 1, 2, 3, \ldots, T$. Instead, the most commonly used specification is the exponential version of equation (3) as follows:

$$\theta(X_{it}) = \exp[\beta_0 + \beta_1 x_{1it} + \beta_2 x_{2it} + \beta_3 x_{3it} + \cdots + \beta_k x_{kit}]$$

$$= \exp[\beta' X_{it}] = \prod_k v_k^k \tag{4}$$

---

[35] A common model in EHA, the Cox partial likelihood model does not specify any particular $q(t)$ or function of time; I discuss this model later in the chapter.

where $i = 1, 2, 3, \ldots, N$ and $t = 1, 2, 3, \ldots, T$. v_k equals the exponential form of the beta parameters $\exp(\beta_k)$. According to this exponential specification, the variable x_k has a multiplying effect on the hazard rate and an additive effect on the logarithm of the hazard rate. When a variable x_k has no effect on the hazard of an event happening, this means that β_k equals 0, or similarly, that the exponential transformation of β_k or v_k equals 1. In a similar manner, if $\beta_k = 0.2$, then $v_k = \exp(0.2) = 1.22$. A common interpretation of this parameter is that a unit increase in x_k increases the hazard rate or risk by 22% (this is, 100% × (1.22 − 1). If $\beta_k = -0.2$, then $v_k = \exp(-0.2) = 0.82$. An interpretation of this result is that a unit increase in x_k diminishes the hazard by 18% (this is, 100% × (0.82 − 1). In general, the estimated parameters are easy to interpret in substantive terms when using the exponential functional specification. In the most general case, one can calculate the percentage of change in the hazard rate that occurs with a one-unit change in any of the exploratory variables x_k using the following formula:

$$100\% \times (v_k - 1)$$

where $v_k = \exp(\beta_k)$; or altogether:

$$100\% \times [\exp(\beta_k) - 1].$$

This formula provides the percentage of change (either increase or decrease) in the hazard rate by a one-unit change in a given independent variable x_k. The units by which the independent variables are measured make no difference in the specification and estimation of the model, but choosing the relevant units can make the interpretation of the model coefficients much easier to understand and report later.

Before proceeding to the specification and estimation of the hazard rate model in which each of the independent variables is assumed to have a constant effect on the logarithm of the hazard rate, one can make use of several exploratory analyses in order to check this important assumption, also called the *proportional hazard rates assumption*. In the exponential specification presented above, variable x_k has an additive effect on the logarithm of the hazard rate. One can then calculate and graph the logarithm of the hazard rate for several values of x_k graphically. When the resulting functional curves are different by a constant amount and the amount appears to be independent of time—so that if the curves were plotted, they would look almost parallel—then the effects of the independent variables on the hazard rate are said to be proportional. In this case, the proportional hazard rate model seems appropriate. If that is not the situation, then it would be beneficial to consider estimating the non-proportional version of the

EHA model that I introduce later in the chapter. In the graph below, I illustrate how the estimated logarithm of the hazard rate should look in the ideal scenario when the effect of an individual independent variable x_1 on the hazard rate is proportional:

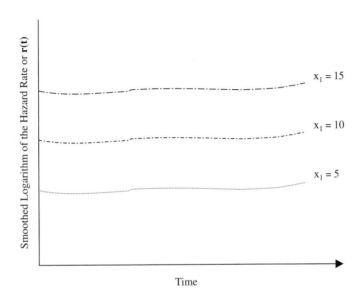

As one can see, the different curves for the logarithm of the hazard rate at three values of x_1 (for the example, values 5, 10, and 25) differ by a constant amount regardless of time. If the effect of x_1 on the hazard was non-proportional, such curves would not appear parallel.

The first step in the multivariate analysis of events consists of specifying the *function of time* or $q(t)$—also referred to as the baseline hazard, because this function gives the value of the hazard rate when all explanatory variables are equal to zero. The different parametric hazard rate models are distinguished by the particular selection of the functional form of $q(t)$. A previous representation of the dependency of the hazard rate on time helps tremendously here because the proper specification of $q(t)$ later aids in obtaining reasonable estimators of the effects of the several independent variables on the hazard rate. Time serves as a *proxy* for the causal factors that are difficult to measure, for example, the effects of individual age or tenure in organizations. In addition, time helps to account for any heterogeneity that cannot be observed but that produces systematic forms of dependency on time. For example, time can account for any variation in the hazard rates that cannot be captured by any of the observed or measured independent variables included in the model.

When specifying the time function $q(t)$, the simplest functional form assumes that $q(t)$ is a constant. This is the popular *exponential hazard rate model* or simply the *exponential model*. Under this model, the main assumption is that duration (represented by T) can be described using an exponential distribution. In other words, conditional to the values of a vector of explanatory variables X_{it}, T has the following exponential distribution:

$$r(t, x) = r_x = \theta(X_{it})\, q = \exp[\beta'X_{it}]\, q \qquad (5a)$$

or if one takes the logarithms of both sides of equation (5a), the following is the log-transformation of equation (5a):

$$\ln r_x = \beta'X_{it} + \ln q \qquad (5b)$$

where q is a positive constant over time. Note that the hazard rate of an event happening is considered constant over time even though it varies depending on the values of a series of explanatory variables X_{it}—again, this is why the notation r_x is used to acknowledge that the rate *only* depends on the values of X_{it}. The density function and the survivor function are easy to obtain in the case of the exponential model; the formulas are the following:

$$S(t) = \exp(-r_x t)$$
$$F(t) = r_x \exp(-r_x t) = r_x S(t).$$

The mean duration (i.e., the mean time at which an event occurs) is computed by $1/r_x$.

Another common specification of the time function is the so-called *Weibull function*. The Weibull function is used to represent a pattern of monotonic temporal variation—meaning that the hazard rate of an event happening either increases or decreases over time, other things being equal. The hazard rate equation in the case of the *Weibull model* can be written as follows:

$$r(t, x) = r_{xt} = \exp[\beta'X_{it}]\, t^{\gamma}. \qquad (6a)$$

Notice that now, the rate, r_{xt}, is used to acknowledge that the rate depends on the values of X_{it} and time. Again, by taking the logarithms on both sides of the equation (6a), it can be re-written as follows:

$$\ln r_{xt} = \beta'X_{it} + \gamma \ln t. \qquad (6b)$$

The only requirement now is that the value of γ must be greater than -1, making this the Weibull function. The value of γ is what determines whether the hazard rate function increases or diminishes over time. When γ equals 0, then the Weibull model is identical to the exponential one. As an illustration, I show, in the table below, the values that the Weibull hazard rate function takes over time when $r_{xt} = 0.5t^\gamma$:

Different Values of Parameter γ

Time (t)	−0.5	0	0.5	1	2	3
1	0.50	0.50	0.50	0.50	0.50	0.50
2	0.35	0.50	0.72	1.00	2.00	4.00
3	0.29	0.50	0.87	1.50	4.50	13.50
4	0.25	0.50	1.00	2.00	8.00	32.00
5	0.22	0.50	1.12	2.50	12.50	62.50
6	0.20	0.50	1.22	3.00	18.00	108.00
7	0.19	0.50	1.32	3.50	24.50	171.50
8	0.18	0.50	1.41	4.00	32.00	256.00
9	0.17	0.50	1.50	4.50	40.50	364.50
10	0.16	0.50	1.58	5.00	50.00	500.00

In each column in this table, I use a different value of the parameter γ, from −0.5 (first column) up to 3 (last column of the table). As a rule, if $\gamma > 0$, then the hazard rate increases over time. When the value of the parameter is greater than −1 but below 0 (i.e., $0 > \gamma > -1$), then the hazard rate diminishes over time. If $\gamma = 0$, then the Weibull model is no different from the exponential model. Later in the estimation part of this chapter, I show how Stata uses the parameter p instead of the parameter γ when referring to the specification of the Weibull model. This is because the Stata program uses a slightly different-looking notation when presenting the Weibull model; this notation is equivalent to the one I presented earlier where $r(t, x)$ is equal to:

$$r(t, x) = pr_x t^{p-1} \tag{7}$$

where $r_x = \exp[\beta' X_{it}]$. In Stata, note that if $p > 1$, the hazard rate function increases over time, whereas if $p < 1$, this hazard rate is decreasing through time. In the case that $p = 1$, then, the Weibull function is identical to the exponential function described earlier.

In Graph 3.5, I show how the density, survivor, and hazard rate functions look for three different Weibull functions as defined by three values of the parameter γ, −0.5, 0, and 3. This graph 3.5 gives an idea of how these functions vary over time, depending on the values of the parameter γ.

The exponential and Weibull are the two simplest models used to describe the distribution of duration T of the event. However, there exist many other

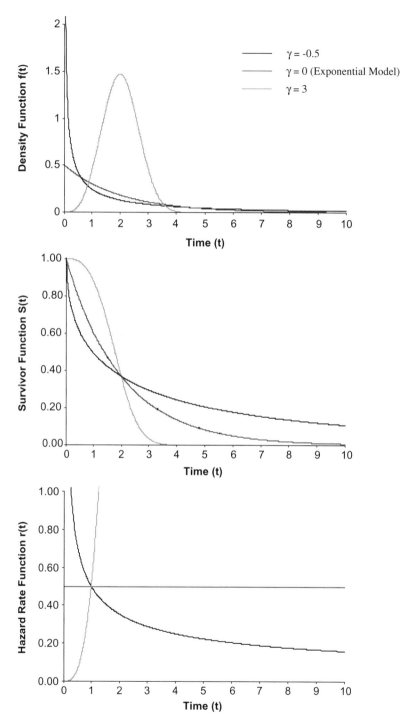

Graph 3.5   Density function $f(t)$, survivor function $S(t)$, and hazard rate function $r(t)$ for three different Weibull models (with different γ values).

Table 3.6
Common specifications of the function of time $q(t)$ and its parameters

Function of Time	$q(t)$	Integral of $q(t)$ $Q(t) = \int_0^t q(u)\, du$
1. Constant Function (or Exponential):	e^α	$e^\alpha t$
2. Monotonic Functions:		
Gompertz	$\alpha + \beta e^{\gamma t}$	$\alpha t + \dfrac{\beta}{\gamma}(e^{\gamma t} - 1)$
Weibull	$\beta(t + \delta)^\gamma$	$\dfrac{\beta}{\gamma + 1}[(t + \delta)^{\gamma+1} - \delta^{\gamma+1}]$
3. Non-Monotonic Functions:		
Log-Logistic	$\dfrac{\alpha t^{\gamma - 1}}{\left(1 + \dfrac{\alpha t^\gamma}{\gamma}\right)}$	$\log\left(1 + \dfrac{\alpha}{\gamma} t^\gamma\right)$
Log-Gaussian[a] or *Log-Normal*	$\dfrac{\dfrac{1}{\sigma t \sqrt{2\pi}} \exp\left[-\dfrac{(\log t - \mu)^2}{2\sigma^2}\right]}{1 - \Phi\left(\dfrac{\log t - \mu}{\sigma}\right)}$	$-\log\left[1 - \Phi\left(\dfrac{\log t - \mu}{\sigma}\right)\right]$

[a]$\Phi(x)$ denotes the cumulative standard Gaussian distribution function, $\int_{-\infty}^x \dfrac{1}{\sqrt{2\pi}} \exp\left(-\dfrac{u^2}{2}\right) du$.

functional possibilities that are not as simple and straightforward as the Weibull or exponential options.[36] Table 3.6 illustrates some of the most frequently used functions of time $q(t)$. Each of the mathematical functions that can be considered in modeling the time function is associated with certain patterns of temporal variation for the hazard rate. A basic distinction is whether the time shape of the hazard rate can be represented by an *increasing monotonic* function (that is, when the hazard increases progressively over time), a *decreasing monotonic* function (when the hazard diminishes progressively over time), or a *non-monotonic* function (when the hazard sometimes increases and sometimes diminishes over time). Two of the most frequent functions in the case of some monotonic temporal variation in the hazard rate are the *Gompertz* function and the *Weibull* function. In the case of non-monotonic functions, it is typical to use the *log-normal* function or the *log-logistic* function.

---

[36] The literature includes a great variety of choices for the time distribution. Two references that contain several specifications are Cox and Oakes [1984] and Lancaster [1990].

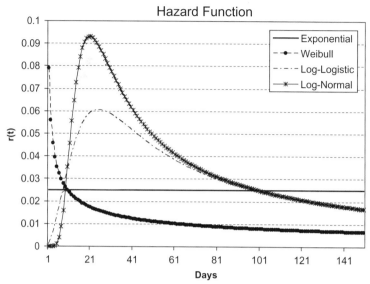

Graph 3.6 Several hazard rate functions.

To give an idea of the different time patterns, Graph 3.6 displays several choices of time distribution to represent the hazard rate function. The hazard function for the exponential distribution is constant. For the Weibull distribution, the hazard function is monotonically decreasing in the Graph 3.6. The hazard rates for log-normal and log-logistic distributions first increase and then decrease over time. The exploratory analysis of the smoothed hazard estimates can help you decide which time distribution to use with your own data.

Suppose that instead of the exponential or Weibull model, I choose the logistic specification for the time function $q(t)$; this will lead to an estimate of the *logistic model*. Placing the specification for $\theta(X_{it}) = \exp[\beta'X_{it}]$ together with the time function $q(t)$, I would be estimating the following hazard rate equation model:

$$r(t, x) = \exp[\beta'X_{it}] \left[\frac{\alpha t^{\gamma-1}}{\left(1 + \dfrac{\alpha t^{\gamma}}{\gamma}\right)} \right]. \tag{8}$$

There exist many different techniques for evaluating whether any given time function is the adequate one for a particular longitudinal data set. For example, one could use several extensions of the more graphical or descriptive methods described in the previous section. I cover some techniques, for example the analysis of pseudo-residuals, later in the chapter when I present the several parametric strategies within EHA.

7.2. *Estimation of the Proportional Hazard Rate Models*

Now that both functional forms have been chosen—the function of explanatory variables and the function of time—one should proceed with the estimation of the multivariate hazard rate model. Estimating this model consists of calculating the value of the different parameters of the function as given below:

$$r(t, x) = \theta(X_{it})\, q(t). \tag{1}$$

A parametric model of the hazard rate facilitates the estimation of the probability of an event for the cases in the sample. In the case of the exponential models, estimating the hazard rate model would imply estimating the following vector of model coefficients β:

$$\beta' = [\beta_0, \beta_1, \beta_2, \beta_3, \ldots, \beta_k]$$

where β_1 is the value of the effect of variable x_1 on the hazard rate, β_2 is the value of the effect of variable x_2 on the hazard function, and so forth. These parametric models are usually estimated by the method of maximum likelihood (ML). The ML estimation consists of finding the values for the parameters in the model that maximize the likelihood function—which of course depends on the functional form of the model. The likelihood function is the product of the values of the density function for each observation in the sample. Assuming a random sample of cases of size N (where the sample observations are independent from each other), the likelihood function of the sample reads as:

$$L = \prod_{i=1}^{N} r(t_i, X_i)^{(1-c_i)} S(t_i, X_i)$$

where c_i is 1 if the individual case i is censored (0 if the case is uncensored); x_i represents the values of all explanatory variables for case i; t_i is the time when the event occurred (for the uncensored cases in the sample) or the end of the observation period (for the censored cases). The statistical power of the multivariable analyses of the EHA is entirely based on the number of events occurring in the sample and not on the number of cases or observations. This might be a bit counterintuitive, but the fact that there might be multiple records for each case or individual in the sample does not affect the statistical power of this multivariate methodology: Only the occurrences of events during the observation period will contribute to the maximization of the likelihood function. This is another reason for the need to be clear in any study about the total number of cases and the number of events occurring in the sample over time.

This routine of choosing the parameters that maximize the likelihood function (L) is easily run using the statistical software programs currently available for the analysis of data sets. Typically, it is easier to maximize the logarithm of the likelihood function or log-likelihood function than to maximize such a function directly. So, the logarithm of the likelihood is expressed as:

$$\mathrm{Log}\, L = \sum_{i=1}^{N}(1 - c_i)\log r(t_i, X_i) + \log S(t_i, X_i).$$

The ML procedure yields the ML estimates, which have desirable statistical properties. These estimators are unbiased and efficient. The log-likelihood function is also used to test central research hypotheses about the significance of certain model parameters (the same way it is done in the case of logit models as shown on pages 154–155). In other words, one can perform log-likelihood ratio tests on nested models to find out whether or not a group of explanatory variables in a model is significant. For example, suppose that L_1 is the value of the likelihood function for a model with $k + g$ parameters or estimated coefficients (model 1), while L_0 is the value of the likelihood function for a model with only k parameters (model 0). Again both models are said to be nested because the k parameters are present in both models 1 and 0 and are the same set of parameters. The parameters in model 1 are thus equal to the parameters in model 0 plus the g new parameters. The null hypothesis or H_0 that could be tested in this case is that all the g parameters (added to model 1) are equal to zero or:

$$H_0: \beta_1 = \beta_2 = \cdots = \beta_g = 0.$$

The alternative hypothesis is that at least one of the g parameters is different from zero and consequently including these g parameters in the model estimation significantly improves the fit of the model. The statistic to use in order to test the hypothesis is calculated as follows:

$$-2[\log L_0 - \log L_1].$$

Again, this statistic is called the *log-likelihood ratio statistic* and follows a χ^2-distribution with g degrees of freedom, where g is the number of additional parameters that are included in the model 1 (this is the model with $k + g$ parameters) in comparison with model 0. g is also the number of parameter constraints applied to model 0 in comparison with the model 1 (the constraints that the g parameters are equal to zero). Thus, model 1 is typically referred to as the full model whereas model 0 is the constraint model. If the null hypothesis is true, then both values of the likelihood (L) for the full model (L_1) and the restrictive model (L_0) are similar. The log-likelihood statistic will be small enough, and its value will be inferior to the value in the χ^2-distribution tables. Consequently, the null hypothesis is not rejected. The opposite will be true if the null hypothesis is false, and the difference between $\log L_0$ and $\log L_1$ will be large enough to obtain a large value for the likelihood ratio statistic. Rejecting the null hypothesis means that *at least one* of the coefficients is different from zero—but not necessarily that each and every one is different from zero.

In addition to the coefficient estimates, the process of ML provides the asymptotic standard errors of the parameters, which consequently facilitates the testing of the significance of each of the different explanatory variables *individually*. The inverse of the matrix of second derivatives provides an estimation of the matrix of covariance of the parameters. For that reason, the square root of the elements of the diagonal is an estimation of the standard error of the parameters of the model. One can also test different hypotheses about the magnitude of each of the parameters in the model. For example, assume one is interested in testing the null hypothesis that β_v equals a constant number h (or using the hypothesis notation: H_0: $\beta_v = h$; where β_v is the ML estimator of the effect of x_v, one of the many independent variables x_k, on the hazard rate). The statistic to compute in order to evaluate whether this null hypothesis is true is calculated as follows:

$$\frac{\beta_v - h}{s_v}$$

where s_v is the estimated standard error of β_v and h is a constant number. The null hypothesis is that the true value of a given parameter β is a constant value h. If one wants to test whether β_v is a significant parameter, then the statistic to compute is much simpler than before. The value of the constant h is now zero (or using the hypothesis notation H_0: $\beta_v = 0$; meaning that the effect of that one of the explanatory variables on the hazard is statistically different from zero). This time the statistic is calculated by dividing the value of the ML estimator by its standard error as shown below:

$$\frac{\beta_v}{s_v}.$$

This ratio is usually referred to as the *Wald statistic*. Many researchers call this ratio the *t*-statistic (or *z*-value), because in sufficiently large samples, it can be treated as if it were the *t*-test in the classic linear regression model. If one wants to perform a two-tailed test with a significance level of 0.05 or 5% (and assuming a large sample), then the null hypothesis (that β_v equals 0) would be rejected if the absolute value of the calculated statistic is greater than 1.96, or if the *p*-value of the *t*-test is lower than 0.05.

In many research situations, one might not be interested in the effect of time *per se* on the hazard rate. This is typically the case when there does not seem to be any theoretical or substantive argument or assumption to be made about the temporal pattern of the social process under study. At other times, one may only want to control for the time component of the hazard rate without having to include any specific time function (at least explicitly). In these kinds of scenarios, when not wanting to make any particular assumption about the functional form of $q(t)$, the *partial likelihood technique* is recommended. This method of estimation was first proposed by Cox (1972; 1975), and has been called *Cox regression*. The Cox regression estimates the coefficients of the explanatory variables in the function $\theta(X_{it})$ without having to specify any functional form for the function of time $q(t)$. These models specify the effect of a set of explanatory variables on the risk of the event—even though this Cox technique operates under the assumption that there is an unspecified time function affecting the risk of events happening in the population of cases. The partial likelihood function is as follows:

$$L_p = \prod_{i=1}^{N} \left[\frac{r(t_i, X_i)}{\sum_{j \geq i} r(t_i, X_j)} \right]^{(1-c_i)}$$

where $r(t_i, X_i)$ is the value of the hazard rate for case i at time t_i, and where t_i is the moment of time when case i had either an event or was censored; and c_i is a dichotomous variable that takes the value of 1 when case i is censored (0 otherwise). The partial likelihood estimators have been shown to be consistent and efficient under general conditions. Consequently, in the cases when one has no interest in modeling the time dependence of the hazard rate and still is confident that the proportionality assumption is valid, this partial likelihood estimation technique is highly recommended.

The proportional hazard rate models assume that the underlying time pattern of the hazard rate model represented by $q(t)$ is identical for all of the cases included in the sample under study. These models also assume that the effect of the different explanatory variables simply multiplies some time-varying baseline hazard rate $q(t)$ so that hazard rates for two different values of x are only function of x and not of $q(t)$. Nevertheless, this proportionality assumption is not always reasonable, and one should verify that that is the case as part of the investigation of events. There are different ways of investigating this issue and checking the veracity of this assumption based on the phenomenon being studied. For example, a simple graph of the hazard rate over time for different groups of cases in the sample can be highly informative because it can easily depict whether or not the time dependence of the hazard rate is similar for the different groups of cases (as illustrated on page 195 in an ideal scenario of proportionality). If that does not seem to be the case, then consider fitting some non-proportional hazard rate models to the observations in the sample (as I will describe later in the Non-Proportional Hazard Rate Models section of this chapter).

In Stata, the command `streg` performs the ML estimation for most of the parametric regression EHA models presented up to this point, namely the exponential, Weibull, Gompertz, log-normal, log-logistic, and generalized gamma models. The syntax is simple: The `streg` command is immediately followed by the list of independent variables to be included in the function of explanatory variables. One of the important options of this command is the `distribution()` option, which specifies the shape of the function of time to be fitted as part of the model. Before running any of these EHA commands in Stata, remember that the `stset` command needs to be used to declare the data to be event history data. This is the way it was done for the employee data set under study here (followed by the Stata output):

```
stset endtime, id(employee) failure(terminat==1)

                id:  employee
     failure event:  terminat == 1
obs. time interval:  (endtime[_n-1], endtime]
 exit on or before:  failure
-------------------------------------------------------------------------
       751  total obs.
         2  obs. begin on or after (first) failure
-------------------------------------------------------------------------
       749  obs. remaining, representing
       325  subjects
       165  failures in single failure-per-subject data
    171381  total analysis time at risk, at risk from t =     0
                              earliest observed entry t =     0
                                 last observed exit t =  1104
```

After declaring the data set to be event history data using the `stset` command, I type the following command line in order to fit the exponential model to the employee termination data:

`streg contact contactt, distribution(exponential)`

Notice that the `distribution( )` option specifies `exponential`. For now, I will only include two variables of interest as independent variables in the model, CONTACT and CONTACTT; I will later introduce all other control variables in the model. By control variables, I refer to the set of independent variables that are needed to be included in the equation models in order to control for their effects on the dependent variable when testing the significance of the effects of a set of independent variables of interest. Methodologically speaking, there is no difference between control variables and independent variables since they all go in the right-hand-side of the equation. Below is the obtained output after the `streg` command presented above:

```
      failure _d:  terminat == 1
analysis time _t:  endtime
             id:  employee

Iteration 0:   log likelihood =   -406.3003
Iteration 1:   log likelihood =  -403.90028
Iteration 2:   log likelihood =  -403.86569
Iteration 3:   log likelihood =  -403.86567

Exponential regression -- log relative-hazard form

No. of subjects =         318          Number of obs  =     734
No. of failures =         161
Time at risk    =      167724
                                       LR chi2(2)     =    4.87
Log likelihood  = -403.86567           Prob > chi2    =  0.0876
```

```
      _t |   Haz. Ratio    Std. Err.      z     P>|z|     [95% Conf. Interval]
---------+------------------------------------------------------------------
 contact |    .7347927     .1242149    -1.82    0.068     .5275593     1.02343
contactt |    1.661876     .4684283     1.80    0.072     .9564735    2.887516
---------+------------------------------------------------------------------
```

Notice that Stata is not reporting the coefficients β_k. Instead, it reports the "hazard ratios" or exponentiated coefficients ($\exp(\beta_k)$ or v_k). So, in the case of the variable CONTACT, the results show that whatever the hazard rate at a particular time for non-referral hires is, the hazard at the same time for referrals is 0.73 times that hazard. This finding that the hazard rate of termination is lower for referrals than non-referrals is significant at the 0.05 level (one-tailed test) without controlling for other important employee characteristics in the model.

Instead, one can use the following distribution options to fit other functional forms of time: distribution(Weibull), distribution(Gompertz), distribution(loglogistic), distribution(lognormal), or distribution(gamma). This is the output obtained for the Weibull model, for example:

streg contact contactt, distribution(weibull)

```
      failure _d:  terminat == 1
analysis time _t:  endtime
            id:  employee

Fitting constant-only model:

Iteration 0:   log likelihood = -406.30034
Iteration 1:   log likelihood = -406.22538
Iteration 2:   log likelihood = -406.22538

Fitting full model:

Iteration 0:   log likelihood = -406.22538
Iteration 1:   log likelihood = -403.75801
Iteration 2:   log likelihood = -403.71981
Iteration 3:   log likelihood = -403.71978

Weibull regression -- log relative-hazard form

No. of subjects =          318              Number of obs  =     734
No. of failures =          161
Time at risk    =       167724
                                            LR chi2(2)     =    5.01
 Log likelihood = -403.71978                Prob > chi2    = 0.0816

---------+------------------------------------------------------------------
      _t |   Haz. Ratio    Std. Err.      z     P>|z|     [95% Conf. Interval]
---------+------------------------------------------------------------------
 contact |    .7334804     .1240166    -1.83    0.067     .5265841    1.021667
contactt |    1.688068     .4783915     1.85    0.065     .9686409    2.941825
---------+------------------------------------------------------------------
   /ln_p | -.0382121      .0713071    -0.54    0.592    -.1779714    .1015473
---------+------------------------------------------------------------------
       p |    .9625088     .0686337                      .8369664    1.106882
     1/p |    1.038952     .0740846                      .9034385    1.194791
---------+------------------------------------------------------------------
```

Note that in addition to the two model coefficients of interest for CONTACT and CONTACTT, the output for the Weibull model in Stata reports a Wald test or z-test for H_0: $\ln p = 0$, for which the z-statistic value is -0.54 with a p-value of -0.59. The null hypothesis cannot be rejected, which means that the $\ln p$-value does seem to be equal to 0. This is equivalent to testing H_0: $p = 1$, and thus, the hypothesis that the hazard is a constant over time cannot be rejected either.[37] In some manuals such as the *TDA* or *Stata Manuals*, the following notation is preferred to represent the Weibull hazard rate function. This notation is equivalent to the one presented earlier in equation (7):

$$r(t, x) = pr_x t^{p-1}$$

where $r_x = \exp[\beta' X_{it}]$.

Now, note that if $p > 1$, the hazard rate function increases, whereas if $p < 1$, the hazard rate decreases over time. In the case that $p = 1$, then, the Weibull function is the exponential function.

Again, observe that the exponentiated coefficients or "hazard ratio" parameters (that is, $\exp(\beta_k)$ or v_k) are displayed on the Stata output. To obtain the unexponentiated coefficients β_k of the last estimated regression model, specify the `streg` command without any list of commands including the `nohr` option right after the `streg` command as follows:

`streg, nohr`

```
Weibull regression -- log relative-hazard form
```

No. of subjects =		318		Number of obs =	734	
No. of failures =		161				
Time at risk =		167724				
				LR chi2(2)	=	5.01
Log likelihood	=	−403.71978		Prob > chi2	=	0.0816

_t	Coef.	Std. Err.	z	P>|z|	[95% Conf. Interval]	
contact	−.3099543	.1690796	−1.83	0.067	−.6413442	.0214356
contactt	.5235845	.2833959	1.85	0.065	−.0318613	1.07903
_cons	−6.598683	.4552185	−14.50	0.000	−7.490895	−5.706471
/ln_p	−.0382121	.0713071	−0.54	0.592	−.1779714	.1015473
p	.9625088	.0686337			.8369664	1.106882
1/p	1.038952	.0740846			.9034385	1.194791

In Stata, all estimation commands such as `streg`, redisplay the estimation results of the last estimated model when the command is typed by itself without any other arguments.

---

[37] Remember that if $\ln x = 0$, then $\exp(\ln x) = \exp(0)$ or $x = 1$; this is because $\exp(\ln x) = x$ and $\exp(0) = 1$.

To illustrate how to interpret some of the coefficients, I estimate the same
Weibull model with some additional independent variables as follows:

```
streg age parttime wage college contact contactt,
distribution(weibull) nohr

      failure _d: terminat == 1
analysis time _t: endtime
            id: employee

Fitting constant-only model:

Iteration 0:   log likelihood = -349.30564
Iteration 1:   log likelihood = -349.28728
Iteration 2:   log likelihood = -349.28728

Fitting full model:

Iteration 0:   log likelihood = -349.28728
Iteration 1:   log likelihood = -342.66335
Iteration 2:   log likelihood = -338.80696
Iteration 3:   log likelihood = -338.78886
Iteration 4:   log likelihood = -338.78886

Weibull regression -- log relative-hazard form

No. of subjects =        277          Number of obs =      658
No. of failures =        138
Time at risk    =     151402
                                      LR chi2(6)    =    21.00
Log likelihood  = -338.78886          Prob > chi2   =   0.0018
---------------------------------------------------------------------
       _t |     Coef.   Std. Err.      z    P>|z|    [95% Conf. Interval]
----------+----------------------------------------------------------
      age | -.0174329   .0108302    -1.61   0.107   -.0386596    .0037939
 parttime |   .990676   .2508433     3.95   0.000    .4990322    1.48232
     wage | -.0637803   .0265751    -2.40   0.016   -.1099212   -.0223607
  college | -.1888893   .2772234    -0.68   0.496   -.7322372    .3544586
  contact | -.3099543   .1690796    -1.83   0.067   -.6413442    .0214356
 contactt |  .5235845   .2833959     1.85   0.065   -.0318613    1.07903
    _cons | -6.235549   .6154788   -10.13   0.000   -7.441866   -5.029233
----------+----------------------------------------------------------
   /ln_p |  .0076693   .0770027     0.10   0.921   -.1432532    .1585919
----------+----------------------------------------------------------
       p |  1.007699   .0775956                     .8665346    1.17186
     1/p |   .99236    .0764144                     .8533445    1.154022
---------------------------------------------------------------------
```

Remember that according to this exponential specification, the variable x_k has
a multiplying effect on the hazard rate and an additive effect on the logarithm of
the hazard rate. Thus, when a variable x_k has no effect on the risk of an event
happening such as AGE (the coefficient is insignificant), then the coefficient for

age or β_{age} equals 0, or similarly, the exponential transformation of β_{age}, v_{age}, equals 1. In the case of the indicator variable PARTTIME, $\beta_{part-time}$ equals 0.99 (significant at the 0.001 level), and consequently $v_{part-time} = \exp(0.99)$ equals 2.69. The hazard of termination is 169% higher for part-time employees compared to full-time employees (that is, 100% × (2.69 − 1). For the case of referral employees (employees who had a contact in the organization prior to applying for the job), the $\beta_{contact} = -0.31$, then $v_k = \exp(-0.31) = 0.73$. An interpretation of this result is that having a personal contact in the organization at the time of application diminishes the termination rate by 27% (that is, 100% × (0.73 − 1). Or, in other words, referrals face a hazard 27% lower than non-referrals (significant at the 0.05 level, one-tailed test). In the case of a continuous independent variable such as WAGE, the coefficient is −0.06, significant at the 0.05 level. This means that a $1 increase in hourly salary decreases the hazard by 6% because $\exp(-0.06)$ is equal to 0.94 and 100% × (0.94 − 1) = 6.

The distribution option facilitates the fitting of a parametric regression model. Such a parametric estimation is more adequate when the shape of the baseline hazard rate is somewhat known. Parametric models help to obtain the most efficient β or vector of regression coefficients and to obtain an estimate of the function of time based on that distribution constraint. However, one can estimate the Cox regression model when looking for a parameterization of the function of time $q(t)$ that has no restrictions on the shape of the baseline hazard. In Stata, the estimation of the Cox regression models can be accomplished with the stcox command. Again, after having declared the data set to be an event history data set using the stset command, type stcox followed by the variable names of the independent variables. Thus:

```
stcox age parttime wage college contact contactt, nohr

         failure _d:  terminat == 1
   analysis time _t:  endtime
                id:   employee

Iteration 0:    log likelihood = -710.70094
Iteration 1:    log likelihood = -702.85463
Iteration 2:    log likelihood = -700.34575
Iteration 3:    log likelihood = -700.30248
Iteration 4:    log likelihood = -700.30246
Refining estimates:
Iteration 0:    log likelihood = -700.30246
Cox regression -- Breslow method for ties

No. of subjects =        277          Number of obs =      658
No. of failures =        138
Time at risk    =     151402
                                      LR chi2(6)    =    20.80
Log likelihood   =  -700.30246        Prob > chi2   =   0.0020
```

```
----------------------------------------------------------------------------
     _t |      Coef.    Std. Err.      z      P>|z|     [95% Conf. Interval]
---------+------------------------------------------------------------------
     age | -.0174559     .0108455    -1.61    0.108    -.0387127     .0038008
parttime |  .9834573     .2516167     3.91    0.000     .4902975     1.476617
    wage | -.0428754     .0337368    -1.27    0.204    -.1089983     .0232476
 college | -.198243      .2773177    -0.71    0.475    -.7417757     .3452897
 contact | -.3099543     .1690796    -1.83    0.067    -.6413442     .0214356
contactt |  .5235845     .2833959     1.85    0.065    -.0318613     1.07903
----------------------------------------------------------------------------
```

Note that Stata does not report any intercept for the Cox regression model. This is because the intercept is subsumed into the baseline hazard or time function $q(t)$, and mathematically speaking, the intercept is unidentifiable from the data. Substantively, in this particular example, results do not seem to change much by fitting the Cox regression model, the Weibull, or the exponential models to the termination employee data set.

7.3. Accelerated Failure Time Models

The so-called *accelerated failure time model* or *log-time model* is a frequently used model for analyzing events. In this case, the natural logarithm of the survival time $\ln t_i$ is expressed as a function of explanatory variables, with the following form:

$$\ln t_i = \beta' X_{it} + \varepsilon_{it} \tag{9a}$$

where X_{it} is the vector of explanatory variables, β is the vector of regression coefficients, and ε_i is the error term with some functional distribution $f(\ )$ to be specified. Here, the functional form of the error is what determines the regression model. In Stata, one only has to add the `time` option to the `streg` command in order to estimate this accelerated failure time models. The following command fits the same model as the one I previously fitted, but now it is presented in the accelerated failure-time metric:

```
streg male age married college wage nights parttime previous
contact contactt, distribution(exponential) time
```

The likelihood function in these models is the same as the proportional hazard rate models. These models can be viewed in either metric (accelerated failure-time or hazard rate), although it should be noted that each one provides its own substantive interpretation of results. When the coefficient β associated to a given explanatory variable is positive, the effect of the variable is to slow down time, and when such coefficient is negative, then the effect is to accelerate. Equivalently, a positive β delays the occurrence of the event, and a negative β accelerates the occurrence of the event. In these accelerated time metric models, the exponentiated coefficients have the interpretation of time ratios for a one-unit change in the corresponding independent variable of interest. For an individual case i, the t_i could be estimated as:

$$t_i = \exp[\beta_0 + \beta_1 x_{1it} + \beta_2 x_{2it} + \beta_3 x_{3it} + \cdots + \beta_k x_{kit}]\tau_i. \qquad (9b)$$

If the individual case had x_1 increased by 1, then:

$$t_{i(x_{1it}+1)} = \exp[\beta_0 + \beta_1 (x_{1it} + 1) + \beta_2 x_{2it} + \beta_3 x_{3it} + \cdots + \beta_k x_{kit}]\tau_i. \qquad (9c)$$

The ratio of $t_{i(x_{1it}+1)}$ to t_i is the exponentiated β_1, or:

$$\frac{t_{i(x_{1it}+1)}}{t_i} = \exp(\beta_1).$$

By default, when fitting the accelerated failure time models, the `streg` command reports the coefficients. If specific `tr` option is specified as part of the `streg` command, the results are shown in terms of time ratios (or exponentiated coefficients). These time ratios are recommended because they are easier to interpret and report to the readers of a study or report.[38]

7.4.   Choosing among Different Proportional Models

Given all the possible parametric specifications of the function of time, the exploratory EHA is a fundamental and very useful tool for getting an idea of the underlying process that generates the hazard rate function in the data. Consequently, it can give a good idea about the shape of the hazard function, which subsequently can aid in the evaluation of which parametric model to choose to fit. From a more statistical viewpoint, there are two other strategies one can follow in order to select a proportional model. The first strategy is used when the parametric models are nested, in which case log-likelihood-ratio tests can discriminate between them. Sometimes a simple Wald test is performed by the statistical package (see below for the case of Stata). The second strategy is used when the models are non-nested; in this case, the Akaike (1973 and 1974) information criterion (AIC) is typically used instead to compare and choose among some non-nested models.

When the models are nested, the log-likelihood-ratio or Wald test is the appropriate test. One can perform this test when comparing the Weibull model versus the exponential model, or gamma versus the Weibull model or the log-normal model, for example. As I detailed earlier in the chapter, the statistic used to contrast this hypothesis is calculated as follows:

$$-2[\log L_0 - \log L_1].$$

---

[38] For more information about these log-time models in Stata, I recommend reading Cleves, Gould, and Gutierrez [2004].

This statistic is called the *log-likelihood ratio statistic* and follows a χ^2-distribution with g degrees of freedom, where g is the number of additional parameters that are included in model 1 (this is the model with $k + g$ parameters) in comparison with model 0. g is also the number of parameter constraints applied to model 0 in comparison with model 1 (the constraints that the g parameters are equal to zero).

When using Stata, Cleves, Gould, and Gutierrez have suggested to start with the generalized gamma model (the most general of the models available using the `streg` command) and then test three hypotheses (here I follow Cleves, Gould, and Gutierrez, 2004):

(a) H_0: $\kappa = 0$; if the null hypothesis is true, then the model is log-normal.
(b) H_0: $\kappa = 1$; if the null hypothesis is true, then the model is Weibull.
(c) H_0: $\kappa = \sigma = 1$; if the null hypothesis is true, than the model is exponential.

So, the strategy here is to start with the estimation of the gamma model and immediately follow by performing the three tests suggested above:

```
streg male age married college wage nights parttime previous
contact contactt, distribution(gamma)

        failure _d:  terminat == 1
analysis time _t:  endtime
              id:  employee

Fitting constant-only model:

Iteration 0:   log likelihood = -367.53793
Iteration 1:   log likelihood = -358.50042
Iteration 2:   log likelihood = -350.26618
Iteration 3:   log likelihood = -350.11259
Iteration 4:   log likelihood = -349.96067
Iteration 5:   log likelihood = -349.66463
Iteration 6:   log likelihood =  -349.4933
Iteration 7:   log likelihood = -348.98776
Iteration 8:   log likelihood = -348.98554
Iteration 9:   log likelihood = -348.98554

Fitting full model:

Iteration 0:   log likelihood = -348.98554
Iteration 1:   log likelihood = -339.91964
Iteration 2:   log likelihood = -333.37402
Iteration 3:   log likelihood = -333.10173
Iteration 4:   log likelihood = -331.83917
Iteration 5:   log likelihood = -331.82223
Iteration 6:   log likelihood = -331.82222
```

```
Gamma regression -- accelerated failure-time form

No. of subjects  =          277             Number of obs  =      658
No. of failures  =          138
Time at risk     =       151402
                                            LR chi2(10)    =    34.33
Log likelihood   = -331.82222               Prob > chi2    =  0.0002
------------------------------------------------------------------------------
       _t |      Coef.    Std. Err.      z     P>|z|    [95% Conf. Interval]
----------+-------------------------------------------------------------------
     male |   -.2354419    .2185291    -1.08   0.281    -.6637511    .1928673
      age |    .0093631    .0111083     0.84   0.399    -.0124088     .031135
  married |    .4890487    .1898546     2.58   0.010     .1169405    .8611569
  college |    .2601119    .2774244     0.94   0.348    -.2836299    .8038537
     wage |    .0511397    .0344205     1.49   0.137    -.0163232    .1186027
   nights |    .6903459    .2878433     2.40   0.016     .1261835    1.254508
 parttime |   -1.091876    .2671011    -4.09   0.000    -1.615385   -.5683676
 previous |    .0259783    .0828621     0.31   0.754    -.1364284     .188385
  contact |   -.3099543    .1690796    -1.83   0.067    -.6413442    .0214356
 contactt |    .5235845    .2833959     1.85   0.065    -.0318613     1.07903
    _cons |    6.030882    .4898968    12.31   0.000     5.070702    6.991063
----------+-------------------------------------------------------------------
  /ln_sig |    .0364474    .1688631     0.22   0.829    -.2945182     .367413
   /kappa |     .880843    .3190909     2.76   0.006     .2554365     1.50625
----------+-------------------------------------------------------------------
    sigma |     1.03712    .1751312                      .7448904    1.443994
------------------------------------------------------------------------------
```

To perform the first test (a), one can read the Wald test directly from the output since `streg` reports that the z-statistic for kappa is 2.76 with a significance level of 0.006. Here, I reject the null hypothesis that the value of kappa is equal to zero. Consequently, the model does not seem to be log-normal. Alternatively, one could perform the Wald test independently by using the `test` command as follows:

```
test [kappa]_b[_cons] = 0

 ( 1)  [kappa]_cons = 0

          chi2( 1) =  7.62
        Prob > chi2 = 0.0058
```

The test statistic looks different because in this case a χ^2-value is reported and `streg` reported a normal test, but these tests are identical (Cleves, Gould, and Gutierrez, 2004). One still should reject the null hypothesis that the model is log-normal.

In addition, a likelihood-ratio test to compare the log-normal model nested with the gamma model can be performed using the following steps. For the first

step, save the estimates of the current model using the command `estimates`
`store` followed by a chosen name to use for the regression equation (in this
case I will use the name "full," which is the name I chose to use for later comparisons of models):

```
quietly streg male age married college wage nights parttime
previous contact contactt, distribution(gamma)

estimates store full
```

Next, fit the constraint model, which in this case is the log-normal model:

```
quietly streg male age married college wage nights parttime
previous contact contactt, distribution(lognormal)
```

Right after the `streg` estimation routine, the command `lrtest` compares
the latest estimated model with the previously saved model using the `esti-`
`mates store` command (remember that I called it "full"):

```
lrtest full, force
```

```
likelihood-ratio test                 LR chi2(1)  =    8.55
(Assumption: . nested in full)    Prob > chi2 = 0.0035
```

The `lrtest` report compares to the asymptotically equivalent Wald test.
Note that I specify the `force` option to `lrtest`. This is because Stata would
otherwise refuse to perform this test. The likelihood test suggests, once again,
that the gamma model significantly improves the fit of the model compared to
the log-normal model.

To test the second hypothesis (b) (so that, if the null hypothesis is true, then
the model is Weibull), the gamma model must be re-fit, and then one should test
whether kappa is equal to 1, just as follows:

```
quietly streg male age married college wage nights parttime
previous contact contactt, distribution(gamma)

test [kappa]_b[_cons] = 1
```

```
 ( 1)   [kappa]_cons = 1

        chi2(  1) =   0.14
      Prob > chi2 =   0.7088
```

Note that the test results do not reject the null hypothesis and, consequently,
one cannot preclude the use of the Weibull model over the gamma model for
fitting the employee termination process in the data. Here, one could also run the
log-likelihood test similar to the way it was done before.

To test the third hypothesis (c) (so that if the null hypothesis is true, then the model is exponential), one can also perform either a likelihood-ratio test or Wald test as below:

```
quietly streg male age married college wage nights parttime
previous contact contactt, distribution (gamma)

test [kappa]_b[_cons] = 1, notest

( 1)   [kappa]_cons = 1

test [ln_sig]_b[_cons] = 0, accum

( 1)   [kappa]_cons   = 1
( 2)   [ln_sig]_cons = 0

          chi2(  2) = 0.21
        Prob > chi2 = 0.9003
```

Note that the `notest` option allows for the use of the `test` command to test two parameters simultaneously. The significance of the Wald test reinforces what is already known about the termination hazard rate: The gamma model does not seem to be the most appropriate model in this employee termination data example. According to this test results, the hazard rate is constant over time and the exponential model is providing a good fit for this data.

However, when the models are not nested the likelihood-ratio and Wald tests are not suitable. Instead, a different goodness-of-fit indicator, the AIC, is recommended for use. This statistic is equal to minus two times the log-likelihood function plus two times the number of free parameters in the model, or:

$$-2 \log L_0 + 2(w + c)$$

where w is now the number of independent variables in the model and c is the number of model-specific distributional parameters. c takes the following values depending of the distribution function:

Distribution	c
Exponential	1
Weibull	2
Gompertz	2
Log-normal	2
Log-logistic	2
Generalized gamma	3

This statistic is attractive to use because it penalizes each model's log-likelihood by the number of parameters being estimated. So, the lower the AIC the better the fit of the model.

To illustrate the AIC, I review the six parametric models that I fit to the employee termination data with the several covariates of interest. The following table gives the summary of the log-likelihood values and AIC values obtained from each model:

Distribution	Log-Likelihood	w	c	AIC
Exponential	−331.92	10	1	685.85
Weibull	−331.89	10	2	687.77
Gompertz	−331.92	10	2	687.84
Log-normal	−336.09	10	2	696.18
Log-logistic	−332.66	10	2	689.32
Generalized gamma	−331.82	10	3	689.64

Following the AIC criterion, the exponential model seems to be the preferred one (i.e., the one with the lowest AIC value). However, in the employee termination process under study, the Weibull model might also be a reasonable model (based on what is known about the termination data in this setting). The Weibull model has virtually the same log-likelihood values and quite a similar AIC score as the exponential model, 685.9 versus 687.8. With that in mind, one could choose to estimate and provide information about both the exponential and the Weibull models. One can also use the `estimates store` and `lrtest` commands to compare the exponential and the Weibull models (as shown on page 214 when the gamma and log-normal models were compared). Later in this chapter, I will run a log-likelihood test to compare both models directly using an example.

7.5.   *Testing the Proportional Hazard Rates Assumption*

Up to now, I have shown how to use the `stcox` command to estimate the most general version of the Cox model (on page 209). The advantage of the Cox model is that one can leave the function of time unestimated, since this model does not make any assumptions about the shape of the hazard over time—it could be constant, increasing, decreasing, or any other temporal pattern imaginable. Whatever shape this baseline hazard rate has, it is assumed to be the same for every individual case in the sample. Right after the `stcox` or the `streg` command, one can use the `stcurve` command for graphing the estimated survival, cumulative hazard, and hazard functions that fit the parametric EHA model. It is especially handy after the `stcox` command because it automates the process of taking estimates at baseline using `stcox` and transforming them to adhere to covariate patterns other than the baseline hazard rate. This is done using the option `basehc(newvar)` in order to obtain and save the baseline hazard values into the `newvar`. The `newvar` refers to the newly created variable that will store the baseline hazard values; in the example below, I named this `newvar` variable h0, as follows:

```
quietly stcox male age married college wage nights parttime
previous contact contactt, basehc(h0)

stcurve, hazard at1(contact=0) at2(contact=1) yscale(log)
```

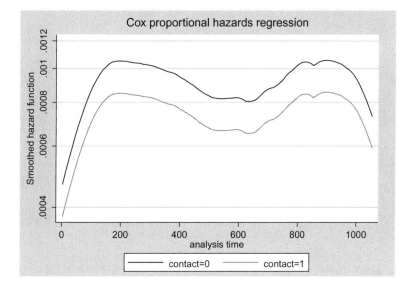

The `stcurve, hazard` command produces a smoothed hazard function estimator (by default, the curve is evaluated at the mean values of all the independent variables). One can request to plot the survival function by using the `stcurve, survival` command instead. The `at1(contact=0)` option requests Stata to plot the hazard rate function when the independent variable CONTACT is equal to zero. The `at2(contact=1)` option does exactly the same when CONTACT equals one. This `at( )` can be generalized to graph the hazard function at different values of any of the independent variables on the same graph.

When using the `yscale(log)` option, the `stcurve` command is particularly helpful because it plots the hazard rate (or survival function) on a log scale. This can help to get a better sense of the proportionality of the hazards assumption under the previously estimated model. If the hazard rates are indeed proportional, when graphed on a log scale they should look parallel or at least close enough to parallel with respect to the smoothing (again as I illustrated on page 195).

One can run other tests to test the proportional hazard rates assumption. The most commonly employed technique to check this assumption is the *analysis of residuals* or Schoenfeld residuals. The idea behind this method is to save the residuals after the estimation of an EHA model, fit a smooth function of time to those residuals, and then test whether there is any significant relationship between the residuals and time. Under the null hypothesis of proportional hazards, one should expect the curve to have a slope of 0, meaning that there is no significant relationship between the residuals and time.[39] Stata makes this analysis of residuals very simple with the `stphtest` command, by automating the process of graphing and testing for individual covariates and globally (that is, for all covariates) the null hypothesis of no slope. Rejecting the null hypothesis, thus, means that the hazard rates are non-proportional.

The `stphtest` command works after the model has been fit using `stcox`. One needs to specify both the `schoenfeld( )` and `scaledsch( )` options when estimating the Cox model in order to save the Schoenfeld and scaled Schoenfeld residuals that later would be used by the `stphtest` command. The Schoenfeld residuals are needed later in order to perform the global test, whereas the scaled Schoenfeld residuals are used to perform the variable-by-variable proportionality tests. The Stata command lines read as follows:

```
quietly stcox male age married college wage nights parttime
previous contact contactt, schoenfeld(sch*) scaledsch(sca*)

stphtest, detail
```

The first command line quietly estimates the Cox model again. By including the `schoenfeld( )` and `scaledsch( )` options in the command line, Stata creates a set of new variables in the data set containing the values for the residuals and scaled residuals. It is convenient in this case to use the `sch*` and `sca*` so that Stata numbers the variables automatically. After estimating and saving these residuals, the `stphtest` with the `detail` option reports the variable-by-variable

---

[39] For a more detailed and comprehensive discussion of residuals, read Wu [1990].

tests along with the overall (or global) test. This is how the Stata output should look:

```
Test of proportional hazards assumption

Time:   Time
-------------+-------------------------------------------------------------
             |       rho        chi2       df       Prob > chi2
-------------+-------------------------------------------------------------
      male   |    -0.14552      2.86        1        0.1908
       age   |    -0.15129      4.01        1        0.1451
   married   |     0.01062      0.02        1        0.8936
   college   |    -0.08656      1.06        1        0.3035
      wage   |    -0.05515      0.44        1        0.5057
    nights   |    -0.02491      0.09        1        0.7668
  parttime   |    -0.06605      0.60        1        0.4400
  previous   |     0.04559      0.27        1        0.6000
   contact   |    -0.08016      0.94        1        0.3319
  contactt   |    -0.04941      0.36        1        0.5471
-------------+-------------------------------------------------------------
global test  |                 12.84       10        0.2329
-------------------------------------------------------------------------
```

In this particular case, I do not find evidence that the model specification violates the proportional hazard rates assumption. That is, all *p*-values are greater than 0.05 and I therefore cannot reject the null hypothesis of proportional hazard rates. The global test has a value of 12.84 (with 10 degrees of freedom) and is not significant either.

7.6. *Non-Proportional Hazard Rate Models*

Often, the hazard rate does not seem to be proportional. This happens when the underlying pattern of the hazard rate over time is not identical for all groups in the population and/or when the effects of certain explanatory variables do not act solely on $q(t)$ multiplicatively. If this is the case, then one must consider estimating a *non-proportional hazard rate model*. Under the most general form, the non-proportional hazard rate model specifies the function of time or baseline hazard rate function also to be a function of explanatory variables (just like the function of explanatory variables). The hazard rate function is written as follows (as specified earlier in this section, on page 193):

$$r(t, x) = \theta(X_{it}) \, q(t, Z_{it}) \tag{2}$$

where X_{it} is the vector of explanatory variables incorporated in the hazard rate model. This EHA model is called the *non-proportional hazard rate model* when the baseline hazard rate $q(t)$ not only depends on time but also on a vector of independent variables Z_{it}. The vector of variables Z_{it} may or may not include the same explanatory variables as the variables included in vector X_{it}. Therefore, in this situation, the hazard rates for two different values of x_k are functions of X_{it} and of Z_{it} via $q(t)$. It is a function of Z_{it} because the time trend of the hazard rate *per se* is now also a function of independent variables.

Let me go back to the data example and examine the factors influencing employees' termination in a given organization. After some careful examination of the survivor and hazard rate functions, assume that the Gompertz model is the most appropriate one to model the rate of employee turnover. By *Gompertz model* I refer to the hazard rate model that incorporates the Gompertz function as a way of specifying the function of time in the hazard rate model. The Gompertz model, without including any vector of explanatory variables affecting the time function, is of the form:

$$r(t) = r \exp[\gamma t]. \tag{10a}$$

An equivalent form is the following (when taking the logarithm of both sides of the previous equation):

$$\log r(t) = \log r + \gamma t \tag{10b}$$

where r is a constant term. In the proportional version of the Gompertz model, r is no longer a constant term. Instead, it is a function of explanatory variables $\theta(X_{it})$. It is frequently assumed that $\theta(X_{it}) = \exp[\beta' X_{it}]$ or similarly that $\log \theta(X_{it}) = \beta' X_{it}$. Therefore, the Gompertz proportional hazard rate model has the following usable formulation:

$$r(t, x) = \exp[\beta' X_{it}] \exp[\gamma t]. \tag{11a}$$

A similar expression of the previous equation is:

$$\log r(t, x) = \beta' X_{it} + \gamma t. \tag{11b}$$

Note that the parameter γ—which measures how much the hazard rate varies with the passage of time t—is a constant and as such does not vary by individual case.

To develop the non-proportional version of the Gompertz hazard rate model, one must still assume that r is a function of independent variables or $\theta(X_{it})$. But now, the parameter γ is assumed not to be constant. Rather, it is a function of several variables as well, as formulated below:

$$\gamma = f(Z_{it}) = \Lambda' Z_{it} \tag{12}$$

where Z_{it} is a vector of explanatory variables that does not necessarily have to be the same as vector X_{it}, even though Z_{it} is typically a subgroup of all variables included in vector X_{it}. With the addition of this functional form for the parameter γ, the time-varying baseline hazard rate $q(t)$ is now specified to also depend on a vector of independent variables Z_{it}. The final formula for the non-proportional version of the Gompertz hazard rate model is:

$$r(t, x) = \exp[\beta'X_{it}] \exp[\Lambda'Z_{it}t] = \exp[\beta'X_{it} + \Lambda'Z_{it}t] \qquad (13a)$$

or similarly, in logarithmic form:

$$\log r(t, x) = \beta'X_{it} + \Lambda'Zt. \qquad (13b)$$

The non-proportional version of the Gompertz model should also be estimated by the ML method.

In Stata, the `ancillary()` and `strata()` options allow the estimation of non-proportional hazard rate models easily. For example, I will now estimate the non-proportional version of the Weibull model using Stata. Recall the Stata Weibull model formula presented earlier in equation (7) (on page 197) where if $p > 1$, the hazard rate function increases, whereas if $p < 1$, the hazard rate decreases over time. In the case that $p = 1$, then, the Weibull function is the exponential function. The exact command line in Stata is as follows:

streg age parttime wage, distribution(weibull) nohr

The following is the obtained relevant output:

```
Weibull regression -- log relative-hazard form

No. of subjects =        281          Number of obs =      670
No. of failures =        140
Time at risk     =     154018
                                      LR chi2(3)    =    18.62
Log likelihood   = -344.27453         Prob > chi2   =   0.0003
-------------------------------------------------------------------------
     _t |    Coef.    Std. Err.     z     P>|z|    [95% Conf. Interval]
--------+----------------------------------------------------------------
    age | -.0207675   .0107921   -1.92   0.054   -.0419196     .0003846
parttime|  1.003233   .2486571    4.03   0.000    .5158744     1.490592
   wage | -.0365953   .0326529   -1.12   0.262   -.1005937     .0274031
  _cons | -6.407754   .6061743  -10.57   0.000   -7.595833    -5.219674
--------+----------------------------------------------------------------
  /ln_p |  .0231866   .0753232    0.31   0.758   -.1244442     .1708174
--------+----------------------------------------------------------------
      p |  1.023458   .0770901                    .8829876     1.186274
    1/p |  .9770801   .0735968                    .8429755     1.132519
-------------------------------------------------------------------------
```

For simplification purposes, this Weibull model only includes AGE, PART-TIME, and WAGE as independent variables this time. Assume one is interested in estimating the non-proportional version of the Weibull model with respect to the independent variable measuring part-time status (or PARTTIME). In the Stata command window, type:

```
streg age parttime wage, distribution(weibull) ancillary(parttime)
nohr
```

where the `ancillary(parttime)` option helps to fit the non-parametric Weibull model that compares differences in the function of time between part-time and full-time employees. In other words, I am comparing simultaneously whether $\theta(X_{it})$ is affected by the variable PARTTIME and whether $q(t, Z_{it})$ is affected by the variable PARTTIME. The output in Stata looks like this now:

```
Weibull regression -- log relative-hazard form

No. of subjects =          281              Number of obs =        670
No. of failures =          140
Time at risk    =       154018
                                            LR chi2(3)    =       7.66
Log likelihood  =  -344.11422              Prob > chi2   =     0.0537
```

	Coef.	Std. Err.	z	P>|z|	[95% Conf.	Interval]
_t						
age	-.0208808	.010791	-1.94	0.053	-.0420308	.0002693
parttime	1.713271	1.250495	1.37	0.171	-.7376536	4.164195
wage	-.0368033	.0326882	-1.13	0.260	-.1008711	.0272644
_cons	-6.530454	.6491478	-10.06	0.000	-7.80276	-5.258148
ln_p						
parttime	-.1169381	.2103373	-0.56	0.578	-.5291916	.2953154
_cons	.0421001	.0818506	0.51	0.607	-.1183241	.2025244

By adding the `ancillary(parttime)` option, the Weibull model is specifying that the parameter p can also be predicted by the variable PARTTIME as follows:

$$p = p_0 + p_1 \times parttime$$

and again, such a model allows for the constant p to differ between part-time and full-time employees. In general, the effect of an independent variable x_k on the time parameter p is represented by p_k. There are two key sections in the Stata output to look at after running the `stset` command. The first section headed by "_t" reports the effects of different independent variables on the hazard rate.

In the example above, the coefficients are for the variables AGE, PARTTIME, and WAGE, just as follows:

```
---------+----------------------------------------------------------------
     _t |
    age |  -.0208808    .010791    -1.94   0.053   -.0420308    .0002693
parttime |   1.713271   1.250495     1.37   0.171   -.7376536    4.164195
   wage |  -.0368033   .0326882    -1.13   0.260   -.1008711    .0272644
   _cons |  -6.530454   .6491478   -10.06   0.000   -7.80276    -5.258148
---------+------------------------------------------------------------
```

The second section is headed by "ln_p" in the case of the Weibull model (it is always the non "_t" section or parameter section for any other functional form) and reports the effects of the different variables on the parameter in the function of time. In the example, this was the parameter section for the Weibull model:

```
---------+----------------------------------------------------------------
   ln_p |
parttime |  -.1169381   .2103373    -0.56   0.578   -.5291916    .2953154
   _cons |   .0421001   .0818506     0.51   0.607   -.1183241    .2025244
---------+-------------------------------------------------------
```

From the estimated results, ln p is equal to 0.04 for full-time employees and $0.08 - 0.12 = -0.04$ for part-time employees. However, given that the Wald test for the part-time parameter is -0.56 with the level of significance 0.578, it does not seem that the shape of the Weibull function of time differs significantly by part-time employment status in this particular case.

The Weibull, Gompertz, log-normal, and log-logistic models can all include the Stata ancillary( ) option. Because the exponential model does not have any additional parameter in the function of time, it does not make sense to include the ancillary( ) option with the exponential model. The generalized gamma model has two time parameters that could be modeled as a function of independent variables. So in the generalized gamma model, one will need to specify the anc2( ) in addition to the ancillary( ) option. The ancillary option allows for the specification of more than one variable affecting the function of time parameters. So it is possible to specify ancillary(parttime contact) in order to explore whether the model is non-proportional with respect to the variables indicating part-time status and referral application.

In general, when specifying the non-parametric version of EHA models, I suggest pursuing the following three strategic steps:

Step 1. Start the EHA modeling exercise by finding the best proportional model that fits the data. This can be accomplished by looking at the exploratory analyses first, before undertaking any multivariate EHA analysis.

Step 2. If one needs to fit a non-proportional model for either methodological or theoretical reasons (i.e., to test the non-proportionality assumption), then add the `ancillary( )` option to the `streg` command, including the name of the variables believed to affect the function of time parameters. Add one variable at a time and evaluate whether their z-values suggest that any particular variable predicts both the main function of independent variables and the function of time. For example, to evaluate the effect of variable x_k on the hazard rate, there are three different scenarios:

1. Only the coefficient for x_k or β_k—when x_k is included in the function of independent variables $\theta(X_{it})$—is significant.

   In practice, if after running the next command line in Stata:

   streg xk, distribution(weibull) ancillary(xk) nohr

   one finds that the x_k is only significant according to the coefficient β_k reported under the "_t" section of the Stata output (where the coefficients for the variables included in $\theta(X_{it})$ are reported) then the model can be assumed to be proportional with respect to variable x_k. Notice that I use xk as the variable name in Stata for any chosen variable x_k of interest.

2. Only the coefficient for x_k or p_k—when x_k is included in the function of time $q(t, Z_{it})$—is significant.

   In practice, if after running the next command line in Stata:

   streg xk, distribution(weibull) ancillary(xk) nohr

   one finds that x_k is significant according to the coefficient p_k reported under the non-"_t" section of the Stata output (where the coefficients for the variables included in $q(t, Z_{it})$ are reported) then the model can be assumed to be non-proportional with respect to variable x_k.

3. The two coefficients for x_k, β_k and p_k—when x_k is included in both $\theta(X_{it})$ and $q(t, Z_{it})$—are significant.

   In practice, if after running the next command line in Stata:

   streg xk, distribution(weibull) ancillary(xk) nohr

   one finds first that x_k is significant according to the coefficient β_k reported under the "_t" section of the Stata output (where the coefficients for the variables included in $\theta(X_{it})$ are reported) and second that x_k is significant according to the coefficient p_k reported under the non-"_t" section of the Stata output (where the coefficients for the variables included in $q(t, Z_{it})$ are reported) then the model can be assumed to be non-proportional with respect to variable x_k.

Step 3. Based on the information above, choose the best parametric specification of the EHA regression model, either the proportional or the non-proportional version. If scenario 1 (in Step 2) is true, choose the proportional version of the EHA model. If either scenario 2 or 3 are true, choose the non-proportional version of the model.

The `strata` option provides another easy way to fit models that assume one and only one independent categorical variable influencing both $\theta(X_{it})$ and $q(t, Z_{it})$. When specifying the `strata(varname)`, `varname` is defined as a categorical variable that identifies the different groups to be compared. Such an option automatically includes a set of indicator or dummy variables (one per category, except the category of reference) created out of the categorical variable included in the command. Technically speaking, if and only if x_s is a dummy variable, then the following command line using `strata`:

`streg xk, distribution(weibull) strata(xs) nohr`

is identical to the following command line using the `ancillary( )` option:

`streg xk xs, distribution(weibull) ancillary(xs) nohr`

where `xs` is the variable name in Stata for any chosen dummy variable x_s. Notice that in the case of `strata`, only the variable of interest is included within the `strata` option. The `streg` routine includes the variable of interest in both the function of independent variables $\theta(X_{it})$ and the function of time $q(t, Z_{it})$.

In addition to the `ancillary` and `strata` options in Stata, there are several user-written commands, that is, commands and programs developed by Stata users to address some of the more advanced EHA estimation procedures. In this case, one could type any of the two command lines below in the Stata command box:

`findit non-proportional hazard rate model`
`findit piecewise hazard rate model`

The `findit` command in Stata will open a new window with a list of links that contain the keywords given after `findit`. Click on the blue clickable hyperlinks and follow the instructions to install these user-written programs. The syntax for these commands can be found in the help files, and their technical support is often supplied by the author of the command. Their contact information is typically at the end of the help file. Scholars continue developing new statistical procedures in Stata and you should look for new possible updates regularly.

Instead of Stata, one can choose to use the TDA program to estimate non-proportional hazard rate models. Use the `rate=` command in order to specify the desired model to estimate, followed by a number indicating the type of model to estimate. The command `rate = 2` is used to request the estimation of the basic exponential model. Below, note what each of the command lines (in ***bold-italics***) will allow the user to estimate in terms of different functional forms:

```
rate =  1; Cox Model
rate =  2; Exponential model
rate =  6; Gompertz model
rate =  7; Weibull model
rate =  9; Log-logistic model
rate = 12; Log-normal model.
```

The model specification can then include two additional commands. The first one is the command to include the list of independent variables in the function of explanatory variables or $\theta(X_{it})$. The TDA command is the following:

```
Xa(0,1) = list of variables
```

where the "`list of variables`" given on the right-hand side of the command refers to the variable names to be listed as part of the model specification.

The `rate` command, together with the `Xa(0,1)` command, estimates the different versions of proportional hazard rate models. Adding the second TDA command `Xb(0,1)` will specify the non-proportional version of any model. This second command allows the user to include the list of explanatory variables in the function of time or $q(t, Z_{it})$; notice that the X is followed by the letter b to indicate to TDA that this is where the vector of independent variables influencing the shape of the function of time is specified:

```
Xb(0,1) = list of variables
```

For example, to fit the non-proportional version of the Weibull model (results are repeated in Table 3.7, under Model C: Non-Proportional Weibull), the following three command lines were needed:

```
rate = 7;
Xa(0,1) = v7, v8, v9, v10, v11, v12, v13, v14, v15, v16;
Xb(0,1) = v7, v8;
```

where the `rate` command indicates to TDA to estimate the Weibull model. The `Xa(0,1)` command is followed by the list of variables to include in the vector of independent variables—in this case, v7, v8, v9, v10, v11, v12, v13, v14, v15, and v16. The `Xb(0,1)` command specifies the list of variables to include in the vector of independent variables in the function of time—in this case only variables v7 and v8 (earlier assigned the values for CONTACT and CONTACTT using the TDA commands `v7(contact)=c7` and `v8(contactt)=c8`).

Table 3.7
Several multivariate termination hazard rate models

Independent Variables	Model A: Exponential		Model B: Proportional Weibull		Model C: Non-Proportional Weibull	
	B	Standard Error	B	Standard Error	B	Standard Error
1. Vector of explanatory variables						
Constant	−5.741***	0.359	−5.743***	0.356	−5.703***	0.349
Male	0.186	0.184	0.184	0.183	0.171	0.181
Age	−0.014	0.010	−0.014	0.010	−0.013	0.010
Married	−0.312*	0.168	−0.309*	0.166	−0.343*	0.170
College degree	−0.221	0.244	−0.218	0.242	−0.206	0.237
Hourly wage	−0.061*	0.030	−0.060*	0.029	−0.063*	0.030
Night shift	−0.717**	0.238	−0.711***	0.239	−0.705***	0.241
Part-time	0.918***	0.220	0.911***	0.222	0.910***	0.220
Previous jobs	−0.060	0.067	−0.060	0.066	−0.068	0.067
Personal contact	−0.296*	0.167	−0.292*	0.167	−0.398*	0.197
Contact is terminated	0.519*	0.255	0.510*	0.259	0.612*	0.288
2. Vector of explanatory variables in the function of time						
Constant			0.011	0.070	0.092	0.093
Personal contact					−0.181	0.143
Contact is terminated					0.072	0.249
χ^2 statistic						
Compared with baseline model[a]		41.980***		41.880***		43.500***
Degrees of freedom		10		10		12.000
Compared with model A				0.020		1.640
Degrees of freedom				1		3
Compared with model B						1.620
Degrees of freedom						2

$N = 325$ employees.
Note: All significant levels are the following: *$p < 0.05$; **$p < 0.01$; ***$p < 0.005$ (one-tailed tests).
[a]The model labeled baseline is a model with no independent variables (only the constant).

To fit the proportional version of the Weibull model (Model B in Table 3.7), the following two lines were used:

```
rate = 7;
Xa(0,1) = v7, v8, v9, v10, v11, v12, v13, v14, v15, v16;
```

7.7.   Piecewise Exponential Models

Of special interest to social scientists is a set of non-proportional hazard rate models that have been widely used in empirical studies due to their simplicity and convenience. These models are called the *piecewise exponential models* (Tuma, Hannan, and Groeneveld, 1979; Blossfeld and Rohwer, 1995). In the most general version of these models, the hazard rate is expressed as follows:

$$r(t, x) = \exp(\beta_p' X_p) \qquad \tau_{p-1} \leq t \leq \tau_p \tag{14a}$$

or alternatively, and in logarithmic fashion, the expression is:

$$\log r(t, x) = \beta_p' X_p \qquad \tau_{p-1} \leq t \leq \tau_p \tag{14b}$$

where p denotes a time interval or time *piece* that goes from τ_{p-1} to τ_p. The basic idea behind these models is to split the time axis into time periods or "pieces" and to assume that the hazard rates are constant within each of those time pieces (allowing them to change across time pieces). Two key features make this piecewise model extremely useful. First, the model allows the hazard rate to vary over time in a flexible way. Second, this model also permits the rate to vary in a non-proportional way with different effects of several explanatory variables within each one of the pre-defined intervals. The model assumes that, given certain values of the independent variables, the hazard rate is constant within each of the time intervals (as specified by the exponential distribution), even though this rate might differ across time intervals. Simultaneously, the hazard rate depends on the vector of explanatory variables X_p in ways that might differ from one time interval to another. Furthermore, the vector of explanatory variables included in the right-hand side of the model X_p does not have to be identical for every time period or interval studied. This model's piecewise features are what makes the model non-proportional.

In Stata, Sorensen (1999) has developed a useful program subroutine called `stpiece` that easily estimates these piecewise exponential hazard models.[40] The subroutine can be installed from within Stata by typing:

ssc install stpiece

The installation command can be immediately followed by the `help` `stpiece` in order to obtain the help document created by Sorensen on how to use `stpiece` to specify the piecewise exponential model in Stata. The data must have been previously declared event history data with the command `stset`

---

[40] Before this program was created, the estimation of the piecewise exponential model usually required following multiple steps, including the definition of the each time interval and splitting the spells using the `stsplit` command. `stpiece` automates this process.

before using the `stpiece` command. In the following, I illustrate this command estimating the piecewise exponential model that includes three independent variables, AGE, PARTTIME, and WAGE. In the Stata command window, type:

```
stpiece age parttime wage, tp(365,730)
```

where the list of variables that follow the `stpiece` command are the independent variables whose effects do not vary between time periods. The `tp( )` option allows the user to define the time periods; this option specifies the exact times at which the records are to be split. In the example above, the records are split into three years, so records are split at $\tau_1 = 365$ and $\tau_2 = 730$. Hence there is a time interval for each of the three years in the sample. The following Stata output was obtained:

```
Invoking stsplit...
Creating time pieces...

        failure _d:  terminat == 1
analysis time _t:    endtime
             id:     employee
Iteration 0:   log likelihood =   -404.8127
Iteration 1:   log likelihood = -357.59547
Iteration 2:   log likelihood =  -344.52404
Iteration 3:   log likelihood =  -344.13876
Iteration 4:   log likelihood =   -344.1381
Iteration 5:   log likelihood =   -344.1381

Exponential regression -- log relative-hazard form

No. of subjects =         281            Number of obs =       955
No. of failures =         140
Time at risk    =      154018
                                         Wald chi2(6)  = 6756.21
Log likelihood  = -344.1381              Prob > chi2   =  0.0000
------------------------------------------------------------------------------
     _t |  Haz. Ratio  Std. Err.      z     P>|z|     [95% Conf. Interval]
--------+---------------------------------------------------------------------
    tp1 |   .0019588   .0007108   -17.18   0.000     .0009618    .0039892
    tp2 |   .0017779   .0006808   -16.54   0.000     .0008394    .0037658
    tp3 |   .0020691   .0009116   -14.03   0.000     .0008725    .0049068
    age |   .9795034   .0105714    -1.92   0.055     .9590014    1.000444
parttime|   2.697156   .6708171     3.99   0.000     1.656538    4.391478
   wage |   .9645605   .0315359    -1.10   0.270     .9046899    1.028393
------------------------------------------------------------------------------
note: no constant term was estimated in the main equation
```

The constant rates for each of the time intervals are listed in the output as "tp#", where # indexes the time intervals sequentially. So that "tp1" is the constant rate for the first year, "tp2" the rate for the second year, and "tp3" the rate for the third year. All three are significant at the 0.001 level. In this particular case, the effect of the variables AGE, PARTTIME, and WAGE are not allowed to vary between time intervals. To test whether the effect of those variables are non-proportional, that is, whether they change between time intervals, add the tv( ) option including those independent variables whose effects are believed to be non-proportional, as follows:

```
stpiece, tp(365,730) tv(age parttime wage)
```

In the previous command line, AGE, PARTTIME, and WAGE are allowed to vary across time intervals. The output would be as follows (I only include the relevant part of the output this time):

```
Exponential regression -- log relative-hazard form

No. of subjects =        281            Number of obs =      955
No. of failures =        140
Time at risk    =     154018
                                        Wald chi2(12)  = 6703.32
Log likelihood  = -339.8583            Prob > chi2   =  0.0000
```

_t	Haz. Ratio	Std. Err.	z	P>|z|	[95% Conf. Interval]	
tp1	.0010142	.0004432	−15.78	0.000	.0004307	.0023883
tp2	.0066083	.0047804	−6.94	0.000	.0016008	.0272796
tp3	.0053366	.0070007	−3.99	0.000	.000408	.0698071
tpage1	.9994835	.0124188	−0.04	0.967	.975437	1.024123
tpage2	.9364903	.0235696	−2.61	0.009	.8914156	.9838442
tpage3	.9489292	.0388979	−1.28	0.201	.8756729	1.028314
tppart1	2.935434	.8387509	3.77	0.000	1.676702	5.139121
tppart2	1.713375	1.026916	0.90	0.369	.529276	5.546545
tppart3	4.587249	4.766702	1.47	0.143	.5984831	35.16032
tpwage1	.9755374	.0383294	−0.63	0.528	.9032329	1.05363
tpwage2	.9523389	.0614322	−0.76	0.449	.8392345	1.080686
tpwage3	.9484291	.1201003	−0.42	0.676	.7399741	1.215607

```
note: no constant term was estimated in the main equation
```

The tv( ) option specifies the list of variables which will appear in the output labeled as "tp*#", where * refers to the first four letters of the variable name followed by # which indexes the time intervals (as described earlier in this section).[41] The effect of AGE only seems to be significant during the second

---

[41] For more information, go to http://econpapers.repec.org and look for Sorensen [1999] and the stpiece command.

year, while the PARTTIME variable only seems to be significant during the first year of tenure at the organization. One has to be warned about one of the problems when splitting observations into time intervals: If the number of observations (and events) in each time period is greatly reduced, this may result in the loss of statistical power when estimating each model for each time period and consequently not meaningful estimated models. So choose the length of time intervals or pieces carefully, ensuring that there are enough observations in each time period.

The statistical program TDA also proves to be of help when it comes to estimating piecewise models. The TDA command to request the piecewise constant exponential model is the following:

```
rate = 3;
```

One must also define the periods by split points on the time axis. The TDA command to define time periods is `tp`:

```
tp = t1, t2, t3,t4 ... te;
```

where `t1` is when the first time period starts, `t2` the second, and so forth, up to `tk` different time periods for which all effects will be calculated. In the case of *e* periods and *c* explanatory variables to be estimated using the `Xa(0,1)` TDA command, a total of $c \times e$ different model coefficients are being estimated. In TDA, other functional forms of time can be easily used (besides the exponential) to formulate the piecewise hazard rate model. For example, a Gompertz model may be specified in order to adjust a non-proportional Gompertz model to each of the time intervals under study. The same approach can be followed for each possible functional specification of the time-varying baseline hazard rate as summarized on Table 3.6.[42]

8. EHA and Some Statistical Software Programs

Given the entire typology of available dynamic models for the study of events presented here in Chapter 3, some models are mathematically easier to work with than others. In addition, only a few statistical programs can estimate many of the non-proportional hazard rate models with many time functional specifications.

---

[42] In Stata, one can also estimate piecewise models using other functional forms of time (besides exponential) by doing some additional programming though (beyond the scope of this book).

Stata and TDA are two of the programs capable of doing so. SAS (for more information, go to www.sas.com) permits the specification of proportional dynamic models with the most common specifications of time, that is, the exponential, Weibull, log-logistic, and log-normal functions.

When it comes to estimating proportional hazard rates, I think Stata is the easiest program to use. One can estimate some of the proportional models, mainly those in which the function of time is exponential or Weibull, with the commands `streg`, followed by the list of explanatory variables. Before running any of these EHA parametric models, again be sure that Stata is aware of the longitudinal nature of the data set by using the command `stset`. This therefore sets the variable that uniquely identifies the individuals or cases in the sample together with the time variable defining the passage of time. One can also estimate nonproportional models by adding the `ancillary( )` and `strata( )` options to the `streg` command.

With the command `stcox`, one can estimate the Cox version of a hazard rate model. The command is also run after the `stset` command to define the longitudinal components of the data set. Stata contains a set of useful commands that begins with the annotation `st` (which references the term *survival time data*): `stdes` provides a description of the longitudinal data loaded in any given Stata session. The commands `sts graph`, `sts list`, and `sts test` generate and plot the survivor and hazard functions (Kaplan–Meier estimates), report its values in a graph over time, and test whether the survivor function differs across several groups of cases in the sample, respectively. Finally, the `ltable` command calculates the life tables for different groups (as I demonstrated in the Exploratory EHA section 6 in this chapter). For more information about survival analysis and tables, I recommend reading the *Survival Analysis and Epidemiological Tables Reference Manual* published by Stata. The most recent version of the manual (at the time of this publication) is Release 9.

SPSS allows for the simple calculation of life tables or the analysis of the Kaplan–Meier survivor function. It also facilitates the estimation of the Cox regression model, with or without time-varying explanatory variables. This can be done when the *Advanced Statistics* module has been installed in SPSS. In the *Statistics/Survival* pull-down menu, one will see that many options are available for analyzing events in a sample study. Below, one can see the SPSS screen of pull down menu that would allow the estimation of the Cox models. In the *Cox Regression Box*, enter the Time variable ENDTIME, followed by the Status variable, TERMINAT in this example. The event is defined by the value of 1 for variable TERMINAT. Then, include the list of independent variables or covariates in the *Covariates Box*. Press OK in order to get the Cox model output. This is what the Cox regression looks like in SPSS:

	employeeid	observation	starttime	endtime	startstate	endstate	terminat	contactt	contact
1	1	1	0	157	0	0	0	0	1
2	1	2	157	162	0	0	0	0	1
3	1								1
4	2								1
5	2								1
6	2								1
7	2								1
8	3								1
9	3								1
10	3								1
11	4								0
12	5								0
13	6								0
14	6								0
15	6								0
16	6								0
17	7								0
18	7								0
19	7								0
20	7								0
21	7								0
22	7	6	74	105	0	1	1	0	0
23	8	1	0	150	0	1	1	0	0
24	9	1	0	1082	0	0	0	0	1

Cox Regression dialog box. Variables: employeeid, observation, starttime, startstate, endstate, contactt, contact, male, married, college, wage, previousjobs, workatapplic, workexperience, parttime, temp, nights. Time: endtime. Status: terminat(1). Define Event... Block 1 of 1. Previous / Next. Covariates: age, parttime, wage, college, contact. Method: Enter. Strata: OK, Paste, Reset, Cancel, Help, Categorical..., Plots..., Save..., Options...

SPSS is easy to use. Unfortunately, the current version of SPSS (Version 14) only permits the estimation of the Cox regression models and not many other advanced multivariate EHA regression models illustrated in this chapter. In addition, the SPSS EHA parametric procedures assume the independence of observations and consequently, they are recommended to use when analyzing events in certain event history data set formats (e.g., where there is only one episode for each individual case in the sample or what is called single-episode data). For these reasons, SPSS survival techniques are recommended when there is only one record per individual case in the sample. For more information, consult some of the *SPSS Reference Manuals*, specifically the one titled *Advanced Statistics*.

9. An Example

Now I will illustrate some of the multivariate models presented in this chapter for the sample of employees in one call center I studied (Castilla, 2005). For this, I use the iterative estimation procedure of ML included in the statistical software program TDA (Blossfeld and Rohwer, 1995). This will aid in learning how to use a program other than Stata. The most recent version of TDA (at the time of this book's publication) is version 6.2, which is still preliminary. This version of the TDA program (like in the examples of files of commands, and notes of the use of the program) can be obtained without charge by visiting the following web page: http://www.stat.ruhr-uni-bochum.de/tda.html (elaborated on by Rohwer and Poetter).

The research question I want to answer is whether social networks matter in the employment history of individuals. In particular, using a sample of 325 employees in a call center, I estimate some multivariate EHA models that explore which variables contribute to the termination of employees at the organization. In doing so, I pay particular attention to the effect of the variable CONTACT or "having a contact or referrer within the organization" at the time of application. I also look at the variable CONTACTT or "whether that referrer is still in the organization or not." These two network variables allow the evaluation of the impact of having connections inside an organization on the hazard rate of termination, after controlling for (and therefore including in the model) several important employee characteristics such gender, age, marital status, education, and wage.

I start by showing the calculation of several termination hazard rate models.[43] Remember that the variable TERMINAT is a dichotomous variable that takes the value of 1 if the employee is terminated during the period of analysis, 0 otherwise. Again, any of the statistical programs that perform EHA require identifying the social process of events under study in two ways. First, a dichotomous "event" variable indicating whether the event happened in a given period of time or record in the sample (the variable is TERMINAT in the example). Second, a "time" variable that measures the time when the event occurred (if the event occurs for a given case or individual) or alternatively, the time when the observation of a given case ceased (if the event does not happen during the observation period for that specific case; this variable then measures the time of censoring). In TDA, remember that four variables are important when defining the termination social process: STARTSTA, ENDSTATE, STARTTIM, and ENDTIME (if necessary, go back to the Practical Example section in this chapter to be reminded of each of these variables measures).

Table 3.7 reports the ML estimates of several parametric EHA models. In all the models, the hazard rate is the function that is being specified and estimated. I also report the values of the *overall log-likelihood χ^2-statistics*—which compare any given multivariate hazard rate model with the baseline hazard rate model, that is the basic hazard model containing no explanatory variables (i.e., the model only including the constant terms). I estimated three models: The exponential hazard rate model, the proportional Weibull model, and the non-proportional Weibull model. All models are statistically significant at the 0.001 level. So in order to compare among the various specifications of the hazard rate function, the *incremental χ^2-statistic* has to be used (given that all three models are nested). The difference between the log-likelihood values for any two nested models is distributed according to the theoretical χ^2-distribution function with g degrees of freedom (where g is the number of variables whose joint significance is being

---

[43] 95% of all dismissals in this organization were voluntary, and given that the absolute number of involuntary dismissals is small enough, I will not estimate competing risk models.

tested). If the difference is significant, one can conclude that the additional variables included in the most complex model contribute to a better explanation of the causes of termination at the organization under study (as reviewed earlier in this chapter).

The CONTACT variable is a significant variable in all hazard rate models. These results confirm that, after controlling for important individual characteristics such as gender, age, marital status, part-time, wage per hour (among many other control variables included in the model), the rate of termination is lower for those hires who were referred to the job than for non-referrals. The coefficient for CONTACT is always negative and significant at the 0.05 significance level. Additionally, the CONTACTT (or "contact is terminated") variable shows that the rate increases considerably once the referrer or contact in the organization leaves the organization. This effect is always positive and significant in the three hazard rate models illustrated on Table 3.7.

By carefully evaluating the incremental χ^2-tests, the best model seems to be the exponential model (which is nested in the proportional Weibull model). The two incremental χ^2-tests comparing the exponential model with the proportional Weibull model and the exponential model with the non-proportional Weibull model are not significant (at the 0.05 level). Consequently, the null hypothesis that the constant parameter γ equals 0 cannot be rejected (in the case of the proportional Weibull model). Nor can it be rejected that the vector of coefficients for the explanatory variables included in the time function (in the case of the non-proportional version of the Weibull model in which I allowed the two social network variables to affect the time function of the hazard rate model) is different from zero. Thus, the simpler exponential model is the preferred model in this particular setting, and the hazard rate is assumed not to vary over time nor do the effects of the variables CONTACT or CONTACTT change over time.

Once again, I show how to interpret the model coefficients. Using the estimates obtained by fitting the exponential model (model A), the coefficient for the variable CONTACT is -0.296; this implies that the logarithm of the hazard rate of termination for referral hires is -0.296 units lower than for non-referral employees, *ceteris paribus* (that is, controlling the other variables of the model). Earlier, I demonstrated a more intuitive interpretation of these estimates that comes with the transformation of these parameters using the following formula:

$$100\% \times [\exp(\beta) - 1]$$

where β is the ML estimate of any of the coefficients in the function of explanatory variables. The formula provides the percentage of change in the hazard rate by a one unit of change in the explanatory variable of interest. According to the exponential model estimates (given that the exponential model is the one that best fits the data), and in the case of hourly wages, the estimated coefficient is -0.061(significant at the level of the 0.05). This coefficient indicates that with each additional dollar earned per hour, the termination hazard rate diminishes by 6% (100% × [exp(−0.061) − 1] = −5.9). Now, when looking at the social network variables of interest in this example study, the termination rate is almost 26% lower for referral employees than non-referral employees (100% × [exp(−0.296) − 1] = −25.62). Notice that when including a dichotomous variable, the exp(β) in itself provides the relative hazard rate for the groups represented by the dichotomous variable (after controlling for the rest of the explanatory variables in the model). Thus, the termination hazard rate for referrals is 74% of the hazard rate for non-referrals (because exp(−0.293) equals 0.744). Alternatively, since 1/0.744 = 1.34, one can claim that the hazard rate of termination is 34% higher for non-referrals than for referral hires in the organization under study.

The second network variable, CONTACTT, provides additional evidence about the social network process behind employee terminations. When the referrer has left the organization, the model shows that termination rate is 25% higher for referral employees than non-referrals—in this case, one must take into consideration both coefficients (CONTACT and CONTACTT), so that the 25% was calculated by adding the two coefficients before applying the formula: (100% × [exp(−0.296 + 0.519) − 1] = 24.98). The coefficient on the hazard rate for referral employees whose referrer is still around is simply −0.296, while the coefficient is −0.296 + 0.519 for those referrals whose referrer is not around.

To test the significance of the model coefficients, one must calculate the ratio of the estimated coefficients and their standard errors. For large samples of observations, these can be treated like the *t*-test in OLS regressions or the *z*-values in the logit/probit regression models. Therefore, if the value of the *t*-test is greater than 1.96, then the coefficient is significantly different from zero for a significant level of 0.05 (if the test is two-tailed; 0.025 if the test is one-tailed). The significance levels for each of the parameters are presented in Table 3.7 using the asterisk (*); this asterisk approach is very common when analysis results are reported in empirical studies. In this example, single asterisk means that the coefficient is significant at the 0.05 level (one-tailed test); double asterisks mean that the coefficient is significant at the 0.01 level; and triple mean that the coefficient is significant at the 0.001 level (one-tailed test). In all models, the variables marital status, hourly wage, and night shift are definitely significant. Moreover, their effects on the hazard rate of termination are in the direction expected according to several theoretical and practical arguments. The same happens with the social network variables CONTACT and CONTACTT (both significant at the 0.05 level).

In Table 3.8, I present the input file I used in TDA to estimate the different EHA parametric models reported on Table 3.7. This file was edited in order to better

Chapter 4

Designing a Study with Longitudinal Data

After reading a research paper published in a top journal, one may think that writing an empirical study is anything but easy. As with any research paper, the authors need to show that their research questions are sufficiently interesting and relevant. They have to demonstrate how the statistical analysis of the available data can address their research question. Finally, they must produce a study that can compete for acceptance and recognition in their chosen academic field of interest. Therefore, authors of research papers have two important obligations. First, they should frame or position their empirical study *vis a vis* previous studies or research papers, showing how their work makes a new and important contribution to their discipline. Second, when writing a study that depends upon the statistical analysis of data, the authors should carefully describe both the data and the methodology used to analyze the data.

This chapter describes how to write a paper that analyzes longitudinal data sets using either event history analysis (EHA) or cross-sectional cross-time analysis of continuous variables. Because much of this discussion applies to other types of papers in which statistical analysis of data is central, I begin by defining the most common elements of a research paper focusing especially on studies that use dynamic methods and techniques to analyze social processes of change. This chapter describes how to compose the multiple sections of a paper, even those that may seem rather obvious (experienced researchers should feel free to skim over elementary material). Recognizing that a research paper may be structured in many different ways, this chapter does not attempt to be definitive or exhaustive. However, its goal is to aid the reader in producing an empirical study that is informative, clear, concise, and self-contained.

To help illustrate my discussion, throughout this chapter I refer to examples drawn from three of my own published articles. The first one examines how national health care systems and other national characteristics affect population health in OECD countries. This paper was published in *International Sociology* in 2004.[44] The second article analyzes the performance implications of using

[44] The full citation of the article is: Castilla, E.J., 2004, Organizing health care: A comparative analysis of national institutions and inequality over time, *International Sociology* 19 (4), 403–435. Copyright © 2004 by Sage Publications.

employee networks as part of the hiring process; the complete paper is available in Chapter 5. A later version of this paper was published in the *American Journal of Sociology* in 2005.[45] In the third article, published in the *International Journal of Technology Management* in 2003, I present the beginnings of a systematic comparative study of social networks of venture capitalists. Specifically, I examine the role of social networks of venture capital firms in the development of industrial regions with both a comparative and a network perspective.[46] I provide more detail about these studies as I use them to illustrate the key components of an empirical research paper.

1.   Writing a Research Paper: The Different Elements

An empirical study using longitudinal data generally sets out to examine a particular research topic and to answer specific research questions. The researcher investigates each question by formulating a series of research hypotheses, by designing a study whose major goal is to reject or accept such hypotheses, and by collecting longitudinal data for analysis. The techniques learned earlier in this book will help to test hypotheses and analyze longitudinal data sets. The following steps are common in most empirical studies using data sets to test theories, including research papers using longitudinal data. Thus, one should begin by stating the questions under investigation, defining hypotheses, assessing the available data, and describing the statistical methods used to test the hypotheses. At the end of a research paper, one should discuss the results and outline possible directions for future research. Finally, the study should provide a bibliography or reference section that lists past studies on the topic and gives sources for the methodology employed. The paper may also contain endnotes and appendices that supplement or clarify important discussions within the paper.

   Fig. 4.1 summarizes the different steps to follow when designing and writing a paper using statistical data analysis. A paper must include a title page listing the author's name, affiliation, and contact information; many journals require a word count on the cover page as well. A succinct *abstract* or *summary* is also recommended at the beginning of the paper. It may be necessary to check the standard length of an abstract in a given specific field of interest since many journals have limits on the word count. I introduce these components briefly before giving a longer explanation of each later in the chapter. The *introduction*

---

[45] The full citation of the article is: Castilla, E.J., 2005, Social networks and employee performance in a call center, *American Journal of Sociology* 110 (5), 1243–1283. Copyright © 2005 by The University of Chicago Press.

[46] The full citation of the article is: Castilla, E.J., 2003, Networks of venture capital firms in Silicon Valley, *International Journal of Technology Management* 25 (1/2), 113–135. Copyright © 2003 by Inderscience Publishers.

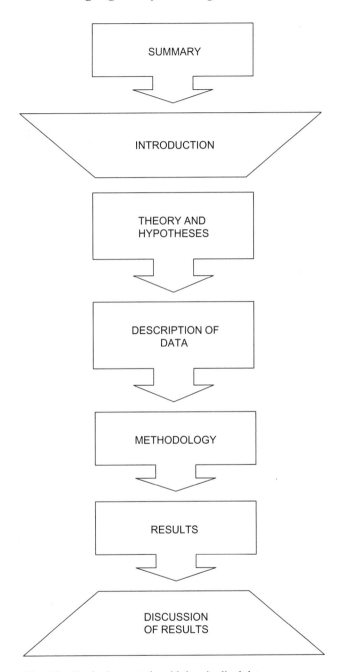

Fig. 4.1   Designing a study with longitudinal data.

provides a justification for the research paper, moving from a general discussion of the topic to the particular questions the paper addresses. An important role of the introduction is to spark readers' interest and to motivate them to read the entire paper. The *theory and hypotheses* section describes the particular questions, theories, and hypotheses the paper investigates and may include a literature review. The *longitudinal data* section describes the data set under analysis. The level of detail provided in this section should depend on the intended audience's familiarity with the data. If the author has collected his or her own data set, it is imperative to describe how it was collected, paying special attention to the variables measured. If the data set is a secondary data set, be sure to refer the readers to the relevant sources in the main text. The next section should briefly describe the *methodology* used in the paper, that is, the models and methods of estimation performed as part of the analysis. When the paper analyzes a longitudinal data set, this section must describe in detail the event history models or panel models used to test the hypotheses. Many scholars combine the longitudinal data and the methodology sections into a *methods* section; one may consider doing that if it makes the paper increasingly concise and clear. The *results* and *discussion of results* sections describe the paper's findings, and the *conclusion* generalizes the study's accomplishments and suggests opportunities for future research. The *list of references* is a key part of a research paper as it provides the most appropriate works for the topic of the investigation and acknowledges scholars who have been working on similar issues.

Finally, research papers may include appendices, tables, figures, and endnotes or footnotes. Appendices are optional, but can provide further useful information or clarification about topics not central to the flow of the paper. Tables and figures can be excellent tools for clarifying and communicating the most important theoretical and empirical ideas in the paper. Some scholars prefer to include tables and figures in the main text, but other may note them in the main text and include them at the end, one per page. I recommend using footnotes sparingly— it is best to incorporate important ideas in the main text and to delete any tangential or irrelevant material.

The remainder of this chapter is devoted to a detailed description of a longitudinal data research paper. This discussion is designed to prepare the reader for the structuring and writing of a research paper.

2.   The Introduction

The introduction orients the reader, providing a framework for understanding the detailed information that appears in later sections. It should first contextualize the study's particular ideas within a broader theory, theme, or topic. To do this, it is important to move from a general discussion of the topic to an outline of specific propositions in the paper. Generally, the introduction takes the form of the inverted trapezoid represented in Fig. 4.1, in which broad discussions are gradually refined into specific points and arguments.

Second, the introduction should summarize previous research on the problem and justify further research in the area. A well-written and exciting introduction can motivate a reader to continue reading the rest of the paper. So aim to be specific and interesting when introducing the topic. For examples of introductions, see the two sample essays included in Chapter 5 or simply pick-up any leading journal and read the introductions to already published papers.

3. Theory and Hypotheses

The goal of the theory and hypotheses section is to articulate the problem or question the paper addresses as well as to review previous relevant work. The section may open with a literature review that gives relevant background information and situates the paper within a larger body of research. The literature review should cite the most pertinent work by other scholars and professionals, identifying monographs, articles, books, and statistical sources by the last name of the author(s), the year of publication, and pagination (when appropriate). Be sure to include the full bibliographical information at the end of the study (I give more information about the format of the references in the *bibliography and references* section). It is best to discuss only literature that bears directly on the research question motivating the paper. When dealing with a large body of literature, cite only representative references and avoid discussions that are overly detailed.

Once the general context of the paper has been presented and previous works have been discussed, one can begin to provide detail about the question the study investigates. Write at least one paragraph to summarize the study's objective, its possible benefits, and its various applications. The principal element of any empirical study, the *research question*, guides how one plans and carries out any scientific investigation. Research questions should be substantive. For example, a researcher interested in studying social networks in labor markets could formulate a question thus: Do social networks help organizations to hire employees with superior performance? The research agenda may involve exploring whether new workers hired using employee networks are better matched to their jobs than socially isolated employees (i.e., employees hired using other recruitment practices such as newspaper ads or internet postings).

After formulating a research question, one should present the strategy for answering it and suggest several possible answers. The statement or set of statements outlining the expected results are called *hypotheses*. In scientific terms, a hypothesis derives from a theory or from existing knowledge about the research subject. Hypotheses are formulated to verify such a theory. For example, in my study of performance and social networks (Castilla, 2005), I wanted to deepen the understanding of social ties by studying whether socially connected employees were more likely to remain in an organization longer and perform better than those who were hired without knowing anyone. Given this research question, I could formulate two testable hypotheses:

Hypothesis 1. *Employees hired via personal contacts are less likely to leave the organization than those hired via other recruitment sources.*

Hypothesis 2. *Employees hired via personal contacts perform better than those hired via other recruitment sources.*

The principal motive of the study is to accept or reject the hypotheses by using the appropriate statistical methodology and data. One should provide a theoretical justification for all hypotheses. In my example, I suggest that individual employees hired via personal contacts perform better than those who applied via other recruitment sources. This is so because social connections provide the candidate and the organization with difficult-to-obtain and more realistic information about each other, which improves the quality of the match. In general, researchers have argued that referral hiring allows the hiring organization to leverage social connections, thereby improving the quality of the match between worker and job—this is known in the economic literature on hiring as the "better match" theory (Castilla, 2005).

Let me emphasize how important it is to explain the hypotheses. After formulating hypotheses, researchers test them against the data that they have collected. These tests usually involve examining the data for patterns and determining whether they match up well with the patterns predicted by the hypotheses. In the end, the data analysis determines whether the research refutes or supports the specific hypotheses. In other words, the hypotheses embody the theoretical concepts, encouraging an evaluation by means of statistical methods, such as the different kinds of longitudinal analysis described in this book. In almost all longitudinal studies, the researcher has several ideas (or working hypotheses) about how independent variables affect a given dependent variable. In the case of EHA, for instance, researchers study the factors that make a certain change more probable (as measured by a qualitative dependent variable) or that make an event more likely to occur at certain times. In addition to rates at which events occur, researchers may also be interested in analyzing duration of events. When conducting analyses with continuous dependent variables, scholars usually study factors that produce a quantitative change in an outcome variable of interest.

Most hypotheses are ultimately about mechanisms or processes (as measured by a vector of variables X that make an event more or less probable in the case of EHA) or that cause a change in a continuous variable (in the case of analysis with continuous dependent variables). In EHA, hypotheses are frequently expressed in terms of explanatory variables affecting the rate (or hazard) of change in a discrete variable. These hypotheses may also be expressed in terms of some explanatory variables affecting the rate of transition from one situation to another (for example, the probability of transition from situation j to situation k; for more information, see Chapter 3). In studies with continuous dependent variables, the hypotheses are expressed in terms of some independent variables increasing or decreasing the value or growth of the continuous dependent variable under study (for more information, see Chapter 2).

After presenting the research questions and hypotheses, one should state a *plan of action*. It is common to begin by specifying the system of concepts or variables that are used, including both dependent and independent variables. Explain the *dependent variable* or outcome variable (an event or a particular process of change under study), indicate the social *unit of analysis* (e.g., a person, a company, or a country), and test the hypotheses with the help of *independent variables*. Sometimes the same independent variable will help testing different hypotheses about different outcome variables. Indicate how different explanatory variables make change more or less likely (in the case of a categorical dependent variable) or tend to alter the magnitude of the change (in the case of a continuous dependent variable). One can also assume that the change can be explained by transforming the explanatory variable (for example, the logarithm of x instead of x). If a variable is transformed, one must explain why.

In my study about the effect of social networks on employee outcomes, examples of outcome variables are the termination of the employee (dichotomous dependent variable) and the change in employee productivity (continuous dependent variable). In the first instance, one might be able to hypothesize whether level of education, experience in the job, previous job experience, salary history, or gender increase or decrease the probability of termination. In the second case, one might be able to hypothesize about how the same independent variables affect productivity growth over time. The main purpose of my social network study was to determine whether networks help organizations to recruit better-matched employees and therefore to test the two hypotheses that I outlined earlier in this section. The first hypothesis is that employees with pre-existing social connections to the organization are less likely to be terminated than socially isolated employees. This hypothesis is supported if the network variable used in the analysis has a negative and significant effect on the hazard rate of termination (i.e., if $\beta_1 < 0$). The second hypothesis—that employees with pre-existing social connections to the organization should be more productive than socially isolated employees—is supported if the network variable has a positive and significant effect on the variable productivity over time (that is, if $\beta_2 > 0$).

Make sure to specify the *units of analysis,* unless they are truly obvious. For example, in labor market and job mobility studies, the units of analysis are usually people. Nevertheless, job mobility studies can take a much bigger social group as the unit of analysis (e.g., the household). The unit of analysis both reflects the researcher's interests and indicates the direction future studies may take.

One should also discuss how underlying patterns of time dependence might affect the change process. Here, it is important to choose among a variety of "clocks." For example, age, historical time (or calendar time), and duration (time length) are different ways of measuring time. In my study, I looked at duration, measured as the length of an employee's tenure in the organization. I examined the hazard rate of termination over employees' tenure on the job. If I had used historical time in my study, I would be examining the hazard rate of termination over a particular historical period of time. It is important to explain the main "clock" in the longitudinal study because the choice of clock can produce different results with distinct theoretical implications. Always specify the chosen *unit of time*—this can be seconds, weeks, months, years, or even decades.

A research paper using longitudinal data should clearly explain the basic concepts underpinning the change process under analysis. In the case of EHA, this conception can be expressed as the probability that individual cases will experience a change or transition over a defined time period. Such a concept can be easily translated into a model of EHA. When dealing with a continuous variable, one must specify whether the estimated model is predicting change in the dependent variable or predicting the value of the dependent variable at time t. Also, one must discuss basic patterns of time dependence in the change process. If one uses EHA, the object of study is presumably the change over time in some discrete variable (categorical or ordinal). If outcomes of a quantitative nature are analyzed, the object of the study is change over time in some continuous variable. Regardless of whether the dependent variable is discrete or continuous, it is important to clearly identify the dependent variable (or the variable that indicates the existence of a process of change over time) as well as the values that this variable can take (in theory as well as in practice). In EHA, if all the transitions between values of the discrete variable are possible, explain changes that can or cannot occur, and note why their occurrence is or is not possible (unless it is an obvious change of status).

Some independent variables that affect the outcome variable may also vary over time (time-varying covariates). In this case, researchers should specify how a particular change depends on current or previous values of the independent variables. The most common assumption is that the probability of change or the change in a continuous dependent variable in time t depends on values of the independent variables in time t or in the previous point of time $t-1$. In response, one might want to lag the independent variables in time; one always has to justify any of these decisions in the main text of the paper. For example, scholars interested in national development often argue that explanatory variables must be delayed by at least one quarter, semester, or year because governments base their economic and other policy decisions on variables of one quarter, semester, or year in advance (such as change in gross domestic product, investment, employment, and inflation).

The theory and hypotheses section should also assess whether the effects of the explanatory variables change over time. For example, many wage growth studies assume that work experience has a proportional effect on salary (i.e., the coefficient of experience on wage growth is invariable during the worker's tenure in the organization). Nevertheless, it is also possible that work experience has a lesser (or greater) effect on salary growth right after the worker is hired and that such effects are smaller (or higher) later on. This probably depends on the job characteristics as well as the state of the labor market, among many other factors. Be specific about this in the hypotheses section when relevant.

Finally, it is necessary to articulate and explain any assumption about effects that do not vary over time. If assuming time-invariant coefficients, justify the assumption in the results section and assess in the discussion section how this may affect the results of the study.

4. Longitudinal Data

Every empirical paper must include a section describing the data used in the analysis. This section describes the sources of the data, details their particular characteristics, and explains how they were collected. The data analysis will either support or refute the hypotheses under study. It will also suggest ways the hypotheses or the research design might be revised in future research studies.

One should describe the variables as instruments that operationalize certain concepts or ideas of concern to the researcher. For example, in my health care paper I used variables such as "average life expectancy" or "infant mortality" as indicators of a nation's population health—I did this because both life expectancy and infant mortality can be measured accurately and because relevant and detailed data were available for the countries under study (Castilla, 2004). Specify the values (or range of values) that the dependent variable can take. If there are a sizeable number of observations, consider including a table or graph charting the dependent variable over time.

Indicate the values of the dependent variable when the variable is discrete. When using continuous variables such as employee salary growth, describe the operationalization of the process of change and indicate the interval or range of values as well as how the variables change over time. If one is studying a process of change, such as a worker finding a job, it would be necessary to identify in the longitudinal data section, which variables are used to represent this process.

It would be possible to use a dichotomous variable that takes the value of 1 starting the day in which a given applicant is hired (0 before that). The dependent variable obviously depends on the characteristics of the available data. The choice of dependent variable also has crucial implications for the formulation and the testing of hypotheses.

Ideally, the researcher can take accurate measurements of the changes or events under study. Often, however, the act of data collection collapses theoretical categories. This happens, for instance, when the data may not have been accurately gathered (in the case of continuous variables). In such a case, one should thoroughly discuss how imperfections in data collection affected the results. One should also be careful about generalizing research and should suggest how the research project might be modified to help others avoid the same pitfalls. Researchers have often contributed to develop the understanding of social phenomena by adding new variables to studies or by better-measuring variables or factors used in previous work.

The longitudinal data section should also describe the measurement of the independent or explanatory variables. As with any empirical analysis, one should explain how different factors or independent variables have been measured and discuss any possible discrepancies between the theoretical constructions and their practical operationalization. Inform the reader about the basic characteristics of these variables (i.e., supply descriptive statistics) and discuss the values of independent variables whose significance is not obvious. Often, researchers describe the independent and dependent variables in tables, giving a brief description of each variable and a theoretical justification for its inclusion in the study. I discuss the presentation of tables further in the *results* section of this chapter (and more examples are included in Chapter 5).

One might also use a table to report possible values of different independent variables. Such a table should include descriptive statistics for independent variables included in any of the longitudinal models as well as information about the average value, range of values, dispersion or variance in values, and number of cases. When independent variables (or covariates) vary over time, it is necessary to discuss how change occurs (for example, whether variables vary abruptly or gradually) and explain how change is measured. Include information about the descriptive statistics and change in explanatory variables over time, presenting the values for the different independent variables at different points within the period of observation. If the explanatory variables that change over time are not continuously measured, discuss how such incomplete measures are handled. For example, one can use the most recently available measurements for these variables or perform some interpolation. Several techniques can be used to interpolate when temporal information is incomplete. Some of them are more appropriate than others depending on the data set and the field of study, but one must always provide an explanation and justification as to why a certain procedure was used.

Finally, state how often the units of analyses were recorded over time, highlighting the points where the dependent (and independent) variables were measured and defining the time frame or observation window. Again, be sure to indicate the unit of time in the study. In EHA, events happen continuously over time; to measure them properly, one must indicate whether the time is measured to the day, month, year, or even decade. When data have been collected at certain regular points in time (every year or every five years, for example), discuss how such aggregation can affect the conclusions of the study. It might even be possible to use a method of estimation that takes into consideration the characteristics of such temporal measurements. State clearly the number of individual cases as well as the number of events occurring in the sample of individual cases.

In general, temporal data derived from dependent variables of discrete nature are right-censored, but in the most unfortunate case, the data can also be censored to the left. The problem of right/left-censored data as well as the reasons behind it should be discussed in the paper. In longitudinal analyses, right-censored data do not normally bias the results of the study. For example, if one randomly excludes cases to the right, the models can still be correctly specified for the results of interest without much bias (although when there are many censored cases, the efficiency of the estimators is reduced, even if the censoring is random). But the right censoring of data frequently happens for reasons closely related to the result of interest. In this case, the results may be both biased and inefficient. For example, assume a study that models the event of firm success using publicly available information such as data about start-up companies' funding from venture capital firms. One might find it difficult to obtain information about start-ups that dissolved relatively quickly or that did not receive funding from venture capital firms. Therefore, neither the existence nor the organization of the data is clear. The sample is truncated, and that can affect the results of the study.

Left-censored data are especially problematic because many change processes of interest depend on the initial conditions. Missing information about the starting conditions of any change affecting the dependent variable generally produces skewed results. When data are left-censored, one should discuss how much left censoring exists, why it exists, and whether it will be ignored or controlled for. When there is temporal dependency over historical time, gaps in data can cause serious problems (this is a type of censoring on the left). One could try to correct for such gaps in several ways (e.g., by trying to predict those variables where no data are available or by evaluating the probable impact of omissions on the results). Regardless of how this problem is solved, one should report and describe the procedure in detail.

In general, it is necessary to include a paragraph reporting the limitations of the data and/or design of the study. Report any problems or gaps in the collection of the longitudinal data, and indicate whether it is believed that there is sample selectivity related to the values of the dependent variable. Give the reader an idea of whether there is selection bias and whether there is left or right truncation of the sample. Because a key aspect of the study is the selection of the sample, one should discuss any selectivity of cases that can be related somehow to the values that the dependent variable can take. After eliminating cases for which the values of the dependent variable are not known, analyze the portion of the sample for which there is sufficient data. For example, collecting information about employee performance means excluding people who were not hired by the organization or who were terminated during the initial training period, even though the results can be biased by this selection mechanism. Similarly, if the analysis examines reasons for an organization's survival, comparisons with other organizations that have also survived for a time period can bias the results, especially if the causal factors represent effects that vary through time.

In addition, consider discussing the methods for dealing with unobserved heterogeneity or the case when relevant independent variables of the social process under study are omitted. Both observed heterogeneity and temporary dependency are partially confounded. There is no simple solution to this problem, so be explicit about how it is handled, even if only to state that the problem is ignored. It is worth explaining the foundations and limitations of the data and outlining how it was collected since they determine the findings in an empirical study and future research.

In this data section (or in later sections when presenting the limitations of the study), one may also want to focus on the different sources of bias and inefficiency associated with the data and the making of causal inferences. Make sure to report any issues of measurement error that could bias the results or make them less efficient. Also, consider possible bias in the causal inferences as the result of omitting explanatory variables that should have been included. The opposite problem can also have a negative impact on the results: One should report attempts to control for irrelevant variables that could reduce the efficiency of the models. Finally, evaluate whether there is any endogeneity problem, which arises when the dependent variable affects the independent variables.

5.   Methodology for Data Analysis

The *methodology* section describes how the longitudinal data has been analyzed. This section especially benefits readers who want to know more about the chosen methodological approach or who are curious to know how the methodology may have influenced the results. The bulk of this section is dedicated to statistical procedures and should draw a fundamental distinction between *exploratory methods* of analysis and *multivariate* (or *explanatory*) *methods* of analyzing longitudinal data. Exploratory methods are useful for describing basic patterns in the data or for evaluating the fundamental assumptions of common multivariate methods.

Explanatory methods are often used to test specific hypotheses. These methods frequently involve estimating multivariate models. Exploratory methods of analyses are currently undervalued. Nonetheless, they are an excellent means of determining what methodology is best suited to multivariate data analysis. These methods give readers a general idea of the change under study (e.g., its frequency and general temporal patterns), but more importantly, they help determine which parametric model to use (e.g., in the case of an event, exploratory methods may aid in the selection of distributions in the hazard rate models).

Regardless of whether or not one uses exploratory or explanatory methods, it is necessary to indicate, which models are chosen and show how they were estimated. It helps tremendously to report the equations that summarize the model's functional form, especially if the intended audience is unfamiliar with it. Use common sense, and do not spend time on unnecessary detail. For example, presenting the equation for a multiple regression is unnecessary since most readers in the social sciences are familiar with the technique (and it is available in any basic statistics book). But when using a Weibull model to calculate the hazard rate, for example, an explanation may be necessary. When using less common models, such as the Gompertz or Gamma models, you may provide an explanation.

Authors of mathematically intensive articles should consider formatting their articles in LaTeX, a program designed for typesetting complex equations. However, recent versions of word processor programs such as Microsoft Word (Windows or Macintosh) or WordPerfect allow users to easily insert equations into the main text. Use these advanced features only to type complex equations.

Informing the audience about the equations used for the estimation and citing some standard references that discuss such equations allow interested readers to learn more about the particular method of analysis. Always mention the properties of the estimators as well as the utility of the method. Report whether the coefficients are consistent, unbiased, or efficient, and note the circumstances under which those properties might be violated. If using standard methods, the discussion should be brief. Mention the statistical software program used in the analysis, especially for new and advanced methodological models. One can use footnotes or even technical appendices to report such technical information.

Below, I show how I wrote the research design section of an early version of my national health care paper (Castilla, 2004).[47] Notice that I combined the data section and the measurement, hypotheses, and methodology sections under the *research design* heading. This should provide an example of how flexible the

---

[47] Again, the full citation of the article is: Castilla, E.J., 2004, Organizing health care: A comparative analysis of national institutions and inequality over time, *International Sociology* 19 (4), 403–435. Copyright © 2004 by Sage Publications.

organization of these sections can really be. In this excerpt, the acronym HCS refers to "health care systems."

RESEARCH DESIGN

Data

The evolution of national institutions is examined using cross-national longitudinal data on national health care structures and population health status across OECD countries during the period 1960–90. The study refers both to public welfare efforts by national governments and to health out-comes as reflected in "quality-of-life" measures. The data analyzed are national level aggregate data for twenty-two OECD countries, mainly obtained from the *OECD Health Data 1995*. This data set provides relevant statistics of health and health care in the OECD region during the 1960–90 period, by single years and by each country. The choice of data set was based on two main factors. First, specific and extensive data on national population health and health care systems (henceforth HCSs) are only avail-able for rich, industrialized countries. Second, the data provide comparable measures across countries, which are available for at least four points in time since 1960, separated by five or ten years. The variables for this study have been selected on the basis of (a) their relevance to the description of the major dimensions of national HCSs and population health status; (b) comparability of measurement across nations and over time; and (c) the availability of time-series observations in a significant number of countries.

Due to the limits set by data availability and sample size, the analysis performed here has several limitations that, if overcome, could improve future research. First, questions about the sampling scheme and problems of selection bias are in order. While general theoretical propositions regarding national HCSs are believed to hold true for all national institu-tions, more research is needed to determine whether these general propo-sitions also operate in developing and newly industrializing countries, as well as other aspects of national institutions. Future analyses of national HCSs should also include some health variables other than life expectancy at birth and mortality rates. More sophisticated performance indicators to capture the variability in population health status (e.g., quality of life) would be of interest to social scientists.

Measurement, hypotheses, and methodology

In the discussion above [omitted in this book but available in Castilla, 2004], I develop a series of general propositions accounting for the most important aspects of national HCSs by the end of the twentieth century. In this section, I present a set of measures and testable hypotheses that derive from my theoretical propositions. I also describe the methodology I use for testing these theoretical explanations. To test propositions regarding similarities and disparities among national health systems, a series of longitudinal analyses of change are performed. Propositions related to the effects of national HCSs on population health are tested using longitudinal regressions and multivariate analysis of variance during the 1960–90 period. Regression analyses are also used to test propositions about the impact of the degree of individualism on the general characteristics that national institutions take, as well as hypotheses referring to the effects of social inequalities on population health status.

In this study, variation is measured using the coefficient of variation (i.e., the ratio of the standard deviation to the mean, expressed as a percent) rather than the more common alternatives such as the standard deviation or variance. The greater the decrease in the coefficient of variation over a specified period of time, the greater the convergence. Since the data were available for different time periods, I also present a measure of the annual rate of change for each of the variables. The annual rate of change in the coefficient of variation (expressed as a percent) is calculated as follows (Williamson and Fleming, 1977):

$$\frac{\left(\dfrac{CV_{T_2} - CV_{T_1}}{CV_{T_1}}\right) \times 100}{(T_2 - T_1)}$$

where CV_{T_1} = coefficient of variation at the earliest year, and CV_{T_2} = coefficient of variation at the latest year. Positive values of the annual rate of change indicate that variation increases across countries over the 1960–90 period. In other words, the more negative the rate of change, the stronger the trend toward convergence.

Proposition 1 asserts that there is convergence in the forms of national HCSs. The form of national HCSs refers to the major characteristics of national HCSs as operationalized in this paper by using health expenditure and social protection: (a) the public and total expenditure on in-patient care, ambulance care, and medical goods; and (b) the level of public coverage against health costs. According to Proposition 1, I derive the following testable hypothesis:

H1. The coefficient of variation in health expenditure and percentage of population covered against health costs decreases over time; that is, the annual rate of change in the variation of these general indicators is negative.

Proposition 2 is about divergence regarding the composition and utilization of national HCSs. The term "composition and utilization of HCSs" here refers to the availability and use of national health care resources, and employment. In particular, it is measured by the following indicators: (a) professional activity such as total health employment, and the number of active physicians; (b) hospital facilities and staff such as available beds, hospital staffing ratios, nursing staffing ratio; and private hospital beds; (c) hospital utilization such as per capita use of in-patient care, admission rates, average length of stay, bed occupancy rates, and hospital turnover rates; and (d) utilization of ambulatory care such as physician's workload and pharmaceutical consumption per person. In agreement with Proposition 2:

H2. The coefficient of variation in the indicators of the composition and utilization of national HCSs—more specifically in the availability and use of national health care resources, as well as health employment— does not decrease over time; that is, the annual rate of change in the variation of such indicators is not negative.

The persistence of different national HCSs and different utilization of these systems is explained by the particular cultural features of individualist and collectivist countries. Propositions 3.1 and 3.2 assert a relationship between the levels of individualism and the major dimensions of public national HCSs. The question is whether individualist countries spend less on health and cover a lower proportion of people against health costs than collectivist ones. Triandis (1995) suggests a variety of attributes that indicate whether a particular country is individualist or collectivist. Triandis points out that collectivist countries tend to favor attitudes that reflect sociability, interdependence, communication, and family integrity. In contrast, individualist countries believe in self-reliance, hedonism, competition, and emotional detachment from in-groups. From Propositions 3.1 and 3.2 and given the observed international trend towards increasing public health expenditure and social coverage over time, I derive the following two testable hypotheses:

H3.1. The level of individualism has a negative impact on the growth rate of public expenditure on health care.

H3.2. The level of individualism has a negative impact on the growth rate of population covered against health care costs.

In order to test these hypotheses, I estimate several multiple regression models. The sample of observations used in the analysis is a panel data set of twenty-two OECD countries for the years 1960, 1970, 1980, and 1990.

To test Hypothesis 3.1, the average annual growth rates of public expenditure on health (as measured by the value per capita) are regressed on GDP per capita and on a measure of the country's degree of individualism. To test Hypothesis 3.2, the regression equations treat the annual growth rate of social protection (as the percentage of population covered against in-patient care cost) as the dependent variable. Given that the data set combines time series and cross sections, ordinary least square (OLS) estimation of regression models is not feasible; instead regression models using a generalized estimating equation approach (GEE) are estimated.[48]

Hypotheses 3.1 and 3.2 predict that the level of individualism should have a significant and negative impact on the growth rate of both public expenditure on health and the percentage of population covered against health care costs. I use the Hofstede individualism index (1980) as a measure of individualism for each country. Despite its limitations, the Hofstede individualism index (HII) is still the most recent continuous index that ranks forty countries in a consistent and systematic manner according to the level of individualism. The index is the result of a factor analysis explaining national differences in employees' answers to fourteen questions.[49] The higher the index, the more individualist the country. The index of individualism ranges from 0.27 (Portugal) to 0.91 (United States). The highest individualism index values are found for the United States, Australia, Great Britain, Canada, and the Netherlands. Among the most collectivist countries are Portugal, Spain, Japan, Austria, and Finland.

---

[48] GEE is a variation of the generalized least square (GLS) method of estimation. Information on the GLS method of estimation can be found in Greene (1997). Since this data set has a large number of panels relative to the number of time periods, the population-averaged panel data model using GEE is better than the standard GLS. The GEE model provides a richer description of the within-group correlation structure for the panels than standard GLS.

[49] Hofstede (1980) provides an explanation of the general societal norms behind the "low HII" and "high HII" manifestations (p. 235). A low individualism index implies the following characteristics: (a) in society, people are born into extended families or clans, which protect them in exchange for loyalty; (b) "We" consciousness; (c) collectivity–orientation; (d) identity is based in the social system; (e) emotional dependence of individual on organizations and institutions; (f) emphasis on belonging to an organization or clan; (g) private life invaded by organizations or clans to which one belongs; (h) expertise, order, duty, security provided by organization, or clan; (i) friendships predetermined by stable social relationships; but need for prestige within these relationships; and (j) belief in group decisions. By contrast, a high individualism index implies the following: (a) in society, everyone is supposed to take care of him- or herself and his or her immediate family; (b) "I" consciousness; (c) self-orientation; (d) identity is based in the individual; (e) emotional independence of the individual from organizations and institutions; (f) emphasis on individual initiative and achievement; leadership ideal; (g) everyone has a right to a private life and opinion; (h) autonomy, variety, pleasure, individual financial security; (i) need for specific friendships; and (j) belief in individual decisions.

Proposition 4 refers to the process of convergence in population health status. The major indicators for measuring the relative health performance of each country available for the 1960–90 period are male and female life expectancy at birth, at age 40, and at age 60, infant mortality (as a percentage of live births), death crude rate (as the number of deaths per 1000 population), the percentage of the population 65 years old and older, and the mean/median population age. Proposition 4 implies that:

H4. The coefficient of variation in the major indicators of population health status declines over time; in other words, their annual rate of change is negative.

In this study, I expect to reject the welfare thesis about the benefits to population health of additional public expenditures and equalizing access to health care. Propositions 5.1 and 5.2 suggest that countries that spend heavily on health and provide universal public coverage against medical costs should have healthier populations than other countries. Thus, I expect to reject that:

H5.1. Public expenditure on health has a positive effect on population health.
H5.2. The level of public coverage against health care costs has a positive effect on population health.

Instead, I want to show that population health depends more on social variables than on the national structure of spending on health itself. Likewise, Proposition 6 suggests that countries with low levels of socio-economic inequality will have healthier populations than countries with high levels of inequality:

H6. The level of socio-economic inequality has a negative effect on population health.

To test Hypotheses 5.1 and 5.2 together with Hypothesis 6, I estimate a set of regression equations using the GEE method of estimation. The level of population health status is regressed on some of the major characteristics of the countries. Again, the panel data set of twenty-two OECD countries for the years 1960, 1970, 1980, and 1990 is analyzed.

Few aggregate measures of population health status exist. To test Hypothesis 6, I use life expectancy at birth for females and for males as well as infant mortality as dependent variables. I control for the level of socio-economic development using GDP per capita (lagged five years). Due to the relatively small sample of countries with available and comparable data, I restrict attention to the two major dimensions of national HCSs: (a) public expenditure and (b) public coverage. I measure public expenditure on health by the value per capita in current prices.[50] One disadvantage of this continuous variable is that it can vary widely from year to year for a given country. To control for possible fluctuations, I averaged per capita public expenditure on health over the 1960–65, 1970–75, and 1980-85 periods. Public protection is measured using the proportion of population covered against in-patient care costs (also lagged by five years).

The socio-economic inequality argument is operationalized by computing the ratio of the percentage of income received by the richest twenty percent of households to the percentage of income received by the poorest twenty percent of households. It is calculated from several issues of the *Human Development Report* during the 1960s, 1970s, 1980s, and 1990s. The lower the ratio, the less the income inequality. On this measure, the United States is found to have more income inequality than most European countries. In the 1980s, OECD countries have an average of 6.3 (compared to 16.8 for developing countries). This means that the percentage of income received by the richest twenty percent of households is 6 times the percent received by the poorest twenty percent of households.

6. Results

The *results* section of the study presents the results and comments briefly upon them. Some social scientists prefer to call this section *results and discussion* and include extensive commentaries alongside their results. Here, however, I consider the *results* and *discussion* sections separately. The first comments on the statistical analysis; the second presents an analysis of the findings.

The results section may be comprised of several components including statistical description or univariate statistics of the variables in the data, exploratory analysis, and multivariate analysis. One should report the univariate statistics on variables in the longitudinal data set if this was not already done in the data section. Then, report exploratory or graphical analyses that help readers understand

---

[50] Total expenditure on health care, as defined in the OECD Health Data, comprises: (a) household final consumption on medical care and health expenses including goods and services purchased at the consumers' own initiative; (b) governmental health services including schools, prisons and the armed forces, as well as specific public programs such as vaccination campaigns, and services for minority groups; (c) investment in clinics and laboratories; (d) administration costs; (e) research and development, excluding outlays by pharmaceutical firms; (f) industrial medicine; and (g) outlays of voluntary organizations, charitable institutions, and non-governmental health plans. The public expenditure on health (as measured by the value per capita in current prices in dollars) measures the quantity of total expenditure on health care financed by national public sectors. This measure of public health funding includes expenditure by government-managed social security institutions and compulsory private health insurance.

the changes in the outcome variable. The goal here is to report the results of the multivariate models, starting with the results for a simple model and gradually increasing in complexity. Readers find it easier to follow a discussion section when this approach is used. Results of multivariate methods of analyses usually appear in tables displaying findings in numerical terms. The accompanying text can then focus the reader's attention to important aspects of the empirical analysis.

Tables communicate information about a set of variables. For any given multivariate model of analysis, a table has three main parts: (1) the title; (2) the body; and (3) a summary of the information. The title of the table should succinctly summarize the table's contents, indicate the main variables, and possibly note the sample under study. Inform the readers about the dependent variable(s), the independent variables, and the source of the information. See some examples of informative titles in Tables 4.1–4.3 (in the next three pages). The style and format of table titles vary depending on the journal. One should closely examine the formatting guidelines before submitting a paper for publication to any journal.

The body of a table displays the information. In designing the table body, one must decide what information to report—typically, the information reported includes means, standard deviations, standardized or unstandardized coefficients, standard errors, and probability values. It is important to decide the best way to report these values and what units to use (e.g., dollars or thousands of dollars); what values of variables to report and how they should be grouped together. Think carefully how to arrange the information on the page; how to label the variables and their values (it is best to identify the variables by words or phrases); and where to draw the lines of demarcation. Avoid vertical lines, since they make tables difficult to read and comprehend. In addition, use horizontal lines in moderation.

Finally, a table should indicate the source of the data, either in the title or in a note to the table. This information about the data source should also appear in the data section of the main text. The table should note the sample size with the number of cases included in the analysis and report how often these units have been observed. Make sure this information appears in the main text as well. When performing EHA, report the number of events that occurred during the period of analysis together with the number of episodes in the tables.

Exploratory analysis tables should include a statistical assessment of the model's adjustment or fit. Those statistics depend entirely on the type of model used (e.g., F-test in the case of OLS or Chi-square in the case of GLS). Report degrees of freedom, levels of probability, the estimated effects of the different covariates, and either their standard errors, z-statistics or t-tests, or significance levels. Make sure that the null hypotheses are clear, and indicate whether one- or two-sided tests were performed. The selection of tests depends fully on how the hypotheses were stated. If the effect of the independent variable is predicted in one direction—i.e., either positive or negative—then, it is required to use a one-tailed test.

Table 4.1
Trends in the characteristics of national health care systems, by OECD country (1960–1990)

Country	Annual Growth Rate of Total Health Expenditure at 1990 Constant Prices (%)			Total Expenditure on Health (as a % of GDP)		Public Expenditure on Health (as a % of GDP)		Per Capita Public Expenditure on Health ($ at PPPs)		Percent of Population Eligible for Coverage Against Total Medical Expenditure	
	1960–1970	1970–1980	1980–1990	1960	1990	1960	1990	1960	1990	1960	1990
Australia	3.7	4.7	4.1	4.9	8.2	2.4	5.6	46	896	100.0	100.0
Austria	3.6	4.7	1.3	4.4	8.4	3.1	5.6	46	922	78.0	99.0
Belgium	5.2	8.1	2.9	3.4	7.6	2.1	6.8	33	1109	58.0	99.0
Canada	6.5	5.1	3.4	5.5	9.4	2.3	6.9	45	1255	68.0	100.0
Denmark	9.3	4.2	1.2	3.6	6.5	3.2	5.3	59	879	95.0	100.0
Finland	11.4	5.4	3.5	3.9	8.0	2.1	6.5	30	1045	55.0	100.0
France	8.7	7.2	5.1	4.2	8.9	2.4	6.6	41	1146	76.3	99.5
Germany	5.1	5.9	1.5	4.8	8.3	3.2	5.9	61	1091	85.0	92.2
Greece	12.1	5.0	4.7	2.9	5.3	1.9	4.5	10	333	30.0	100.0
Iceland	9.0	6.8	5.5	3.3	7.9	2.5	6.9	39	1190	100.0	100.0
Ireland	8.7	10.8	−1.2	3.8	6.7	2.9	5.0	27	559	85.0	100.0
Italy	8.5	6.7	3.4	3.6	8.1	3.0	6.3	41	1029	87.0	100.0
Japan	13.7	9.0	3.7	3.0	6.8	1.8	4.8	16	841	88.0	100.0
Netherlands	8.4	2.2	2.0	3.8	8.0	1.3	5.7	23	914	71.0	69.1
New Zealand	—	2.4	0.9	4.3	7.4	3.5	6.1	74	819	100.0	100.0
Norway	5.8	6.9	2.9	3.3	7.5	2.6	7.1	37	1137	100.0	100.0
Portugal	—	10.9	4.2	—	6.6	0.8	3.6	5	336	18.0	100.0
Spain	16.1	7.8	5.4	1.5	6.9	0.9	5.4	8	640	54.0	99.0
Sweden	9.7	3.4	1.7	4.7	8.6	3.4	7.7	66	1312	100.0	100.0
Switzerland	8.1	2.3	3.6	3.3	8.4	2.0	5.7	57	1205	74.0	99.5
United Kingdom	5.2	4.8	2.0	3.9	6.0	3.3	5.1	64	803	100.0	100.0
United States	6.5	4.8	3.6	5.3	12.7	1.3	5.2	35	1105	20.0	44.0
Mean	8.3	5.9	3.0	3.9	7.8	2.4	5.8	39.2	934.8	74.7	95.5
Standard deviation	3.3	2.5	1.7	0.9	1.5	0.8	1.0	19.5	273.1	26.0	13.3
Coefficient of variation (%)	39.7	41.8	56.3	23.5	18.8	34.1	16.6	49.6	29.2	34.8	13.9

Note: The expenditures in constant prices are calculated from current price spending data deflated by the private consumption price index. The mean is the unweighted arithmetical average of the country rates which are shown.
Sources: OECD Health Data File (1995).

Table 4.2
Trends in national population health, by OECD country (1960–1990)

Country	Life Expectancy						Mortality				Demography			
	Overall		Females at Birth		Males at Birth		Infant Mortality (as a % of Live Births)		Death Crude Rate (per 1000 Population)		Population 65 Years Old and Over		Mean Age	
	1960	1990	1960	1990	1960	1990	1960	1990	1960	1990	1960	1990	1960	1980
Australia	70.7	76.5	73.9	80.1	67.9	73.9	2.0	0.8	8.5	7.0	8.4	11.2	28.0	32.1
Austria	68.7	74.8	71.9	79.0	65.4	72.5	3.8	0.8	12.7	10.7	11.9	15.1	33.0	35.0
Belgium	70.3	75.2	73.5	79.1	67.7	72.4	3.1	0.8	12.4	10.5	12.0	15.0	32.0	36.0
Canada	71.0	77.0	74.3	80.4	68.4	73.8	2.7	0.7	7.8	7.2	7.6	11.0	28.0	32.0
Denmark	72.1	75.8	74.1	77.7	72.3	72.0	2.2	0.8	9.6	11.9	10.6	15.6	32.0	37.0
Finland	68.5	75.5	71.6	78.9	64.9	70.9	2.1	0.6	9.0	10.0	7.5	13.4	28.0	37.0
France	70.3	76.4	73.6	80.9	67.0	72.7	2.7	0.7	11.3	9.4	11.6	14.0	32.5	34.7
Germany	69.7	75.2	72.4	79.1	66.9	72.7	3.4	0.7	11.6	11.3	10.2	—	32.0	37.0
Greece	68.7	76.1	70.4	79.4	67.3	74.1	4.0	1.0	7.3	9.3	8.1	14.0	34.0	38.0
Iceland	73.2	77.8	75.0	80.3	70.7	75.7	1.3	0.6	6.6	6.7	8.1	10.6	23.5	29.1
Ireland	69.6	74.6	71.8	77.5	68.5	72.0	2.9	0.8	11.5	9.1	11.1	11.4	28.0	28.0
Italy	69.2	76.0	72.3	80.2	67.2	73.6	4.4	0.8	9.6	9.5	9.2	—	—	36.0
Japan	67.9	78.6	70.2	81.9	65.5	75.9	3.1	0.5	7.6	6.7	5.7	11.9	28.0	37.0
Netherlands	73.2	77.2	75.5	80.1	71.6	73.8	1.8	0.7	7.5	8.6	9.0	12.8	28.0	37.0
New Zealand	70.9	75.2	73.9	78.3	68.7	72.4	2.3	0.8	8.8	7.9	8.7	11.0	24.6	31.1
Norway	73.4	77.1	75.9	79.8	71.4	73.4	1.9	0.7	9.2	10.9	11.1	16.3	32.0	37.0
Portugal	63.3	74.0	67.2	77.9	61.7	70.9	7.8	1.1	10.5	10.4	8.1	—	—	—
Spain	69.0	77.0	72.2	80.5	67.4	73.4	4.4	0.8	8.6	8.6	8.1	13.4	32.0	32.0
Sweden	73.1	77.4	74.9	80.4	71.2	74.8	1.7	0.6	10.0	11.1	11.8	17.8	—	37.0
Switzerland	71.2	77.4	74.1	80.9	68.7	74.0	2.1	0.7	9.8	9.5	11.1	14.9	32.0	37.0
United Kingdom	70.6	75.7	74.2	78.6	68.3	72.9	2.3	0.8	11.5	11.2	11.7	15.7	32.0	37.0
United States	69.9	75.9	73.1	78.8	66.6	71.8	2.6	0.9	9.5	8.6	8.8	12.2	28.0	32.0
Mean	70.2	76.2	73.0	79.5	68.0	73.2	2.9	0.8	9.6	9.4	9.6	13.5	29.9	34.7
Standard deviation	2.2	1.2	2.0	1.1	2.5	1.3	1.4	0.1	1.7	1.6	1.8	2.1	3.0	3.0
Coefficient of variation (%)	3.2	1.5	2.7	1.4	3.7	1.8	47.2	18.6	17.9	16.8	18.9	15.4	9.9	8.6

Note: The mean is the unweighted arithmetical average of the country rates which are shown.
Sources: OECD Health Data File (1995).

Table 4.3

Multivariate regression models of some population's health variables on national health care variables, level of national development, and socio-economic inequality using OLS estimation. Pooled sample of 22 OECD countries in 1970, 1980, and 1990 ($N = 66$)

Independent Variables	OLS Regression Coefficients of Population's Health Indicators			
	Life Expectancy at Birth	Female Life Expectancy at Birth	Male Life Expectancy at Birth	Infant Mortality (as a % of Live Births)
Constant	72.90***	73.65***	68.07***	2.13***
Lagged GDP per capita in thousands ($T - 5$)	0.03***	0.02**	0.02*	−0.01***
Lagged per capita public expenditure on health in thousands ($T - 5$)	−0.03	0.05	0.03	0.00
Social protection against medical care costs ($T - 5$) (as % of population covered under public scheme)	−0.28	−0.38	−0.43*	0.04
Social protection against in-patient care costs ($T - 5$) (as % of population covered under public scheme)	0.30	0.40*	0.45*	−0.05
Household income inequalities ($T - 10$)[a]	−0.23***	−0.10	−0.26***	0.09***
Year 1980	−1.37***	0.69	0.16	−0.19
Year 1990	−0.70	0.82	0.65	0.09
R^2	0.71	0.74	0.68	0.73
F-statistic	13.71***	16.26***	12.29***	15.37***
N	66	66	66	66

*Significant at the 0.15 level.
**Significant at the 0.10 level.
***Significant at the 0.05 level.

[a]As measured by the ratio of the percentage of income received by the richest 20% of households to the percentage of income received by the poorest 20% of households. It was calculated from the Human Development Report (various issues).

Sources: OECD Health Data File (1995); and Human Development Report (various issues).

The notes to the table give additional information and clarify the body of the table. Notes inform about the meaning of special symbols, such as those used to indicate significance levels; give explanations of how special calculations were made; and detail sources of data (if not given in the title). Note whether one- or two-sided hypotheses tests were performed, and define variables and their values. Remember to index notes using numbers or letters.

Returning to my study of national HCSs, before the results section, I wrote a section describing trends in OECD social expenditure on health, and another section describing population health trends in OECD countries. Table 4.1 describes general trends in national HCSs by OECD country. In an earlier version of my article, I describe the table in the main text as follows:

For OECD countries, the share of total health spending in GDP rose from almost 4 percent in 1960 to nearly 8.5 percent in 1993. This represents a growth rate twice that of GDP over the 1960–90 period. The public share of total health expenditure increased even faster (from 2.4 percent in 1960 to 6.2 percent in 1990), and, on average, is nearly 75 percent of total health care expenses in 1993. Much of the increase is associated with the growth in public sector health care services prior to the mid-1970s, by which time both physical and financial access to public hospitals was almost universal throughout OECD countries, and public provision of ambulatory and primary services was almost equally widespread.

Since 1975, there has been a slowdown both in the rate of growth in total expenditures and in the public component of expenditure (Table 4.1: columns 1–3). This has been associated with the slower economic growth since the mid-1970s and with conscious policies to curtail health care costs, with both to prices and to utilization or intensity of care. The average annual growth rates are 8.3 and 5.9 percent during the 1960–70 and 1970–80 periods, respectively. The rate drops to 3 percent in the 1990s. There are considerable differences between countries, both in the share of GDP devoted to health care (Table 4.1: columns 4–7) and in per capita levels of expenditure (columns 8 and 9). Traditionally, researchers attribute these variances to different standards of living (e.g., expenditure on health typically rises more than proportionally as personal incomes increase). Access to health services, both primary and hospital care, has become nearly universal throughout OECD countries, and we have seen a trend to increase the percentage of population eligible for public coverage. In the 1980s, 75 percent of the OECD countries had quasi-universal public coverage against medical care benefits in comparison with only 58 percent of OECD in the 1970s. By 1990, the average percent of population eligible for coverage against total medical expenditure in the OECD area is 95.5 percent (Table 4.1: column 11).

In Table 4.2, I present the trends on population health for all OECD countries in the sample and chose to describe it in the main text as follows:

> Overall, there is some improvement in the quality of life in good health in industrial countries. Infant mortality has been reduced. In 1960, on average, across the twenty-two OECD countries under analysis, three babies in a hundred died before their first birthday. The figure is now less than 1 in a hundred (Table 4.2: columns 7 and 8). Life expectancy has increased, including for those in the older age groups. Females born in 1990 are expected to live almost 7 years longer than those born in 1960, while males will live 5 years longer (Table 4.2: columns 1–6). A 60-year-old woman today lives almost 4 years longer than women who were aged 60 in 1960, while a 60-year-old man lives 2 years longer than his 1960 counterpart. The improvement has been greater for women than men. Over the last thirty years, most OECD countries have experienced first an increase and subsequently a marked decline in total fertility rates, to the point where, with very few exceptions, fertility rates in almost all OECD countries are now below the level needed to ensure the replacement of the population (OECD Health Data, 1995). As a consequence of both the longer-term trends towards lower birth rates and greater longevity, almost all OECD countries will experience a rapid and large aging of population by the end of the century. The proportion of the population aged 65 or over increased from 9.6 percent to almost 14 percent of the population between 1960 and 1990 (Table 4.2: columns 11 and 12). By the year 2040, 20 to 29 percent of most OECD populations will be over the age of 65. This population aging will be one of the most significant events of the next century and is likely to have a strong impact on the evolution of social programs.

Consider including *figures* in a research paper. Graphs and diagrams communicate ideas with clarity, precision, and efficiency. Label the figure carefully, and be sure to explain it in the main text. Longitudinal studies with a continuous dependent variable often employ time-series plots that represent time on the *x*-axis and the dependent variable on the *y*-axis (as described in Chapter 1). For example, my study displays the performance trajectory of employees in a firm in one chart. Performance changes likely reflect learning and skills acquisition within the organization. According to human capital theory, skills can be acquired in a number of ways: Formal education is perhaps primary, but the second most important method of acquiring human capital is on-the-job training, or "learning by doing" (Arrow, 1962). On-the-job training makes a worker more productive at the current firm as well as at other firms in the same industry. One could identify different hypothetical functional relationships between performance and tenure within the organization and represent those relationships in a figure. Fig. 4.2 illustrates some of these hypothetical relationships between tenure and performance. The traditional perspective in labor economics and in the learning literature assumes that a new employee's performance is initially low, then improves, and eventually reaches a plateau, so that the new employee performance curve is an *S*-shaped function. However, an inverse *U*-shape performance curve is more likely for stressful jobs or physically demanding tasks. Some jobs, especially those that are service-oriented, may be characterized by good early performance and subsequent "burn-out." Only by paying attention to the tenure–performance relationship can the most appropriate performance growth model be identified.

Whereas Fig. 4.2 illustrated theoretical performance curves over tenure, Fig. 4.3 charts the actual average productivity curve for a group of employees in one real organization. The graph is from my study of Phone Customer Service Representatives (hereafter "CSR") at a phone center in the United States (Castilla, 2005). The CSR is a full-time, hourly-paid employee whose duties consist of answering customers' telephone inquiries about their credit card accounts. Trained for accuracy, speed, and efficiency, CSRs handle up to 5000 phone calls per month. Managers often monitor phone calls to insure that the CSRs achieve their courtesy and accuracy goals. The phone center measures performance by calculating the average number of calls a CSR answers per hour in any given month. The performance curve shows that the average productivity of a new employee starts low but improves over time. Eventually, after 12 months in the company, the level of productivity worsens slightly, although variance in productivity increases and the number of employee survivors decreases.

When dealing with longitudinal analysis of events over time, consider using a diagram to illustrate processes of social change. Diagrams generally use boxes and arrows to explain the causal process by which the change occurs. For example, Fig. 4.1 diagrams the process of designing and writing a research paper. In diagramming a change process, each box represents a state or value of the discrete entity being studied; each arrow represents a possible path of change. If one is interested in examining some processes of change within the set of all the possible changes, clearly differentiate those that are studied from those that are not studied by using different line styles. Always delimit the object of study at the beginning of an essay. For example, one could be interested in the transition process from being a student to getting a first job but not in transition processes after that (e.g., going back to school or getting a second job).

Fig. 4.4 diagrams the employment history of a hypothetical member of a given population sample, illustrating several events or changes that can take place. The *x*-axis represents the person's age; the *y*-axis represents her labor status at a

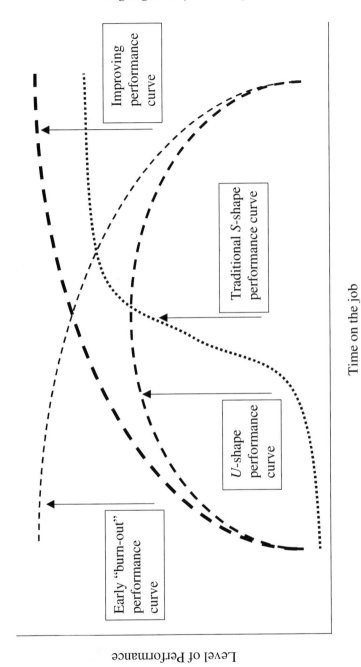

Fig. 4.2 Hypothetical performance curves.

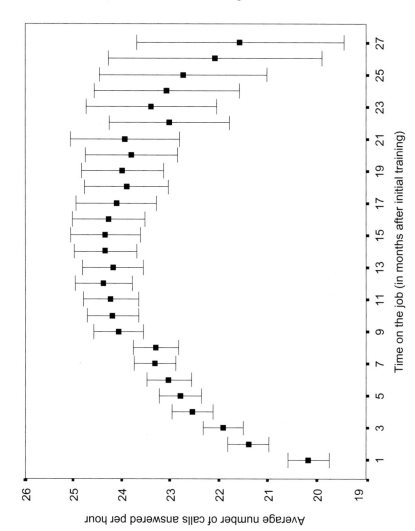

Fig. 4.3 Error bar for number of calls answered per employee per hour over time on the job. *Notes:* (a) For all employees (without correcting for turnover). (b) The graph includes the 95% confidence interval for the number of calls.

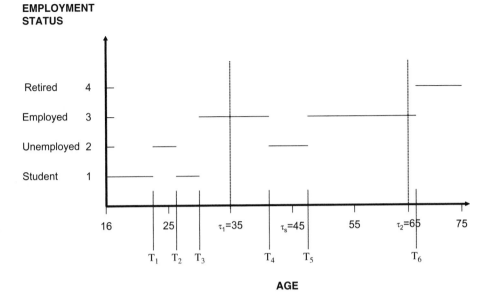

Fig. 4.4 Employment history of a hypothetical person.

particular moment in her life, $y(t)$. In this hypothetical case, the person becomes part of the active population when she turns 16 years old, at the moment T_0. She is a student until the moment T_1 when she drops out of school to search for a job. She is then unemployed for a few years. At time T_2 she decides to go back to school. She finds a job at time T_3, but at time T_4 she becomes unemployed again. After a period of time T_4–T_5 of unemployment, she goes back to work, but retires shortly afterward at the age of 65 years (T_6) and becomes part of the inactive population. The person remains in the inactive population group until she dies at the age of 75. This study could analyze five different events relating to the individual's changing status in the labor market: (a) the transition from being a student to becoming unemployed; (b) the transition from being unemployed to becoming a student; (c) the transition from being a student to being employed; (d) the transition from being employed to being unemployed; and (e) the transition from employment to retirement. The type of event and its succession in time varies for each individual case in the sample, so that some people can have more events than others as well as different combinations of events at different moments of their lives. A graph can illustrate all events under consideration. This is especially useful when multiple events are analyzed and when the graph helps to highlight the particular events under study.

Many software programs can help plot variables and create several graphs easily. Microsoft Excel is particularly noteworthy for its flexibility and ease of use.

The program's Excel Chart Wizard can easily create column, bar, line, area, surface, and XY scatter charts. For example, Figs. 5.3 and 5.4 in Chapter 5 (the second paper example) were created using Excel. They plot different specifications of the dependent variable (memberships and cumulative memberships) for different groups of countries. Statistical software packages such as SPSS also allow the creation of tables that can be easily copied into a word-processing program. SPSS offers many options for graphing different sets of variables. Other software programs can make other types of figures and pictures. If interested in network analysis, consider using visualization programs such as Pajek or Mage.[51]

In a recent study, I presented the beginnings of a systematic comparative study of social networks of venture capitalists in two different regions of the United States (Castilla, 2003).[52] Specifically, I examined the role of social networks of venture capital firms (VC henceforth) in the development of industrial regions like Silicon Valley, California, from both a comparative and a network perspective. By analyzing a sample of investments from VC firms in Silicon Valley, I showed how a uniquely interconnected group of firms has dominated the structure of VC in Silicon Valley from the beginning. In other words, how Silicon Valley's entrepreneurial efforts and VC activities are embedded in densely connected networks. Researchers have attributed Silicon Valley's relatively high growth and development rates to the historical development and the particular structure of the valley's social networks (Saxenian, 1990, 1994; Florida and Kenney, 1987; Nohria, 1992). This is what I wrote in my research paper when I was describing Fig. 4.5 (Castilla, 2003):

> To examine the most recent networks of VC firms in Silicon Valley (from 1995 up to the first quarter of 1998) I used the data from Pricewaterhouse Coopers' Money-Tree-Historical Data of VC funds and investee companies. I use Mage (Richardson and Richardson, 1992), a computer graphic program, to explore and evaluate the social structure of VC firms using dynamic three-dimensional color images. A graph can give us a representation of a social network as a model of a social system consisting of a set of actors (in this case, VC firms) and the existing ties among them. The graph for the venture capital social network in Silicon Valley is given in Fig. 4.5. In the graph, the points (called nodes) are used to represent each

---

[51] To learn more about one of these visualization programs, visit Pajek's website at http://vlado.fmf.uni-lj.si/pub/networks/pajek/. The program can be downloaded for free.

[52] Again, the full citation of the article is: Castilla, E.J., 2003, Networks of venture capital firms in Silicon Valley, *International Journal of Technology Management* 25 (1/2), 113–135. Copyright © 2003 by Inderscience Publishers.

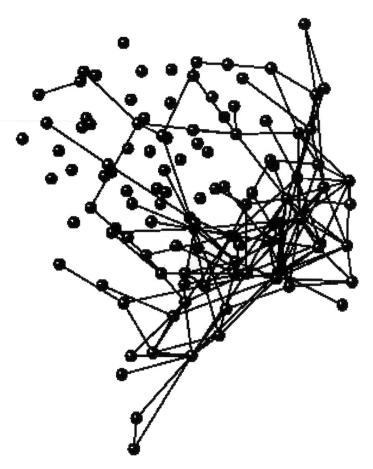

Fig. 4.5 Network of venture capital firms in Silicon Valley.
Source: PriceWaterhouseCoopers (1995–1998).

of the existing VC firms in Silicon Valley during the 1995–1998 period. The lines connecting the points represent the ties between those organizations. In this case, the presence of a tie between any two VC firms indicates that they have at least one investee company in common; that is, that both VC firms cooperated to finance the same company or companies. In this sense, a tie is either present or absent between each pair of venture firms. In addition, the length of the tie between two VC firms is inversely proportional to the number of investee companies in common: The shorter the tie (i.e., the closer the two VC firms in the 3-dimensional space), the higher the number of investee companies in common. In Silicon Valley, during the 1995–1998 period, there were 111 firms, and 312 lines between the pairs of firms; the sum of the values of such ties is 550, indicating the number of occasions on which at least two different VCs cooperated to co-invest money in the same company. The graph looks rather complicated or dense since the level of connection among VC firms in Silicon Valley is high (the density of the network is .025). The five most prominent VC firms, *Accel Partners, Kleiner, Perkins, Caufield & Byers, Crosspoint Venture Partners, Sequoia Capital,* and *Hambrecht & Quist Venture Capital* are located in the center of the plot; whereas firms like *Alta Communications, Alpha Partners, Robertson, Stephens & Company,* and *Softbank Technology Ventures Fund* are all on the side of the plot, but still weakly connected to the main network component.

Finally, consider putting some additional material in *appendices*. Located at the end of the paper, appendices allow the author to provide details about the research that are not strictly essential to understanding the main points and do not belong in short notes or footnotes. A common strategy is to include in appendices all materials that reviewers may want to know but that editors may choose to omit if the paper exceeds the journal's page limitations. While appendices should be succinct, they should also contain all information required for clarity purposes.

Technical appendices can provide extended descriptions of sampling procedures or samples, details about variable constructions, special features of the model or the method of analysis that may not be generally known, and anything else peripheral to the paper's argument but essential to a full understanding of the analyses. For example, a technical appendix could include an alternative set of longitudinal models to those presented in the study. These alternative models could be designed to assure the reader that the findings reported in the main text are robust and do not depend on any specific longitudinal methodology used. The models reported in the appendix are useful but not essential, and the editor of a journal might decide to omit such an appendix before publishing the paper.

7. Discussion of Results

Although one may think that tables and diagrams speak for themselves, different readers often "see" different things in the same tables and numbers. The purpose of the *discussion of results* section is to tell readers what the key results are and what the numbers mean to the author(s). Do not simply recite numbers that also appear in the tables. Instead, help the reader navigate through the table and interpret the results. Explain what the findings imply. By the end of the paper, the reader must be clear on whether or not the data analyses support the main theoretical arguments of the paper. In other words, one should explicitly accept or reject the hypotheses of the study in this section. Later, this should help propose new hypotheses or assess whether more and/or better data are needed to test the hypotheses of the study.

The *discussion of results* section needs to emphasize the estimated effects of the independent variables, stating whether those effects support the hypotheses under study. For the key explanatory variables, one should explain the meaning of the model coefficients. For example, if the estimated effect of years of schooling on the hazard rate of getting a job is 0.2 with a standard deviation of 0.005, one should inform that an additional year of education increases the rate of getting a job by 22% (this is, $100\% \times [\exp (0.2) - 1]$) and that such increase is significantly different than 0 at the level of significance of the 0.05 (using a one-sided test). One may wish to report and discuss the standardized effects in order to establish, which explanatory variables have the greatest predictive power of the events under study. The preceding chapters in this book can help interpret the coefficients in longitudinal models. The discussion sections of the two research papers included in the next chapter exemplify how to report those effects.

As in any other multivariate analyses, if one suspects that multicolinearity among the independent variables might be a problem, one should highlight this and provide some statistics to help readers assess whether the degree of multicolinearity significantly put the results into question. If there is an interest in time dependence, discuss the findings pertaining to this topic. Depending on the questions motivating the investigation, one may choose to discuss time dependence before discussing the effects of the variables; at other times, it may be better to discuss time dependence later. Any analysis of longitudinal data should also answer two basic questions. First, it should provide information about how the dependent variable varies over time. If the dependent variable does not vary as predicted, one should explain why that may have happened. Second, the analysis should assess the effects of the independent variables on the dependent variable and identify whether these effects vary over time. If these effects differ from what the hypotheses predicted, explain why that might be the case.

In my paper about OECD health care systems and their impact on population health, the results section explored the extent to which countries with low levels of inequality in the distribution of income and education have healthier populations over time (Castilla, 2004). Table 4.3 reports the panel OLS regression estimates of the coefficients obtained for five of the regression models including the pooled sample of 22 countries in the 1960s, 1970s, 1980s, and 1990s. Each column of the table lists a separate equation, with the slope associated with several independent variables. I decided to discuss the results as follows:

In column 1 [of Table 4.3], life expectancy at birth is regressed on national health care variables and the level of national development and socio-economic inequality. Columns 2, 3, and 4 report the regression results for female life expectancy, male life expectancy, and infant mortality one at a time. Not surprisingly, in all models, the level of socio-economic development as measured by the GDP per capita has a positive and significant effect (at the 0.1 level, one-tailed test) on the propensity of a country to have healthier populations. However, the amount of public expenditure on health per capita appears to have an insignificant effect on the health status of their populations. In the case of overall life expectancy at birth, expenditure on health has a negative but insignificant effect. The relationship between public expenditure and health status appears to be opposite to the one hypothesized by welfare theorists (Hypothesis 5.1). A possible explanation for this result might be that since OECD countries have a high level of development, additional spending on health does not greatly increase the health status of the population since such populations already live relatively longer. Along such lines, health economists have stated recently that greater reductions in mortality rates and improvements in life quality are better achieved through additional expenditure on formal education rather than through additional expenditure on medical care (Auster *et al.*, 1969; Fuchs, 1979).

None of the models strongly supports the thesis that adopting universal public coverage of medical care benefits improves population health (Hypothesis 5.2). On the contrary, the sign of the coefficients is opposite to the one predicted by welfare theorists in the case of social protection against medical costs. However, although the significance level of the estimates differs slightly among models, all these coefficients are never significant at the 0.1 level.

In conclusion, either the "health care" variables considered in this study do not adequately capture the benefits of efficient and universal HCSs, or HCSs do not necessarily improve the health status of citizens. Moreover, HCSs may even have a negative impact on national health outcomes unless they are combined with and supplemented by other public policies regarding education, housing, etc. Social scientists interested in comparative analyses need more data on lifestyles, family structures, income, and education differentials that could account for such differences in health. However, national stratification systems play a significant role in affecting population health. The analysis gives strong support to hypothesis 6—that public policies adopted in order to reduce socio-economic inequalities have a positive effect on the propensity of a country to have a healthier population. Three of the four regression models predict that the higher the levels of socio-economic inequality, the higher the differentials in health, and therefore the lower overall population health status. These negative effects of inequalities on health are significant (at the 0.05 level) except in the case of female life expectancy at birth. The latter suggests the importance of race and gender when examining differentials in health outcomes. The four models here presented are the most appropriate of all previous ones, accounting for at least 70 percent of the variation in health outcomes.

8. Conclusion

The *conclusion* should briefly and clearly restate the main arguments and findings of the article. As in the introduction, researchers use the conclusion to situate their work in a larger context. This section thus moves readers from the specific information offered in previous sections to a more general vision of why the results should matter and contribute to the knowledge in a certain discipline. This is exactly the opposite trajectory to that of the introduction. As a rule, do not introduce new topics or subjects in the conclusion. The conclusion serves as a way of wrapping up an empirical analysis.

The conclusion must note the implications of the results explaining to the reader what he or she has learned from the analyses. It should also present some recommendations for future research, outlining theoretical or empirical avenues for further inquiry. The final few pages of the paper are the last opportunity to convince readers that the paper makes a key and novel contribution to a field of study and is therefore worth remembering and quoting.

Following is how I conclude my systematic comparative study of social networks of venture capitalists in two different regions of the United States (Silicon Valley and Route 128) (Castilla, 2003):

In the present study, I have shown some of the most important differences between the network structure of venture capital firms in Silicon Valley and in Route 128. First, collaboration among venture capital firms in Silicon Valley is more pronounced and dense than in Route 128. The network analyses clearly show that the network of VC firms in Silicon Valley is

denser overall, which illustrates in substantive terms the culture of cooperation among local firms. The average number of ties in Silicon Valley is
twice the average number in Route 128. In addition, the network of firms
in Silicon Valley scores higher in all centralization measures. The network
analyses of components and cliques in both regional networks of VCs confirm that the Silicon Valley social network is much more cohesive and connected. Not only is the structure of VCs denser, but firm participation in the
network is higher. In Silicon Valley, 70 percent of all VCs cooperate with
other VCs to fund companies (only 50 percent of Route 128 VCs engage
in such co-operative behavior).

Second, Silicon Valley VC firms make more local investments than VC
firms in other prominent regions such as Route 128. In Silicon Valley, local
investee firms account for almost 70 percent of funded companies and
70 percent of the money invested. This figure drops to 40 percent in the case
of VC firms in Route 128. Differences in investment patterns between VCs
in Silicon Valley and VCs in Route 128 are therefore quite significant. Over
50 percent of Silicon Valley VC portfolios consist of investee companies in
their first, second, and third stages of development. Route 128 VC firms
tend to invest much more in buy-outs and follow-on companies. Both firms
in Silicon Valley and Route 128 invested substantially in the Software &
Information and Communication industries during the three-year period of
analysis.

The present structure of the social networks in Silicon Valley and its historical development can explain the higher growth and development of the
region in comparison with other regions in the world. This network structure
explanation challenges the market-centered theory of regional development.
Both Silicon Valley and Route 128 have comparable concentrations of skills
and technology, and even similar numbers of companies and institutions.
However, important network differences emerge when only examining the
venture capital sector—the most critical source of capital for many start-up
companies since the 1950s. Venture capitalism in Silicon Valley is central to
the region's social and professional development; VC firms in other regions
tend to play a less active role, even though venture capital networks rely on
major inflows of technical entrepreneurs, venture capitalists, management
talent, and supporting services.

A network approach to regional study that explores the network structure
among a region's important actors can account satisfactorily for Silicon
Valley's superior performance. My network analysis of venture capital firms
and the differences across two U.S. regions is only the beginning of a
regional network approach to regional development. Entrepreneurial efforts
and venture capital activities are indeed more successful when embedded in
a densely connected network such as Silicon Valley, mainly because such

embeddedness shapes and facilitates the allocation of resources, support, and information necessary to develop new local organizations and ensure the success of the region over the long run.

In general, social networks matter because trust, information, action, and cooperation all operate through social relations. To develop better models for explaining how these crucial network factors affect regional development first requires that we understand how they are shaped by networks and how those networks are re-shaped in turn. Regional development, in this sense, may rely more on networks of social actors since social actors ultimately influence economic and organizational practices as well as the institutional infrastructure of a region. In this sense, a theoretical network perspective on regional development should operate under at least five basic premises: (1) organizations are connected to one another, and they are therefore members of regional social networks; (2) an organization's environment consists of a network of other organizations and institutions (such as universities, VC firms, law firms, trade associations); (3) the actions (attitudes and behaviors) and their outcomes (performance, survival, and legitimacy) of organizations can be best explained in terms of their position in networks of relationships; (4) certain networks promote certain actions, but in turn constrain other actions; (5) networks of organizations change over time. Those five premises are important to consider when comparing regions and/or industrial districts. A comparative analysis of regional development should therefore pay attention to at least some of these network premises. We also need more knowledge about what kinds of regional networks promote superior economic and social development.

The analyses in my paper also have important implications for practice and policy. Any attempt to replicate Silicon Valley is unlikely to succeed at the level that Silicon Valley has unless dense social networks among actors that promote trust and cooperation are simultaneously developed and supported over time. Replicating Silicon Valley's infrastructure has been the goal of many governments in recent years; however these governments have not necessarily worked to create and maintain beneficial networks between firms in different institutional settings. The case of Taiwan is a successful story of regional development. During the 1970s and 1980s, Taiwanese government agencies and policy makers set out to improve the region's position in the international economy by creating a technology park (the Hsinchu Science-based Industrial Park) and a venture capital industry. Moreover, they recruited Taiwanese and Chinese engineers and entrepreneurs working in the United States and they forged connections to the U.S. market (Kraemer *et al.,* 1996; Mathews, 1997). The Taiwanese case suggests that regions aiming to develop in the information technology era need to create an infrastructure of institutions that fund and support new firms as well as to facilitate and promote financial, technical, and technology connections among Taiwanese firms and between Taiwanese firms and institutions in other regional communities like Silicon Valley.

In my paper, I have attempted to study the venture capital sector by asking how differences in the network cooperation among VC firms in two regions can explain differences in regional development. Future analyses should extend the network analyses to other economic sectors and institutions. A network approach can help social scientists and policy-makers understand the relationship between social networks of actors and regional development. Such network analyses are indispensable steps for understanding industrial and regional economies.

9. Bibliography and References

The list of references constitutes an important, often neglected, part of a research project. In this section, researchers need to select readings and references to include, giving equal weight to theoretical, empirical, and methodological work. The bibliography should be brief and useful. One should never use it to demonstrate how much one has read or what one knows. A good references section selects the most appropriate works for the topic of the investigation and acknowledges scholars who have been working on similar issues. In a good bibliography, less is more. The bibliography needs to be useful to the reader who wants to continue learning some more about the research of a paper. Strategically speaking, the list of references is often used by journal editors to understand who the audience is and select reviewers of the work accordingly. So spend time in preparing and thinking about whose work to cite in a paper.

The format for the references section depends upon the specific journal one is targeting. Always list all items, including data files and statistical sources, alphabetically by author (providing the full list of multiple authors) and, within author(s), by year of publication. Do not use abbreviations. Many journals discourage writers from using italics, bold, or other formatting options. Check the guidelines of journals of interest on this. Many software programs can help an author to be systematic in creating an electronic bibliographic database and in citing of articles and books. An example of such a program is Endnote (available for Windows and Macintosh).

When citing publications in the main text, remember that all references to monographs, articles, and/or statistical sources need to be identified parenthetically by last name of the author, year of publication, and page number (when appropriate). When the author's name appears in the text, report it thus: Castilla (2003). But when the author's name does not appear in the text, cite it thus: (Castilla, 2003). Page numbers follow the year of publication: (Castilla, 2004: pp. 34–36). When multiple references to an author appear in the same year, distinguish them by attaching a letter to the year of publication: 2003a, 2003b, and 2003c. When a piece of research has more than three authors, use "*et al.*" when referring to the publication in the main text but report the full names of all authors in the list of references.

10. Abstract of a Study

In general, the abstract of a paper is the part of a study that is read first. Although I discuss it at the end of this chapter, the abstract is normally the first section of the study, appearing after the title but before the introduction. The abstract gives the reader an overview of the study and summarizes the information presented in other sections. The majority of readers use the abstract to decide whether they wish to read the study. A good abstract is essential, and it should be one of the last parts of the study to be written.

Abstracts for empirical studies are usually written in a generic way. The type of information to include is quite conventional: (1) general information about the subject and object of study of the research paper; (2) main objectives of the study; (3) information about the data and methodology used to address the research question; (4) the most important results of the study; and (5) a brief conclusion or recommendation. The summary must be as brief and concise as possible. Most of the leading journals in the field impose a maximum length in the 100–300 word range. Do not confuse the introduction with an abstract. The best summaries focus on the study's innovations and emphasize its results.

The following are the abstracts of two of the research papers I have introduced in this chapter. The abstract for the health care article (Castilla, 2004) is:

Research has not satisfactorily answered the question of whether institutions are becoming more similar across countries despite the existence of nationally distinctive social, cultural, political, and economic factors. This study presents an institutional account for the process of change in national institutions over time, and specifically examines one of the most important national institutions, health care systems (HCSs). Longitudinal data on national health care systems and population health in the OECD countries show that there has been increasing convergence in the general characteristics of national HCSs since 1960, mainly in both the expenditure on health and the level of public coverage of medical costs. At the same time, the composition and utilization of such HCSs have diverged, particularly concerning the national availability and utilization of health care resources. Despite these differences in national HCSs due to differences in national cultural traits like the degree of individualism, countries have healthier populations, and improvements in health appear to converge all over the industrialized world. Results from this study indicate that population health is ultimately affected by social variables such as the level of socio-economic inequality rather than by the way countries organize health care.

The abstract for the social networks and employee performance article (Castilla, 2005) is the following:

Much research in sociology and labor economics studies proxies for productivity; consequently, little is known about the relationship between personal contacts and worker performance. This study addresses, for the first time, the role of referral contacts on workers' performance. Using employees' hiring and performance data in a call center, the author examines the performance implications over time of hiring new workers via employee referrals. When assessing whether referrals are more productive than non-referrals, the author also considers the relationship between employee productivity and turnover. This study finds that referrals are initially more productive than non-referrals, but longitudinal analyses emphasize post-hire social processes among socially connected employees. This article demonstrates that the effect of referral ties continues beyond the hiring process, having long-term effects on employee attachment to the firm and on performance.

11. Final Recommendations for Writing a Research Paper

The information presented up to now may seem overwhelming, but it sounds more complicated than it actually is. Just be sure to translate any theoretical propositions into a set of testable hypotheses, describe the studied data and its content, accompany the presentation with informative tables containing descriptive statistics over time, and explain the methodology, using equations when necessary. This chapter would be incomplete without some final suggestions that apply to all sections of any research paper. Many of my suggestions focus on how to give a paper a polished and professional appearance.

1) *Always keep your audience in mind.* The ultimate goal of a paper is to communicate ideas and findings to other people. Consider what the audience already knows, figure out what the audience will be interested in discovering and what the audience will find hard to understand, and write accordingly. One should always keep in mind a few specific readers who are representative of the audience when designing and writing a research paper.

2) *The art of good writing comes with rewriting.* No paper is final until publication, and rewriting a paper will only make it clearer and more concise. The more rewriting, the more focused and tightly structured the paper will

become. If possible, ask colleagues to read drafts of a paper and incorporate their feedback and suggestions. No one can write a quality research paper in one draft.

3) *Organize your paper carefully.* One's ideas can be communicated more effectively both by breaking a paper into the major sections discussed in this chapter and by dividing those sections into additional sub-sections as needed. Be sure to label all sections and subsections with underlined, bold, or italic type. If the paper is to be submitted to a particular journal, check a recent issue for formatting conventions and follow the journal's recommendations for labeling sections. Topics should be presented in a logical, hierarchical order. First, introduce a general thought and then get more specific. Finally, keep an eye on the section length, remain paced, and be balanced throughout the text. Do not spend 15 pages on the theory section and 3 pages reporting the findings. The findings section is important, and the amount space devoted to it should reflect this.

4) *Write well.* Be clear. This is obvious, but it is often difficult to become self-aware about the obfuscation and ambiguity that may affect one's writing. For a start, avoid overusing the passive voice. Try to use the active voice as often as possible. First person ("I") is perfectly legitimate (and highly advisable) in scientific writing. Good writing is extremely important. Journals frequently reject manuscripts simply because they are poorly written. A paper should be well argued and the prose should be clear and accessible.

5) *Be concise.* Keep most sentences and paragraphs moderate in length. Very short sentences and paragraphs can make a paper appear choppy and can make the logic hard to follow. However, one's paper can lose the reader if there are too many long and convoluted sentences. The best writers vary the length of their sentences through the paper.

6) *Spell correctly; use the right words; be careful with using variable names and abbreviations.* Since modern word processors automatically check spelling and grammar, I recommend using spell checkers regularly and checking drafts carefully for errors. Do not try to impress readers by using words that are not common in everyday speech or in the field. Be sure to know the meaning and connotation of a particular word before using it. Finally, identify variables and values by words or sentences. Do not use the names of variables as they appear in the computer programs (not even in tables). Make sure to use the complete name of the variable in the text and tables so that the reader understands. Mnemonics may be used, but only when those mnemonics are obvious to the reader. One should also code metric variables so that higher values mean more of what the variable measures, and one should label variables to reflect this. Describe models in an equation or in words and try to avoid computer software terminology.

7) *Pay attention to detail.* Bad word choices, formatting errors, and blotched papers are likely to leave readers and editors with a poor opinion of the author. Readers may conclude that an author who is careless about the appearance of his or her document may have also been careless in collecting and analyzing data. They may even question the value and validity of the author's work.

8) *Use an appropriate format and layout for the paper.* Make the paper physically attractive and easy to read. Double-spaced text and appropriate margins (at least one inch on all sides) help tremendously with readability. Twelve-point type in one of the standard fonts (Times, Helvetica, Arial, or Courier) is recommended. One should avoid using "designer" fonts in a research paper. Use underlining or italic for emphasis, but avoid using both. Number the pages; label and highlight each main section, subsection, and sub-subsection.

9) *Use figures and tables to present visual information.* Produce clear and informative tables and figures. Errors in tables and figures both confuse the reader and make the writer appear sloppy. Remember, too, that no table or figure is self-interpreting; consequently, accompany all tables and figures with an interpretation and discussion in the main text.

10) *Think strategically about the title of the paper, and be thoughtful when writing an informative short abstract for the paper.* The title and abstract are the two most important ways to attract scholars to a paper. The research paper will be known by its title, and a good title will attract readers just as much as a poorly worded title will discourage them. The title should explicitly describe the topic of the study, should indicate the scope of the study (neither overstating nor understating its significance), and should be self-explanatory to readers in the chosen area. The expected length of a title varies by discipline. In the social sciences, one normally writes a short title followed by a longer descriptive subtitle.

11) *Be interesting and novel.* This is easier said than done. Since most scientific writing is highly structured and follows a given format, the paper may become boring. Some great social scientists who write well suggest that a paper should begin with an intriguing problem that is then developed and finally solved. The paper should have a story line that is organized logically and presented in an interesting way. Pick any leading journal and have a look at their abstracts and articles before beginning to write.

12) *Spend time writing the conclusion.* Include a brief but concrete summary; one paragraph is usually enough. Make a conclusion about the substantive problem in light of the findings. Avoid introducing a new finding or thought in the conclusion.

12. Preparing a Research Paper for Publication

Once the paper has been circulated among colleagues, its arguments have been revised and clarified, and it has been checked for typos or misspellings, the paper is ready to be submitted for publication in a top journal. When targeting a journal, check its submission guidelines in advance. The most general requirement is the paper length and format. Journals normally recommend a maximum number of pages of double-spaced text in 12-point font, excluding references, tables, figures, and appendices.

Editors of most journals require an author to submit several copies of the manuscript and abstract on white paper. One should also enclose a self-addressed stamped postcard to obtain acknowledgment of manuscript receipt. Nowadays, most journals require the article to be submitted in an electronic format. Some journals prefer to receive files via CD-Rom, FTP, electronic mail, or by some other means. When submitting material electronically, always label disks and files appropriately (including the author's name in the name of the file). Also, one should inform the editors about the name and version number of the word-processing program used. A single file that includes all parts of the paper should be submitted. Never submit files that contain "links" to other files; the reliability of such links cannot be controlled after a submission. Tables and figures created from spreadsheets or other database software programs can be converted to modern word processor format. Refereed journals send papers anonymously to readers. The journal editors typically rely on the judgments of those readers. To protect anonymity, most journals ask that only the title and abstract be included on the first page of a manuscript. One should attach a cover page with the title of the manuscript, author's name, affiliation, and contact information, but all identifying references and footnotes should appear on a separate page.

The research paper can now be sent for publication. Be patient, and follow the advice of the famous science fiction novelist and scholar Isaac Asimov: "You must keep sending work out; you must never let a manuscript do nothing but eat its head off in a drawer. You send that work out again and again, while you're working on another one. If you have talent, you will receive some measure of success—but only if you persist." In the next chapter, I present two examples of full research papers using longitudinal analyses. One of them is my complete study about the impact of social networks on employee productivity over time (referred to in some of the previous sections of this chapter). The second paper included in Chapter 5 is a study that uses EHA to determine which factors influenced national science activity during the twentieth century.

Chapter 5

Two Applications of Longitudinal Analysis

In this chapter, I provide two examples of papers written using longitudinal data sets and methods. The *first paper*, titled "Social Networks and Employee Performance" illustrates some of the methods covered in Chapters 2 and 3, namely event history analysis and cross-section cross-time analysis. This is an earlier version of the paper that was later published in the *American Journal of Sociology* in 2005.[53] It examines the role of social networks on workers' post-hire outcomes in a large call center in the United States. Using longitudinal data on the call center pool of 4165 job applicants, which include information about ties between employees and about their demographic characteristics, I identify and test the mechanisms by which the hiring of new workers using employee referrals shapes employees' productivity and turnover over time. This study demonstrates that the effect of referral ties continue well beyond the hiring process, having considerable long-term effects on employee attachment to the firm and on performance. I structure my argument as follows. First, I test the central prediction of the "better-match" theory common in labor economics. The proposition here is that if referrers help to select better-matched employees, one would expect that after controlling for observable human capital characteristics, workers hired via employee referrals should be more productive than non-referrals at hire. Second, and independently of any superior initial performance, I examine whether referrals' performance advantages are manifested in a steeper post-hire performance improvement curve. In this sense, if productivity improvement is a reflection of learning, network ties might affect both the potential levels of performance, as well as the rate at which employees learn. Third, and irrespective of any differences in performance trajectories between referrals and non-referrals, referrals may exhibit lower turnover than non-referrals.

More importantly, I study how interdependence between referrals and referrers affects work performance. I look at the performance implications of what has been called the "social enrichment process," according to which the interdependence between referrals and referrers shapes employee performance. Even if one assumes that referrals and non-referrals have equivalent résumé credentials, perform equally well in the interview, or even exhibit similar performance trajectories, employers may still hire referrals at a higher rate simply to accrue the benefits of the social integration phenomenon in the workplace. Referrals might

[53] The full citation of the article is: Castilla, E.J., 2005, Social networks and employee performance in a call center, *American Journal of Sociology* 110 (5), 1243–1283. Copyright © 2005 by The University of Chicago Press.

be coached and trained by their referrers. At the same time, networks might provide the support that minimizes turnover and increases positive work attitude and job satisfaction (and therefore productivity) in an organization.

The *second paper*, titled "Institutional Aspects of National Science Activity," illustrates the use of the event history methodology. An earlier version of this paper was presented at the American Sociological Association Annual Meeting in 1997 (Toronto). This article examines the institutional factors influencing national science activity during the 20th century. In particular, it investigates the conditions under which nation-states are more likely to join any of the unions or organizations comprising the ICSU, the primary and oldest international science institution in the world. Using historical and quantitative data, I develop an event history analysis to account for repeating events. I seek to test two of the major theoretical arguments concerning the expansion of national science activity over time. First, I test variations in the impact of socio-economic development. Second, I explore the role of the worldwide system's linkages in the process of national adherence to scientific organizations (institutional arguments). I restrict my attention to the effect of (1) the degree of economic openness of a country (economic linkages) and (2) the degree of rise and linkages to the worldwide system, specifically the expansion of the international discourse and the proliferation of a "scientific consciousness."

According to the dynamic model proposed here, institutional theories have the greatest power to predict the rate at which nation-states join scientific organizations. The rate increases more quickly following World War II with increasing globalization. In addition, the number of ICSU organizations' memberships of a country increases over time, while the hazard rate at which a nation-state joins an additional ICSU organization or union decreases. This inverted U-shaped relationship between the rate and the cumulative number of unions' memberships of a country holds for both core and peripheral countries. Furthermore, the increasing dimensions of the world-system as well as the impact of its discourse on nation-state behavior have been shown to have a strong positive effect on the hazard rate for both core and peripheral countries. On the contrary, development and modernization arguments do not seem to offer significant explanations for the ICSU join rate over time. Thus, the model does not support the hypothesis that socio-economic development has a strong positive effect on the rate at which countries join ICSU Scientific Unions. Development arguments explaining the national process of joining any ICSU Union are of importance early in the "science diffusion" process. After 1945, institutional factors better account for nations' decisions to join scientific organizations. Finally, although core countries have significantly higher hazard rates than peripheral countries, there is evidence of certain convergence in joining rates over time. Differences between developed and developing countries' entry into scientific organizations seem to have decreased slowly over time. The study also finds no substantial variability in the effects of the different development and institutional factors for core and peripheral countries.

1. Social Networks and Employee Performance (Paper 1)

Abstract: Much research in sociology and labor economics studies proxies for productivity; consequently, we know very little about the relationship between personal contacts and worker performance. This study addresses, for the first time, the role of referral contacts on workers' performance. Using unique employees' hiring and performance data in a call center in the United States, I examine the performance implications of hiring new workers using employee referrals over time. When assessing whether referrals are more productive than non-referrals, I also consider the relationship between employee productivity and turnover. Consistent with the better-match argument in economics, I find that referrals are initially more productive than non-referrals. In the long run, however, my longitudinal analyses support a more sociological explanation that emphasizes post-hire social processes among socially connected employees. This article demonstrates that the effect of referral ties continues beyond the hiring process, having considerable long-term effects on employee attachment to the firm and on performance.[54]

1.1. Introduction

For decades, we have seen a stream of theoretical and empirical studies in economic sociology and labor economics examining how recruitment sources relate to employees' outcomes such as turnover and tenure, starting wages, and wage growth (for a detailed review of these studies, see Granovetter 1995; and more recently, Petersen *et al.*, 2000; Fernandez *et al.*, 2000). Although many of these studies have sought to determine whether hires made through personal contacts are better matched than those made through other channels, none has focused specifically on the performance implications of hiring new employees by using current employees' connections.

Examining information on employee productivity promises to constructively advance research in this area. Such information is crucial because social relations and productivity can be related in complicated ways (and more importantly, they are likely to be confounded). In general, economists believe that social networks are not independent of productivity, and are, therefore, valuable proxy variables when performance data are not available. For instance, the better-match argument common in labor economics argues that social connections provide high-quality information that will improve the match between the job and

---

[54] Again, this is an earlier version of the following article: Castilla, E.J., 2005, Social networks and employee performance in a call center, *American Journal of Sociology* 110 (5), 1243–1283. Copyright © 2005 by The University of Chicago Press.

the person. Under this theory, social relations act as a proxy for information about the job candidate that is difficult and expensive to measure or observe directly, such as employee productivity. However, a more sociological explanation suggests that regardless of whether personal connections reliably *predict* future employee performance, connections among employees can still *produce* more productive employees even after they have been screened and hired. Social interactions that occur among socially connected employees at the new job setting may enrich the match between the new hire and the job, and may thus affect employee performance over time. This "embeddedness" account emphasizes how the presence of personal contacts and their departure at times facilitates and at times lessens employees' productivity and attachment to the firm. However, the field's lack of direct measures of employee productivity renders us incapable of adjudicating between these two competing theoretical accounts.

One way to make progress on this subject is to directly examine the relationship between personal networks and worker performance. Using comprehensive employees' hiring and productivity data from a large call center in the United States, I examine, for the first time, the performance implications of hiring new workers via employee referrals, using referrals as indicators of preexisting social connections. My study also provides a further understanding of how worker interdependence impacts their performance. I structure my argument as follows. First, I test the central prediction of the "better-match" theory in economics. The proposition here is that if referrers help to select better-matched employees, one would expect that, after controlling for observable human capital characteristics, workers hired via employee referrals should be more productive than non-referrals at hire. Second, I examine whether referrals' performance advantages are manifested in a steeper post-hire performance improvement curve. If productivity improvements occur as a result of employees acquiring knowledge and skills, network ties might affect both potential levels of performance as well as the rate at which employees learn. Third, since turnover and performance are likely to be related, I consider the process of turnover when assessing whether referrals are better than non-referrals (i.e., the fact that referrals might exhibit lower turnover than non-referrals). Finally, I test a more sociological proposition, which presumes that interaction between the referral and referrer at the workplace enriches the match between the new hire and the job.

My present study uses a direct measurement of what constitutes a "better" employee, one that is an *objective* measure of productivity. Thus, this is an exceptional opportunity to tackle an important research question that has never been addressed before. Consistent with the better-match argument, I find that employee referrals are initially more productive than non-referrals. In the long run, however, my analyses do not seem to support the better-match explanation. Instead, I find support for the more sociological argument that stresses how post-hire dynamics of social relations among socially connected employees' influences employee productivity over time. Given the results of my analyses, I suggest that the better-match mechanism should be complemented by the social interaction and embeddedness arguments in sociology. Even if one assumes that referrals and non-referrals have equivalent work abilities, perform equally well in the interview, or even exhibit similar performance trajectories, employers may still prefer to hire referrals at a higher rate simply because of the benefits of social integration in the workplace. Referrers might mentor and train their referrals. At the same time, the social support provided through networks might also increase positive work attitude and job satisfaction (and therefore productivity) and minimize turnover in an organization. In this study I show that the effect of referral ties goes beyond the hiring process, having significant long-term effects on employee attachment to the firm and on performance. I find that the effect of being a referral on performance is contingent on the referrer's continued presence in the firm. The departure of the referrer has a negative impact on the performance of the referral, even after the referral has been working in the organization for some time.

1.2. Hypotheses

1.2.1. Better Match Implies Better Performance

It has been argued that the social connections inherent in referral hiring benefit the hiring organization by improving the quality of the match between worker and job. In the economic literature on referral hiring, this argument is known as the "better-match" account. It proposes that personal contact hires perform better than isolated hires because social connections may help to obtain difficult and more realistic information about the job and the candidate.[55] Wanous (1978, 1980), for example, posits that individuals who possess more accurate and complete information about a job will be more productive and satisfied with the organization than will individuals who have less accurate and complete information. This is mainly because job candidates who have more complete, relevant, and accurate information will have a clearer understanding of what the job entails and will thus be more likely to perform well on the job than will candidates lacking such information.

---

[55] Previous theoretical accounts of the role of networks in screening and hiring discuss in detail the different mechanisms that could be producing the better match (see Fernandez *et al.* [2000] for a review).

Ultimately, the "better-match" theory posits that employers may benefit from referral hiring because referrals simply exhibit superior performance and are therefore better workers than non-referrals. However, the traditional post-hire indicators of employees' better matches used in existing empirical studies have been anything but direct measures of productivity; and consequently, evidence for the better-match hypothesis is quite mixed.[56] Perhaps, the main reason for the inconclusive nature of these studies is that none has satisfactorily analyzed performance, the most important indicator of whether a referral employee is a better worker than a non-referral employee. Therefore, it is difficult to claim to have examined the match quality in depth without having measured productivity, one of the bases upon which employees are evaluated and compensated. The present study uses a direct measurement of what constitutes "better": an objective measure of employee productivity.[57] Thus, I can provide a strong test of whether referrals are better matched than non-referrals. If referrals are better matched to the job than non-referrals, one would then expect some performance advantage associated with referrals at hire:

Hypothesis 1. *Referrals initially perform better than non-referrals.*

This hypothesis could be questioned on the grounds that all hires (referred and non-referred) have been screened on performance-based criteria. Employees are selected on observable individual characteristics gathered from their résumés or observed during the interview. Nonetheless, if one does not take into account the selection process—the fact that employers hire the survivors of the organization's screening process—the effect of the referral variable on initial performance might be biased (Berk, 1983; Heckman, 1979). For this reason, previous studies analyzing only hires when relating recruitment source and employee's outcomes are likely to be biased (Breaugh, 1981; Breaugh and Mann, 1984; Quaglieri, 1982; Taylor and Schmidt, 1983; for an exception, see Fernandez and Weinberg, 1997). The present study tests Hypothesis 1, correcting for the selection of hires in pre-hire screening. This correction will help to perform the mental experiment of what the initial performance of all applicants would have been had they been hired without screening, and to determine whether there

---

[56] The traditional post-hire indicators of employees' better matches used in this literature have been higher starting wages and slower wage growth (Quaglieri, 1982; Simon and Warner, 1992), lower turnover (Corcoran *et al.*, 1980; Datcher, 1983; Decker and Cornelius, 1979; Quaglieri, 1982; Gannon, 1971; Simon and Warner, 1992; Sicilian, 1995; Wanous, 1980), different time path of turnover (Fernandez *et al.*, 2000), and even better work attitudes and lower absenteeism (Breaugh, 1981; Taylor and Schmidt, 1983).

[57] Admittedly, some studies have shown that people hired through social contacts received better *subjective* performance evaluations (Breaugh, 1981; Breaugh and Mann, 1984; Caldwell and Spivey, 1983; Medoff and Abraham, 1980, 1981; Swaroff *et al.*, 1985).

exists any difference in initial performance between referrals and non-referrals at the time of hire.

Hypothesis 1 emphasizes referrals' advantages over non-referrals at the beginning of their work contract with the organization. However, these accounts of the better-match story are still incomplete because they ignore the tendency for networks to recruit employees with superior performance careers. In this sense, cross-sectional analyses may miss the role of personal contacts in building such a performance career. If the benefits of good early jobs found through contacts later translate into labor market advantages, the effect attributable to social networks is attenuated in the cross section. The possibility that network ties themselves influence productivity over time needs to be further explored with longitudinal data on employee performance. Following the better-match predictions, if referrals are better matched to the job than non-referrals, they might not only perform better right after being hired, as suggested in Hypothesis 1—they should also perform better than non-referral hires in the long run. Even if Hypothesis 1 was not supported, the advantages of social ties could be manifested over the tenure of the newly hired employee in two ways. First, referrers may provide information that helps employers choose recruits who can potentially reach a higher level of performance than non-referrals. Second, referral hires might be able to learn the job and adjust to its requirements more quickly than non-referrals. These two propositions imply:

Hypothesis 2. *Referrals have better performance trajectories than non-referrals.*

When assessing whether referrals perform better than non-referrals, I will consider the issue of turnover. The obvious relationship between turnover and performance has not been explored in empirical studies.[58] Generally, productivity can appear to improve due to two separate processes. Under the first process, particular individuals show true improvement in performance over time. This process is consistent with the learning theory (Arrow, 1962). However, there is a second process whereby performance growth is affected by turnover. Since turnover may change the composition of the workplace, the observed positive correlation between tenure and performance when measured across the cohort of workers (not for any particular individual) could be entirely due to population heterogeneity. If low-productivity performers are leaving first (Tuma, 1976; Price, 1977; Jovanovic, 1979), then what looks like productivity improvement is

---

[58] A number of authors began a conceptual exploration of the positive organizational consequences of turnover (Dalton and Todor, 1979; Mobley, 1980, 1982; Staw, 1980).

actually due to a selectivity effect.[59] Thus, the average worker's productivity will improve as long as low-productivity employees leave the organization at a higher rate than good employees. The net effect of the different rates at which low- and high-productivity employees leave the firm could *look* like productivity improvement over time when measured across the cohort of workers. But this is not *true* longitudinal productivity improvement because of the change in the composition of the workforce.[60] Any attempt to assess whether referrals are better matched in this dynamic context requires separating these two processes. Therefore, Hypothesis 2 will be tested controlling for the risk of turnover.

1.2.2.   Post-Hire Interdependence in Performance

The last mechanism by which referral hiring might affect performance is sociological; it emphasizes post-hire social processes that occur among socially connected employees. This proposition presumes that interaction between the referral and referrer at the new job setting enriches the match between the new hire and the job. The experience of the referral hire might simply be richer and more gratifying because the referrer is around and available to help, answer questions, provide feedback, and participate in non-work related social activities. In addition, referring employees can serve as informal mentors and enhance training and performance in the workplace. This process, termed the "social integration" or "social enrichment" process (Fernandez *et al.*, 2000), is distinct from the better-match argument because it occurs post-hire. Thus, social relations between referrals and referrers affect new hires' attachment to and performance in the organization.

Fernandez *et al.* (2000) present evidence for interdependence of referrals' and referrers' turnover patterns. They show that referral ties affect employees' attachment to their firm, but suggest that referrer turnover may have some implications for referral performance, even after the referral has been in the organization for some time. For example, the departure of the referrer may in itself prompt the referral to re-evaluate her own satisfaction with the current job; such a re-evaluation may subsequently lower her commitment to the job, and consequently

---

[59] Bartel and Borjas (1981) already introduced the question about the effect of labor turnover on wage growth within the job. They argue that the observed positive relationship between tenure and wage growth could be entirely due to population heterogeneity. There exist some unobserved individual characteristics that lead to low wages and high turnover rates for some workers, and to high wages and low turnover rates for others.

[60] Few researchers have conceptually examined this individual performance–turnover relationship in depth (Porter and Steers, 1973; Price, 1977). In general, the findings of such studies are quite mixed. For example, Bassett (1967) found that high-productivity performers were more likely to leave the organization; Seybol *et al.* (1978) found higher performers less likely to leave; and Martin *et al.* (1981) found no relationship between performance and turnover.

her performance. Another mechanism could be that the referrer's exit reduces the quality of the work setting to an unsatisfactory level, again lowering the referral's performance and possibly leading her to quit. Even if the referrer's employment termination does not affect the referral's performance, it may still increase her likelihood to quit; referrers who leave the organization may convey information about external job opportunities back to friends and colleagues, increasing the chance that the referral herself will be lured away to another company (Fernandez *et al.*, 2000). One final possibility is that the referral employee may feel a sense of obligation not to embarrass the referrer; such sense of obligation might decline or even disappear after the referrer departs, lowering the referral's performance.[61] All previous scenarios suggest that the referrer turnover has negative consequences for referral performance.

However, the Fernandez *et al.* study does not focus on the fact that referrer turnover could also have some positive impact on the attitudes and performance of those referrals who remain in the organization. Krackhardt and Porter (1985) found in their study of three fast-food restaurants that the closer the employee was to those who left the restaurants, the more satisfied and committed she would become. This observation has support in dissonance studies. If a person observes a friend leaving and attributes dissatisfaction to the friend's decision to quit, that person's decision to stay may require more justification. One way the person could justify her decision to stay is to develop stronger positive attitudes towards the job and the workplace.

Clearly, it is difficult to universally predict the effects of referrer turnover on the performance of the referral. In this study, I examine what happens to the performance curve of employees whose referrer leaves; this involves comparing the performance curves among non-referrals, referrals whose referrer leaves, and referrals whose referrer stays. If the referrer's decision to quit has a negative impact on the productivity of the referral, this leads to:

Hypothesis 3a. *The turnover of the referrer worsens the referral's performance improvement trajectory.*

Assessing the "social enrichment" effect requires analyzing whether or not it is the presence of the referrer that improves referrals' performance. Thus, all the previous hypotheses about the better-match argument could be complemented as being about social enrichment. For instance, if the referrer teaches the referral the ins and outs of the job at the beginning of the job contract or during the training, it is the presence of the referrer that accounts for any performance differential

---

[61] I thank one anonymous reviewer for pointing this mechanism out.

between referrals and non-referrals—even between referrals whose referrer is present and referrals whose referrer is not present during the first months in the organization. Similarly, the referrer could help the referral along a quicker performance improvement trajectory. In fact, one could argue that if the referrer were to influence the referral, this influence should be strongest at the very beginning. Non-referrals might subsequently build a social network that dissipates the referral's initial advantage. This leads to an alternative hypothesis:

Hypothesis 3b. *The presence of the referrer improves the referral's performance improvement trajectory.*

Finally, I explore whether the positive effect of workplace interaction between referrer and referral is enhanced when the referrer's level of performance is taken into account. After careful analysis of the classic Hawthorne plant data, Jones (1990) demonstrated that workers' productivity levels were highly interdependent. In my setting, Jones' finding suggests that there might be a relationship between the performance of the referred and referring employees: If referrals are exposed to high-performance referrers, their performance should be much higher than the performance of non-referrals or individuals referred by low-productivity referrers. Conversely, social interactions with a low-productivity referrer at work might have a negative effect on referrals' productivity. If referrals' performance is affected by the amount of exposure to the referrer, the difference in results of exposure to a high-performance referrer as compared to that of a low-performance referrer should be explored.

1.3.   Research Setting

The job I study is the Phone Customer Service Representative (hereafter the "CSR"), an entry-level job at a large phone center (the "Phone Center"), within a large international financial services organization in the United States. CSRs are full-time employees, paid by the hour, whose duties consist of answering customers' telephone inquiries about credit card accounts. New hires are given approximately two months of classroom and on-the-job training (OJT) before working on the phone. CSRs are trained in order to improve their accuracy, speed, and efficiency while processing phone calls. Managers often monitor phone calls to ensure that CSRs achieve the Phone Center's courtesy and accuracy goals.

I wish to highlight two important features of the organization under study. First, the Phone Center is a single site with a centralized human resources function. It keeps particularly clean and orderly databases, which allow every phase of the CSR hiring process to be identified. A second feature of the Phone Center particularly relevant to this study is that in addition to recording supervisors' subjective ratings of employee performance, the Phone Center collects objective and precise measures of productivity for the CSRs. This should greatly improve the estimates of the impact of recruitment source on employee productivity. I have also been able to learn about the Phone Center's screening criteria and performance expectations. Consequently, I can more precisely specify the set of appropriate individual control variables that affect labor market matching. In addition, I can consider the extent to which an applicant's referral status is a proxy for other characteristics that might make the applicant desirable to the recruiters at the Phone Center.

In the remainder of this section, I describe the employment process at the Phone Center as illustrated in Fig. 5.1. I start with the records of the Phone Center's hiring activities during the two years from January 1995 until December 1996. The Phone Center's human resources (PCHR) department tracked 4165 external employment inquiries for CSR jobs over this two-year period. Only 8% (336) of the original applicants were hired. I tracked 334 of these employees from the time of their hire until June 1997, when I ended the performance data collection. Around 290 hires completed the two-month training period at the Phone Center. For those hired CSRs, I examine two of the most relevant post-hire outcomes at the Phone Center: turnover and productivity. Whenever possible,

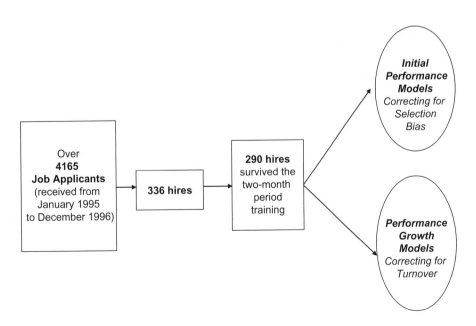

Fig. 5.1 Employment process under study.

I incorporate evidence that I gained through observation and interviews of the different professionals at the Phone Center (mainly from the PCHR staff).

1.3.1. Recruitment and Training

As a part of their standard operating procedures, PCHR professionals record the recruitment source for every employment inquiry. This information is recorded when a potential employee makes her initial contact with the Phone Center. I interviewed PCHR recruiters to determine the screening criteria and perform-ance evaluations that were used to recommend candidates for employment. They informed me that the CSR job involves significant customer interaction, PCHR screens applicants based on verbal and interpersonal skills from brief interviews or phone interactions. PCHR recruiters also look for people with prior customer service experience. They also tend to look for applicants who they believe will be reliable employees, preferring applicants who are currently employed, and who have had some previous work experience. In addition, they look for evi-dence of basic keyboarding and computer skills on the application. Very relevant to this study, the PCHR personnel emphasize that "referrals are treated the same as everyone else."

The Phone Center runs the training session for a cohort or "class" of about 15–20 new hires. The training consists of about six weeks of classes and two weeks out of class working in a controlled area (what they refer to as "on-the-job training" or OJT). The OJT takes place on the first floor of the Phone Center and not in the main area where the CSR would work (the main call area is located on the second floor of the building). The OJT period is important in this study for various reasons. First, during the training, performance is never mea-sured or evaluated; and most importantly, both referrals and non-referrals go through an identical hiring and training process. As one of the PCHR managers put it: "Nothing will prevent a new hire [whether referral or non-referral] from having to go through all the weeks of training. Our training is designed to bet-ter prepare our new hires to perform their duties as excellent CSRs." Second, during the training, recent hires learn about the CSR job, getting very clear information about what the job entails, including job content and responsibili-ties. Simultaneously, the employer learns whether the new employee is well suited for the job. As one PCHR member put it: "We can get a good idea of who is not going to be a good match to the CSR job; some hires do not even bother to show up to complete their training!" Quite a few new hires quit during the training period—during the period of my analysis, over 10% (44 of 334) of the hires left the firm during training. Therefore, I coded performance histories only for those 290 employees who were hired and completed the training process.

1.3.2. Turnover and Performance

PCHR personnel are concerned about the performance of their hires. However, recruiters do not seem to learn much about the quality of the employee once she has passed through the hiring process, even though, as one of the PCHR managers joked: "It is all a selection issue; if recruiters were doing their job right, we would not see so much turnover in our Phone Center" (field quotation from a PCHR manager). Regardless of whether recruiters are screening candidates using the "right information" or not (quoting one PCHR member), PCHR professionals are aware of the limitations associated with the screening of candidates.

PCHR personnel are also concerned about the costs of high employee turnover (although turnover is low by call center standards). Answering phone call after phone call in a high-pressure, highly structured environment demands a set of skills for which it is difficult to screen. As the PCHR director put it: "People leave their jobs because of the working environment. The job burns you out!" The numbers confirm statements like these; almost half of the CSR terminations at the Phone Center were due to job abandonment or job dissatisfaction (45.7% in 2000). One of the PCHR managers remarked that: "People do not want to be in [the Phone Center] all day. So the question is how we can change the bonus structure so that we can help reduce turnover." In previous years, PCHR professionals have dealt with the issue by hiring more people when turnover is high. Although this hiring practice might help to keep a stable number of CSRs answering the phones, it does not solve the problem of the high costs associated with employee turnover.

In terms of performance, unit supervisors at the Phone Center pay close attention to the average handle time, that is, the amount of time in seconds that it takes a CSR to complete a phone call with a client. A PCHR manager stated that: "Average handle time is the ultimate variable we are looking at, always controlling for the quality of the call." Every year, one of the PCHR managers computes simple statistics (i.e., means) for the whole year, taking into account the tenure of the employee. This manager prepares a report that is presented and discussed at PCHR meetings. Henceforth, PCHR managers try hard to understand the main predictors of handle time and quality so they can better screen for well-performing employees. PCHR staff, however, do not seem to have any clear idea about the observable characteristics that could help them identify and hire individuals with higher productivity potential. As the head of the PCHR department put it: "Based on our experience of thirty years, what you see in the résumé or during the interview is not a predictor of performance at all." Although PCHR closely monitors performance, CSRs are hardly ever fired because of low productivity: Only slightly more than 1% of all hires are terminated each year for performance issues. Still, supervisors intervene with poorly performing employees by rebuking them about their low productivity and/or low quality; they are also in charge of helping these CSRs to improve their performance. In addition to this monitoring and control, the Phone Center has a basic incentive plan where those employees with the highest performance ratings (i.e., highest average number of calls answered per hour) in a given time period get an increase of 5–10% in their salaries. Normally such salary revision occurs once or twice a year, and very few employees—less than 10%—get a raise.

1.3.3. *Measuring Tenure and Employee Performance*

For hires, I coded the two main dependent variables in the post-hire analysis: duration in the organization and performance during their tenure with the Phone Center from hire up until June 1997 when I ended the collection of performance data. Objective and subjective performance measures were examined monthly (at the beginning of each month). Because hires go through a training period of about two months, I have a maximum of 27 months of performance observations per employee hire.[62] Almost half of the hires were still with the organization at the end of my study. For hires, the days of tenure with the Phone Center range from a minimum of 3 days up to a maximum of 1104 days, with a median of 480 and a mean of 528 days.

I measure performance using the average number of calls a CSR answers per hour in any given month (corrected for call quality).[63] This measure is calculated using handle time. The Phone Center computer automatically calculates the average time a CSR takes to complete a phone call. Compared with other available performance measurements, the average handle time provides a good measurement of how efficient a CSR is. This measure is exceptionally accurate: It is measured across a large number of calls which are randomly routed to CSRs by the Phone Center computer (over 5000 calls per month for the typical CSR at

---

[62] The maximum number of performance observations is for those employees who were hired at the very beginning of my hiring window, January 1995, and who stayed in the organization until my last month of performance observation, June 1997. Because performance measures are available at the beginning of each month, the performance of hires is not available for the first three months (i.e., the two months of training, plus the first month after their training).

[63] Unit managers listen to a sample of calls for each CSR and rate the quality of their calls, evaluating each CSR on a monthly basis across courtesy and accuracy. Both measures of quality are typically at ceiling and exhibit little variance across people and/or over time. The evaluation scale ranges from 0 up to 1 (when all monitored calls are of maximum accuracy and/or courtesy) for the whole sample during the months of observation. Because of the lack of variation across observations, I do not use such measures of employee productivity as dependent variables in this study. Instead, I divide the average number of calls answered per hour by the product of both quality measures to compute a quality-adjusted average handle time for each employee. This calculates the number of calls answered per hour, adjusted for quality. As expected, both measures (number of calls quality- and non-quality adjusted) are highly correlated (with a correlation coefficient of 0.99).

about two-and-a-half minutes per call), and thus equates the difficulty of tasks across CSRs. In addition, it is measured automatically, and therefore is not subject to the normal problems of subjective performance ratings (e.g., supervisor evaluations). The maximum value observed in any month of tenure was approximately 26.5 calls answered in an hour, and the minimum was 19.5, with a mean of 20.3 phone calls per hour and a standard deviation of 3.63. The number of phone calls answered per hour is on average initially low, but tends to improve over the first year in the job, peaking at the fifteenth month, when an average CSR answers over 24 phone calls per hour. After the fifteenth month, the level of productivity worsens slightly—although variance in productivity also widens and the number of employee survivors decreases.

1.3.4. Independent Variables

Two different sets of variables are used in this study to predict an employee's performance trajectory. The first set of variables includes human capital variables that are believed to influence not only screening decisions but also an individual's productivity. Years of education and previous job experience are two of the most important variables. Experience includes variables such as months of bank experience, months of non-bank experience, number of previous jobs, whether the hire was working at time of application, and tenure and wage in the last job (as a proxy for job status prior to the job at the bank site; these variables are coded as zero for people who had not had a previous job). Since work in the human capital tradition argues that the value of human experience declines over time, I captured this effect in the analyses by entering a squared term for months of non-bank experience. In the analysis, I also include measures of different individual skills and capabilities such as having some computer knowledge or speaking another language (both are dummy variables). I also include a dummy variable to distinguish repeat applicants from first time applicants (1 for repeat applicants, 0 otherwise). The maximum number of applications from individuals is three. An important demographic variable is gender (coded 1 when male, 0 when female). Finally, I also control for the state of the market, i.e., the number of job openings and the number of applications on the date the candidate applied.[64]

---

[64] One might expect that the higher the supply of jobs in the organization, the less selective the organization can be. This may possibly worsen the employee–job match, leading to increased turnover of the hires and a worsening of the hires' performance. The demand side of the state of the labor market economic argument implies that the higher the demand for jobs in the organization, the more selective the organization can be. This increased selectivity should be reflected in an overall better match of hires to their jobs, in lower employee turnover, and in improved performance.

The second set of variables includes those measuring the availability as well as the characteristics of referrers, not only at time of the referral's application but also during her employment at the Phone Center. The first network variable included is a dummy variable indicating whether the respondent is a referral. My analyses are conservative tests (given the fact that I have only one of an employee's network ties) of the effects of social embeddedness of workers on productivity. The second set of variables measures the characteristics of the referrer, including variables such as wage, education, tenure in the firm, and performance rating in the organization. I also include variables about the referrer's structural accessibility to successful referrals, such as previous employment as a CSR.[65] All the referrer's characteristic variables are allowed to change over time except for education, which is considered constant. For non-referrals, all these variables are coded zero. Hence, the effects of referrers' characteristics are conditional on the applicant's being a referral. I also coded a dummy variable to distinguish those referrers who received a good subjective evaluation, as recorded in the Phone Center's computer files.[66] Finally, a time-varying dummy variable is coded as 1 once the referrer has left the organization.

Table 5.1 presents descriptive statistics for the independent variables included in the performance and selection models for applicants and hires. The job is female dominated—only 22% of the hires are male. Hires on average have about 13.6 years of education, with about 3 months of bank experience, 71 months of non-bank experience, and 48 months of customer service experience. Less than 14% have a bachelor's degree, and 74% have some computing experience. Sixty-seven percent of the hires were working at time of application; their number of previous jobs is 3 on average, with approximately two years of tenure in their last job. Half of the hires were referred by an employee in the firm. The table also includes the initial performance variables for the new hires. The average number of calls answered per hour is initially 20 calls, with courtesy and accuracy levels close to 1.

1.4.   Methods

My hypotheses pertain to the performance implications of hiring new employees using referral programs. Accordingly, my methodological approach is to break down the post-hire employment process into individual components and to model

---

[65] Studies show that when employees find their jobs through contacts with high rank and prestige, they tend to get better jobs themselves (Lin, 1999; Marsden and Hurlbert, 1988). Referrers may also vary in their accessibility to successful referrals (Fernandez and Castilla, 2001).

[66] In preliminary analyses, I also used the firm's information about bad evaluations. But this bad evaluation dummy variable is almost always 0; only 16 out of the 4165 applications (0.39%) were referred by employees who got bad evaluations. One of those candidates was hired and completed the training. In the case of good evaluations, 31 out of 350 applications made by "good" referrers were hired and completed the training.

Table 5.1
Means and standard deviations for variables in the performance models

	Applicants		Hires[a]		Hires Surviving One Year[b]	
	Mean	SD	Mean	SD	Mean	SD
Independent variables						
Gender (1 = male)	0.337	0.473	0.224	0.418	0.216	0.413
Repeat application (1 = yes)	0.096	0.295	0.072	0.260	0.025	0.157
Marital status (1 = married)	0.423	0.494	0.420	0.494	0.443	0.497
Skills						
Computer	0.731	0.443	0.741	0.439	0.784	0.413
Language	0.196	0.397	0.138	0.345	0.131	0.338
Years of education	13.750	1.866	13.607	1.723	13.528	1.768
Bachelor's degree (1 = yes)	0.185	0.388	0.138	0.345	0.146	0.354
Experience						
Works at time of application	0.550	0.498	0.679	0.468	0.714	0.453
Months of bank experience	2.102	14.402	3.335	14.649	2.374	10.561
Months of non-bank experience	64.629	62.361	71.721	60.136	72.631	55.334
Non-bank experience, squared	8064.834	28859.890	8746.934	20540.150	8321.410	14600.030
Months of customer service	34.195	44.877	48.085	53.577	47.272	47.023
Number of previous jobs	3.224	1.139	3.048	1.240	3.031	1.226
Tenure in last job (in days)	576.968	1032.603	776.493	1264.242	815.091	1234.666
Salary in last job	6.431	3.659	6.086	3.450	6.426	3.198
Application behavior						
Number of applications	19.937	16.783	18.534	15.246	18.126	13.624
Number of jobs openings	18.198	10.672	18.469	10.477	19.131	10.602
Application source						
External referral	0.374	0.484	0.510	0.501	0.548	0.499
Referrer's characteristics at time of application[c]						
Tenure (in years)	4.019	3.984	3.478	3.523	3.819	3.622
Wage	9.749	5.635	9.536	4.816	9.821	5.257
Years of education	12.430	1.218	12.432	1.202	12.422	1.189
Performance (1 = good evaluation)	0.228	0.419	0.209	0.408	0.211	0.410
Ever worked as a CSR (1 = yes)	0.297	0.457	0.351	0.479	0.330	0.472
Terminated	0.034	0.182	0.041	0.198	0.028	0.164
Dependent variables						
Number of calls answered per hour			20.291	3.629	20.312	3.501
Number of calls answered per hour (quality adjusted)			20.044	3.594	20.059	3.477
Maximum level of performance			26.497		27.949	
Minimum level of performance			19.458		19.319	
Courtesy (worst level = 0; best level = 1)			0.998	0.007	0.998	0.008
Accuracy (worst level = 0; best level = 1)			0.983	0.024	0.983	0.022
Number of cases	4114		290		199	

[a] Hires who survived the initial training period (approximately two months).
[b] Hires who stayed in the company for at least 12 months after the start date.
[c] The mean and standard deviations for these characteristics are calculated only for referrals (these referrer's characteristics are coded as zero for people who were not referred).

each of these pieces to understand performance careers within organizations (see Fig. 5.1). For those hired CSRs, I examine turnover and productivity, the two most relevant post-hire outcomes at the Phone Center. I estimate (1) models for initial performance and (2) models for performance growth. The initial performance models are estimated controlling for the screening of employees. The performance growth models are corrected for the turnover propensity of employees.

1.4.1. Initial Performance Models

In order to analyze the determinants of starting performance, the dependent variable I use is the starting average number of phone calls answered per hour (adjusted for call quality) immediately after the initial two-month training period.[67] I estimate the parameters of models of the form:

$$Y_0 = BX + \varepsilon \tag{1}$$

where Y_0 is the first available performance measure in the job as a CSR after training, X is a vector of covariates that contains characteristics of the individual at the time of entry into the Phone Center, as coded from their job applications, and ε is the disturbance term assumed to be normally distributed and well-behaved (uncorrelated with the covariates).

The performance equation proposed above has been traditionally estimated for the hires using OLS, mainly due to the lack of information about job applicants. As the result of observing performance only for the applicants who got hired, these past models do not correct for selection bias. To correct for such selection bias, I use the Heckman selection model (Gronau, 1974; Lewis, 1974; Heckman, 1976). This model assumes a regression like the one described in equation (1). However, the dependent variable, performance, is not observed for all applicants or hires who were terminated during the two-month training. So there is a selection equation, and the applicant is hired and completes the initial training period in the organization if:

$$Y'Z + \mu > 0 \tag{2}$$

where Z is a vector of covariates that affects the chances of observation of performance for a given applicant and μ is normally distributed with mean 0 and standard deviation of 1.[68] Presumably, firms hire those applicants who, based on

---

[67] The performance measure is normally distributed and no logarithm transformation was therefore required. Nevertheless, in addition to the modeling of starting performance, I also modeled the logarithm of starting performance and obtained very similar results (available upon request). I also used the number of phone calls per hour (without adjusting for the quality of the call) and obtained similar results.

[68] Following Stolzenberg and Relles (1997) and Winship and Mare (1992), I run several tests using different sample selection models to ensure my results are robust.

available information from their résumés, are expected to be most productive. But firms may also take into consideration the state of the labor market: that is, the number of job openings that needs to be staffed and the number of available applications (demand for jobs). Z is therefore a vector of covariates that contains the characteristics of the job applicant (X) plus two variables controlling for the state of the market at the time of application. The correlation between ε and μ is some parameter ρ; so that when ρ is different than zero, only the Heckman selection model provides consistent, asymptotically efficient estimates for the parameters in equation (1).

1.4.2. Performance Growth Models

For the study of performance growth, I analyze longitudinal data using regression models of change. While a comparison of cross-sectional analyses at different points in time provides some insight into this process, models of change represent it explicitly. The performance data structure is a pooled cross section and time series. The data are unbalanced: The number of observations varies among employees because some individuals leave the organization earlier than others (while many workers opt to stay in the organization). Research studies typically model such data with fixed-effect estimators, which analyze only the within-individual over-time variation. This choice is unappealing in this context because the majority of the independent variables (i.e., those variables coded from the application) do not vary over time.

To test my hypotheses about the determinants of change in productivity, I estimate various cross-sectional time-series linear models using the method of generalized estimating equations (GEE). These models allow estimating general linear models with the specification of the within-group correlation structure for the panels. I report the robust estimators that analyze both between-individual and within-individual variation. Specifically, I use the method of GEE developed by Liang and Zeger (1986). This methodology requires the inclusion of a correlation structure when estimating these models.[69] Any of these estimated longitudinal models will be corrected for the turnover process. So following Lee (1979, 1983), Lee *et al.* (1980) and Lee and Maddala (1985), I control for the retention of employees over time by including the previously estimated turnover hazard when I estimate such longitudinal models. This results in a two-stage estimation procedure. So I estimate GLM correcting for turnover as follows:

$$Y_{i,t} = \alpha Y_{i,t-1} + BX_{i,t} + \delta \overline{\pi}(t, Z_{i,t}) + \varepsilon_{i,t} \qquad (3)$$

---

[69] Under mild regularity conditions, GEE estimators are consistent and asymptotically normal, and they are therefore more appropriate for cross-section time-series data structures.

where Y_t is the performance measure in the job at time t and $\bar{\pi}$ the estimated turnover hazard rate (using event history analysis):

$$\pi(t, Z_{i,t}) = \exp[\Gamma' Z_{i,t}]\, q(t) \tag{4}$$

where π is the instantaneous turnover rate. This rate is commonly specified as an exponential function of covariates multiplied by some function of time, $q(t)$. The log-linear form for the covariates is chosen to ensure that predicted rates are non-negative. Z is a vector of covariates that affects the hazard rate of turnover for any given hire. Z is indexed by i to indicate heterogeneity by case and by t to make clear that the values of explanatory variables may change over time.[70] I estimate the effect of the explanatory variables in model (4) using the Cox model that does not require any particular assumption about the functional form of $q(t)$ (Cox, 1972, 1975). I tested for the effect of turnover across different specifications, functional forms, and measures of turnover and always found similar results.

1.5. Results

1.5.1. Initial Performance

Table 5.2 shows the differences in the levels of initial productivity by application source. The performance of referral workers appears superior to the performance of non-referrals, especially if we look at both quality-adjusted and non-quality-adjusted number of calls: Referrals' average number of calls answered is higher than non-referrals' (although the difference of a half-call is barely statistically significant at the 0.1 level). The table shows little difference between referrals and non-referrals on other dimensions of performance such as courtesy or accuracy. In Table 5.2, the exploration of performance differences between referrals and non-referrals does not control for other individual variables or the screening process that CSR applicants undergo before starting to work at the Phone Center. In the next tables, multivariate regression models are used to further explore this difference in initial performance between referrals and non-referrals.

Table 5.3 provides the results of the initial performance regressions correcting for both the selection of candidates from a pool of applicants and their retention during the training.[71] Employees are selected on observable individual

---

[70] The model in (4) is the most general form of any parametric models that are generally distinguished by the different choices of $q(t)$.

[71] OLS multivariate regressions were also estimated to examine the impact of the referral variable on initial performance (Castilla, 2002). From the results of these traditional models, the only significant variable in the prediction of employee performance is non-bank experience, which has a negative impact. These results are available upon request; although I argue that these traditional OLS results might be biased because these traditional OLS models do not take into account the selection process, that is, the fact that employers hire the survivors of their screening process.

Table 5.2
Means and standard deviations for measures of initial performance by application source

	All Hires[a]		Referrals		Non-Referrals		Performance Differences between Referrals and Non-Referrals	
	Mean	SD	Mean	SD	Mean	SD	Difference	t-Test[b]
Dependent Variables								
Number of calls answered per hour	20.291	3.629	20.561	3.584	20.009	3.666	0.552	1.296[†]
Number of calls answered per hour (quality adjusted)	20.044	3.594	20.290	3.478	19.787	3.706	0.502	1.191[†]
Courtesy (worst level = 0; best level = 1)	0.998	0.007	0.998	0.007	0.999	0.008	0.001	0.600
Accuracy (worst level = 0; best level = 1)	0.983	0.024	0.983	0.023	0.984	0.025	0.002	0.579
Number of cases	290		148		142		290	

[a] Hires who survived the initial training period (approximately two months).
[b] [†] $p < 0.1$, *$p < 0.05$, **$p < 0.01$ (one-tailed tests).

Table 5.3

Initial performance regression models correcting for screening and completion of training (OLS models with sample selection)[a]

	Only Referral		Referral and All Controls		Referral and Some Controls	
	Coefficient	SE	Coefficient	SE	Coefficient	SE
Main Model						
Constant	18.551***	2.091	21.410**	11.491	19.481***	2.002
External referral	0.745*	0.435	0.703[†]	0.465	0.737*	0.436
Gender (1 = male)			0.598	1.413		
Repeat application (1 = yes)			−0.404	1.058		
Computer			−0.090	1.047		
Language			−1.030	0.777	−1.076[†]	0.624
Years of education			−0.046	0.162		
Works at time of application			0.118	1.673		
Months of bank experience			−0.016	0.015		
Months of non-bank experience			−0.018*	0.009	−0.010**	0.004
Non-bank experience, squared			0.000	0.000		
Months of customer service			0.002	0.014		
Number of previous jobs			−0.047	0.433		
Tenure in last job (in days)			0.000	0.000		
Salary in last job			0.077	0.106		
Number of applications			0.001	0.021		
Number of openings			0.039	0.020		
Selection Model						
Constant	−1.457***	0.257	−1.449***	0.257	−1.443***	0.257
External referral	0.250***	0.064	0.248***	0.064	0.248***	0.064
Gender (1 = male)	−0.256***	0.074	−0.254***	0.074	−0.257***	0.074
Repeat application (1 = yes)	−0.130	0.112	−0.133	0.112	−0.132	0.112
Computer	0.176*	0.080	0.174*	0.080	0.175*	0.080
Language	−0.189*	0.087	−0.200*	0.086	−0.200*	0.086
Years of education	−0.012	0.019	−0.013	0.019	−0.013	0.019
Works at time of application	0.310***	0.069	0.314***	0.069	0.311***	0.069
Months of bank experience	0.000	0.002	0.000	0.002	0.000	0.002
Months of non-bank experience	0.002	0.001	0.002	0.001	0.002	0.001
Non-bank experience, squared	0.000*	0.000	0.000*	0.000	0.000*	0.000
Months of customer service	0.003***	0.001	0.003***	0.001	0.003***	0.001
Number of previous jobs	−0.075*	0.031	−0.076*	0.031	−0.076*	0.031
Tenure in last job (in days)	0.000*	0.000	0.000*	0.000	0.000*	0.000
Salary in last job	−0.015	0.010	−0.015	0.010	−0.015	0.010
Number of applications	−0.003	0.002	−0.003	0.002	−0.003	0.002
Number of openings	−0.001	0.003	0.000	0.003	−0.001	0.003
Wald χ^2 statistic	3.20		21.88		14.26**	
$\rho > \chi^2$	0.074		0.147		0.003	
Rho	−0.202	0.236	−0.231	0.373	−0.183	0.226
Test of independence of equations	0.66		0.27		0.57	
$\rho > \chi^2 (1)$	0.417		0.602		0.451	
Number of job applicants (Hires)	3972 (272)		3972 (272)		3972 (272)	

[a] Performance is measured as the average number of calls answered per hour (quality adjusted).

[†]$p < 0.1$, *$p < 0.05$, **$p < 0.01$, ***$p < 0.001$ (two-tailed tests; except the z-test for the effect of the "external referral" variable which is a one-tailed test).

characteristics from their résumés or during the interview. One needs to account for the fact that employers hire people who passed the organization's screening process. In Table 5.3, I correct for the selection of hires in pre-hire screening by including the factors for which PCHR says they screen and the control variables that affect labor market matching in the selection equation of the Heckman model.[72] The first column includes only the referral variable in the performance equation (Model 1). The second column includes all controls (Model 2); the third and final model (third column of the table) includes only those individual controls with a z-value more than 1 in Model 2 (i.e., language and months of non-bank experience).[73]

Referral appears to be an important variable at both the pre-hire and post-hire stages of the employment process. First, referrals are more likely to be hired and to complete the initial training. More importantly, once I control for the other "observable characteristics," referrals show a better level of performance, as measured by a higher quality-adjusted average number of calls answered. Referrals answer, on average, an additional phone call per hour when compared to non-referrals—the difference in the number of calls is slightly over 0.7. This is true for any of the three models of performance. The referral effect is significant at the 0.05 level (one-tailed) in the model without controls or with those controls with a z-value higher than 1 (i.e., language and non-bank experience). When all control variables are included in the model (column 2 of the table), the referral effect is still about 0.7 quality-controlled calls; but the difference is now barely significant (at the 0.1 level).[74]

The results in the selection part of the Heckman model show that applicants are more likely to get hired and complete training if they are employed at the time of application, if they have more months of customer service experience and work experience outside the financial services sector, or if they report longer tenure on their last job. In the study, the number of previous jobs that candidates report has the expected negative sign (significantly related to being selected). This is consistent with the recruiter's preference for candidates with a lower number of previous jobs (since those who change jobs a lot during their work

---

[72] My analyses reported in Table 5.3 do not change much when I exclude the two variables measuring the state of the market in the outcome equation.

[73] The effect of any independent variable with a z-value below one can be considered very insignificant, and therefore negligible.

[74] Even when applicants with foreign language skills are less likely to be selected than applicants without such skills, foreign language skills seem to have a negative effect on CSR performance. Candidates with foreign language skills answer fewer phone calls per hour than candidates without such skills (the difference is not significant at the 0.05 level, though). Work experience outside the financial services sector worsens the CSR average number of calls answered (significant at the 0.01 level); customer service experience is never significant, although it has the expected sign.

histories might be more likely to leave). While education did not emerge as a significant predictor, the dummy variable reporting some computer experience is a significant negative predictor of being selected. However, applicants with foreign language skills are less likely to be selected than applicants without such skills. The final human capital variable—candidate's last salary on the job—is not significant, although its coefficient has the expected sign. Applicants who report a higher wage on their last job seem less likely to be selected. This is consistent with recruiters' concerns that such candidates might be overqualified and more likely to leave the firm.

Controlling for other factors, males are less likely to be selected than females. PCHR recruiters speculate that females have a better sense of how to conduct customer service interactions, even though the substantive part of the equation demonstrates that females do not seem to perform any differently from males. The recruiters' preference for female employees may be an effect of gender stereotyping in the service industry. Even when there are no objective reasons showing that women perform better in any given service job, women are more often recruited for such positions. None of the variables controlling for the state of the market at the time of application has any significant impact on the likelihood of the candidate to be hired and complete the training period. Finally, the dummy variable distinguishing between referrals and non-referrals is statistically significant. Referrals are more likely to be hired and to complete their training than non-referrals.[75] Although referrals appear to be more appropriate candidates for the CSR job, referrals' advantages at the interview and training stage cannot be explained by the individual background control variables alone.[76]

There are clear productivity advantages in the hiring of employees using referral programs (and the bonus associated with the referral program). In this setting, the referral program seems to bring measurable post-hire advantages in the hiring of employees, even though certain PCHR representatives express their belief that the referral bonuses are counterproductive since people refer others for money. Only after completing my analyses on initial performance that control for the selection of hires, could I determine that these perceptions do not reflect reality. In actuality, the use of the referral program seems to provide two

---

[75] I estimated the same Heckman regression models excluding those 44 hires who were terminated during their training. The selection coefficient for referral is still statistically significant and has about the same magnitude. Moreover, the deletion of those 44 cases does not change the pattern of the effects of all the variables in both the selection and the substantive equations in the Heckman model.

[76] To complete my test of the "better-match" argument at the time of hire, I analyzed terminations during the training period (Castilla, 2002). There were 44 terminations during this period. My probit models with sample selection suggest that referrals seem less likely than non-referrals to leave their job during training. In termination models beyond training, however, the effect is not significant (consistent with Fernandez et al., 2000).

advantages in this particular Phone Center. First, there is evidence that the referral program increases the quantity of job applicants. Second, the referral program seems to recruit better-matched employees; there are definite initial performance advantages in the hiring of employees using the referral program. I show a net productivity gain of 0.7 phone calls per hour. Over an 8-hour day, this corresponds to 5–6 calls a day; and over a 40-hour week to about 28 calls, about an hour's worth of work has been saved. So I estimate an initial productivity increase of about 2.5%.

Finally, there is another way in which referring employees can signal information about the performance quality of referral candidates. Employees can access "upstream" information that could be available because of the tendency of people to refer others like themselves. According to this homophily mechanism, referrals are more likely to be like referrers, and since referrers have already survived a prior screening process, the homophily mechanism would lead the applicants referred by employees to be better performers than non-referred applicants (see Montgomery, 1991; Ullman, 1966). To address whether the referrer's characteristics and level of performance might influence the initial performance of the employee, I ran several Heckman regression models adding referrer's characteristics to the main performance equation. I find no evidence that additional information about referrers improves the fit of the initial performance model.[77] The fact that recruiters never contact referrers or look up their information explains why I see no support for these hypotheses in this Phone Center.

1.5.2. Performance Trajectory

I now analyze whether referrers help to recruit better performing employees over time. Table 5.4 presents the results of the several performance growth regression models correcting for the turnover rate. I estimate the Turnover Hazard Rate Selection Regression Models to test Hypothesis 2—whether referrals have better performance trajectories than non-referrals once we control for the risk of turnover. There does not, therefore, seem to be any support for the better-match theory in the long run. In addition, most of the appropriate individual human capital characteristics have insignificant effects on employee performance over time. By looking at the results on Table 5.4, it seems that the three significant variables in the prediction of employee performance growth are non-bank experience (which has a negative impact, significant at the 0.05 level), customer service experience (which has a positive impact, significant at the 0.001 level), and

---

[77] These results are available in Castilla (2002). Referrers' characteristics do not seem to provide any additional information about the future performance of the referral ($p < 0.776$; incremental $\chi^2 = 1.78$; df = 4); nor does information about the referrer's evaluation by the firm ($p < 0.727$; incremental $\chi^2 = 0.12$; df = 1).

Table 5.4
Turnover hazard rate selection regression models predicting performance correcting for turnover rate

Main model	Only Referral		Adding Controls		Adding Some Controls		Turnover Rate Model (Cox Regression Model)	
	Coefficient	SE	Coefficient	SE	Coefficient	SE	Coefficient	SE
Constant	5.890***	0.340	6.148***	0.545	5.844***	0.351		
Performance, month T-1[a]	0.762***	0.015	0.748***	0.015	0.768***	0.014	0.012	0.048
Tenure in months	-0.054***	0.019	-0.060***	0.020	-0.055***	0.019		
External referral	0.037	0.111	-0.009	0.113	-0.015	0.107	-0.435	0.320
Gender (1 = male)			0.103	0.146			0.254	0.340
Repeat application (1 = yes)			-0.225	0.194	-0.179	0.174	0.541	0.565
Computer			-0.074	0.138			0.168	0.456
Language			-0.167	0.157	-0.161	0.139	0.510	0.440
Years of education			0.013	0.039			0.052	0.093
Works at time of application			-0.009	0.131			-0.075	0.369
Months of bank experience			-0.002	0.004			0.018***	0.006
Months of non-bank experience			-0.005*	0.002	-0.004***	0.001	-0.009*	0.005
Non-bank experience, squared			0.000	0.000			0.000*	0.000
Months of customer service			0.004***	0.001	0.003**	0.001	0.004	0.004
Number of previous jobs			-0.013	0.051			-0.211	0.157
Tenure in last job (in days)			0.000	0.000			0.000	0.000
Salary in last job			0.024	0.019			-0.055	0.049
Number of applications			0.006*	0.003	0.006*	0.003	-0.001	0.013
Number of openings			-0.001	0.006			-0.014	0.019
Turnover hazard rate[b]	1.767†	1.199	1.762†	1.028	1.417†	1.011		
Wald χ^2 statistic	3298***		3732***		3680***		39.11***	
$p > \chi^2$	0.000		0.000		0.000		0.002	
Person–month observations (employees)	2983 (257)		2983 (257)		2983 (257)		3188 (260)	

[a] Performance is measured as the average number of calls answered per hour (quality adjusted).
[b] The turnover hazard rate is estimated from the turnover Cox Survival Model reported in the table.
†$p < 0.1$ *$p < 0.05$ **$p < 0.01$ ***$p < 0.001$ (two-tailed tests; except the z-test for the effect of the "external referral" variable which is a one-tailed test).

the number of applications at time of application (which has a positive impact, significant at the 0.05 level). Looking at the coefficient for the estimated hazard rate in the performance growth model, one can see that the likelihood of turnover is associated with higher performance growth over time. In other words, the model seems to suggest that those employees who are more likely to leave the Phone Center are those whose performance improves over time.

By examining the results of the Turnover Hazard Rate part of these Selection Regression models (reported in the last two columns of Table 5.4), I do not find statistically reliable evidence of referrals having a lower turnover rate.[78] As in the performance models, very few of the appropriate individual human capital and other control variables seem to have any significant effect on turnover. Additional months of bank experience seem to increase the rate of turnover whereas months of customer service experience improve an employee's chances of staying in the Phone Center.

The model of employee performance reported in Table 5.4 represents an important step towards correcting for a lack of empirical research concerning both the evolution of productivity and turnover decisions of hires within organizations. My model attempts to control for the process of turnover. I find that neither poor performers nor good performers are more likely to leave the Phone Center; the effect of the performance at time $t - 1$ on the turnover rate is not significant. This finding seems to suggest that the performance of the employee is not a good predictor of turnover in this research setting. However, as pointed out above, the coefficient for the estimated hazard rate in the performance growth model shows that those employees who are more likely to leave the Phone Center tend to significantly improve their performance over time (significant at the 0.1 level).

The initial performance models show clear productivity and early-turnover advantages when employees are hired using referral programs. The analyses of performance growth, however, show that referrals do not perform any better than non-referrals over time at the Phone Center. Moreover, the path of performance estimated for the employees at the Phone Center does not seem to reflect any improvement and/or skill acquisition (i.e., "learning by doing"; see Arrow, 1962). In this Phone Center, on-the-job tenure does not seem to make a worker more productive, especially in the long run. My findings suggest that an inverse U-shape curve is very descriptive for the CSR performance curve (consistent with the findings in many other studies of service-oriented jobs; see Staw, 1980).

---

[78] Results do not change across a variety of parametric transition rate models (the Cox model presented in the table, the proportional exponential model, or the proportional Weibull model). I also performed these tests separately for voluntary (quitting) versus involuntary (fired or laid off) turnover and found no reliable differences. Most of the turnover was voluntary, though; only 18 (11%) of the overall job terminations were involuntary.

The performance improvement mostly occurs during the first three months at the job (including the training). After that, the performance tends to decrease (the coefficient for tenure is negative and significant in all estimated models; also the coefficient for average performance in the previous month is below one and significant in all models).

To address whether referrer's characteristics and level of performance might influence the performance trajectory of the employee referral, I also ran several models adding referrer's characteristics to the main performance equation. The results were similar to those in the initial performance models; I found no evidence that information about the referrer improves the fit of any performance improvement model.[79]

1.5.3. Interdependence at Work

The dynamic models in Table 5.5a include the "referrer leaves" dummy variable in order to evaluate whether the departure of the referrer has a negative impact on the referral performance curve. All four sets of models show that the referral variable does not seem to have a significant effect on employee performance. Overall, referrals and non-referrals do not differ in their productivity trajectories. These results again do not support the "better-match" theory in the long run. Instead, I find statistically reliable evidence of interdependence between referrals' productivity and referrers' turnover. Consistent with Hypothesis 3a, referrals whose referrer has left show a worse performance trajectory than those whose referrer has stayed (and even than non-referrals). The "referrer leaves" coefficient in the dynamic regression models—including the most significant controls only—is -0.28 (significant at the 0.05 level, two-tailed test). I therefore find support for Hypothesis 3a.[80]

---

[79] Referrers' characteristics do not seem to predict or provide any information about the future performance of the referral ($p < 0.575$; incremental $\chi^2 = 2.90$; df $= 4$); nor does information about the referrer's evaluation by the firm ($p < 0.255$; incremental $\chi^2 = 1.29$; df $= 1$). These results are available in Castilla (2002).

[80] I also tested the reciprocal effect that referrals had upstream on their referrers' turnover and performance. Consistent with the social enrichment argument, one can easily imagine that the referrers might also be affected when their referrals depart. Thus, I explored the data for evidence of whether the referral turnover decreases referrers' performance and/or increases the chances of referrer turnover. Unfortunately, the study design did not allow me performing a robust test of this effect. First, in order to avoid problems of left censoring, I only looked at referrers who were themselves hired in the two-year hiring window ($N = 82$). Of the 119 referrals these people made, only 18 were hired, and of these, only 7 terminated. Second, given that not all employee referrers were (or had worked as) CSRs before (or during the period under study), their performance was not measured as number of calls per hour. Instead I used the organization's yearly subjective employee evaluations, which exhibit little variance across employees or over time. Despite the lack of statistical power, the effect of the referral terminating in both the turnover and the performance models is as predicted (the results are never statistically significant, though). This is consistent with Fernandez *et al.* (2000).

Table 5.5a
Cross-sectional cross-time regression models predicting performance correcting for turnover rate

	Only Referral		Adding Social Interaction		Adding Controls		Adding Some Controls		Turnover Rate Model (Cox Regression Model)	
	Coefficient	SE	Coefficient	SE	Coefficient	SE	Coefficient	SE	Coefficient	SE
Main model										
Constant	5.890***	0.340	5.859***	0.341	6.149***	0.546	5.823***	0.352		
Performance, month T-1[a]	0.762***	0.015	0.763***	0.015	0.748***	0.015	0.769***	0.014		
Tenure in months	-0.054***	0.019	-0.053***	0.019	-0.058***	0.019	-0.053***	0.019	0.015	0.048
External referral	0.037	0.111	0.061	0.114	0.020	0.117	0.012	0.110	-0.654*	0.354
Referrer leaves (1 = yes)			-0.279*	0.159	-0.275*	0.155	-0.281*	0.157	0.966*	0.504
Gender (1 = male)					0.105	0.147			0.271	0.340
Repeat application (1 = yes)					-0.220	0.192	-0.174	0.172	0.537	0.562
Computer					-0.065	0.139			0.105	0.452
Language					-0.171	0.156	-0.162	0.138	0.537	0.445
Years of education					0.013	0.039			0.051	0.092
Works at time of application					-0.007	0.131			-0.078	0.362
Months of bank experience					-0.002	0.004			0.019***	0.005
Months of non-bank experience					-0.005*	0.002	-0.004***	0.001	-0.009†	0.005
Non-bank experience, squared					0.000	0.000			0.000*	0.000
Months of customer service					0.004***	0.001	0.003***	0.001	0.004	0.004
Number of previous jobs					-0.015	0.051			-0.186	0.160
Tenure in last job (in days)					0.000	0.000			0.000	0.000
Salary in last job					0.022	0.019			-0.053	0.049
Number of applications					0.006*	0.003	0.006*	0.003	0.001	0.013
Number of openings					-0.001	0.006			-0.013	0.019
Turnover hazard rate[b]	1.767†	1.199	1.746†	1.204	2.076†	1.221	1.644†	1.199		
Wald χ^2 statistic	3298***		2098***		3754***		3713***		41.33***	
$p > \chi^2$	0.000		0.000		0.000		0.000		0.001	
Person–month observations (employees)	2983 (257)		2983 (257)		2983 (257)		2983 (257)		3188 (260)	

[a] Performance is measured as the average number of calls answered per hour (quality adjusted).
[b] The turnover hazard rate is estimated from the turnover Cox Survival Model reported in the table.
†$p < 0.1$ *$p < 0.05$ **$p < 0.01$ ***$p < 0.001$ (two-tailed tests; except the z-test for the effect of the "external referral" variable which is a one-tailed test).

This finding suggests that the effects of referral ties continue beyond the hiring process, having later effects not only on attachment to the firm but also on performance. The model shows that there are no statistically significant differences in productivity trajectories between non-referrals and referrals whose referrer stays in the organization. I find, however, that the "breaking of the tie" between referrer and referral has important negative consequences for the productivity of the referred employee—to the extent that referrals perform worse than non-referred employees if their referrer leaves the organization. In the last column of Table 5.5a, I report the event history analysis results of the turnover process. Now the referral variable appears to have a significant negative impact on the likelihood of an employee leaving the organization. However, referrals whose referrer has left show a lower survival rate than referrals whose referrer has stayed.

Hypothesis 3a is about examining the impact of the exit of the referrer on the referral's level of performance. Assessing the social enrichment effect also requires analyzing whether the presence of the referrer is what improves referrals' performance (Hypothesis 3b). I was unable to test whether the presence of the referrer during training accounts for any initial performance advantage of referrals in comparison with non-referrals. By the time referred hires complete their training, most of their referrers are still with the company, and therefore the referral variable and the referrer presence variable are almost identical.[81] I can, however, test whether the referrer can help the referral along a quicker performance improvement trajectory. To test Hypothesis 3b, I run a similar model (to the one in Table 5.5a) including now the variables "referrer stays" and "referrer leaves." The results of such analyses are displayed on Table 5.5b. Since the effects of the control variables on performance are almost identical to the effects reported in Table 5.5a, those effects are omitted in Table 5.5b. The referrer's continued presence does not seem to help the referral along a quicker performance improvement trajectory.

As some evidence supporting the social enrichment in this Phone Center, I learned from my interviews with PCHR personnel about what they thought were important post-hire determinants of retention and productivity. In a high pressure, highly structured environment where CSRs are answering up to 5000 phone calls a month and where work is closely scrutinized, the PCHR staff have indeed thought about the potential benefits of the social enrichment process. As mentioned earlier, one trainer claimed: "People leave their jobs because of the working environment. The job burns you out!" Thus, it is not surprising that prior to my study, the staff had been considering introducing a formal "buddy" system, where long-time employees would be paired with new hires as a means of reducing turnover and increasing job satisfaction. The underlying theory is

---

[81] Only 11 referrers left the organization during the training period of their respective referrals.

Table 5.5b
The presence of the referrer in predicting the performance of the referral

	Coefficient	SE	Turnover Rate Model (Cox Regression Model)	
			Coefficient	SE
Main model				
Constant	6.149***	0.546		
Performance, month T-1[a]	0.748***	0.015	0.015	0.048
Tenure in months	−0.058***	0.019		
Referrer stays (1 = yes)	0.020	0.117	−0.654*	0.354
Referrer leaves (1 = yes)	−0.275*	0.155	0.966*	0.504
Control variables[b]				
Turnover hazard rate[c]	2.076[†]	1.221		
Wald χ^2 statistic	3753.50***		41.33***	
$p > \chi^2$	0.000		0.001	
Person–month observations (employees)	2983 (257)		3188 (260)	

[a] Performance is measured as the average number of calls answered per hour (quality adjusted).
[b] The effects of the control variables on performance are very similar to those effects reported in Table 5.5a. They are not reported in this table.
[c] The turnover hazard rate is estimated from the turnover Cox Survival Model reported in the table.
[†]$p < 0.1$ *$p < 0.05$ **$p < 0.01$ ***$p < 0.001$ (two-tailed tests; except the z-test for the effect of the "referrer stays" and "referrer leaves" variables which are one-tailed tests).

identical to that of the social enrichment process. One trainer highlighted the importance of having a friend on the job: "It really helps in making the job more comfortable." This trainer did not specifically say that the referrer might play an important role in this process. But when probed about the possibility that the referrer might be acting as an informal "buddy," she said that this was "probably right" (Fernandez *et al.*, 2000). Similarly, one hiring manager pointed out that "I once had two employees who were dating each other come together to let me know about their decision to leave the floor." When I probed about the possibility of them being referral and referrer, she said that this was "very possible given that many employees refer relatives and friends for the CSR job."

Finally, I also test whether referrers' characteristics and level of performance influence the performance trajectory of the referred employee. Again, adding the information about the productivity of the referrer who leaves the Phone Center does not improve the fit of any performance growth model.[82]

---

[82] Once again, referrers' characteristics do not seem to predict or provide additional information about the future performance of the referral ($p < 0.620$; incremental $\chi^2 = 2.64$; df = 4); nor does information about the referrer's evaluation by the firm ($p < 0.308$; incremental $\chi^2 = 1.04$; df = 1). Results are available in Castilla (2002).

1.5.4. Limitations and Future Research

This paper expands the scope of previous recruitment source studies by analyzing unique objective measures of performance. As with the results of any study
based on one sample drawn from a single organization, caution needs to be exer-
cised when generalizing the results of this study. Thus, I believe that this research
can be extended in several interesting directions. The first and most obvious area
to be explored involves developing studies to continue testing the social integration and embeddedness arguments in more comprehensive and detailed ways.
I have explored whether the effect of referral ties on employees' performance
continues beyond the hiring process. A dummy variable for whether the respondent is a referral is included as a network variable. I focus on referrals as a
matter of necessity. In particular, one can believe in social enrichment without
believing that it has to come through a referral process. After all, once an
employee enters the organization, she presumably develops a much wider
network of employee connections and friendships. Studies need to look at the
social enrichment process by including dynamic information about employees'
multiple networks before and after they come to work at an organization. It is
important to collect information about new workplace connections and personal relationships as they evolve over employees' tenure on the job. Additional
research dealing with multiple ties over time (and their effect on individual
performance) is much needed.

The second extension is closely related to the previous one. Although the
current and previous studies have documented important recruitment source
differences in several post-hire indicators of employees' better matches, most
of them exclusively focus on the bilateral nature of ties and their consequences. In my study, I only look at how the tie between two actors (referrer/
referral) affects the performance for one of these actors. One point often
emphasized in the "post-hire interdependence" literature, however, is that the
social relationships that affect outcomes are decidedly multilateral. In this
regard, more research is required to understand how multilateral networks
affect outcomes and how an employee's degree of integration into multilateral
social networks might itself be affected by the referrer (perhaps by the referrer's status or simply by the referrer's position in the formal or informal
network of the workplace). Without studies that collect and analyze data on
more complex relationships, many of the intervening mechanisms are left
open to speculation.

As a sociologist, my own bias is to be more interested in the "post-hire interdependence in performance" aspect of this study. In order to better understand
how post-hire social integration lies at the root of any differential in performance, I suggest more qualitative data studies which can give us a better sense of

the mechanisms at work. Despite my limited presentation of what I learned from interviews and observation at the Phone Center, there is no way I could have captured these more complex relationships in my data since I did not have information about these social networks over time within the Phone Center. In the absence of such data, many of the intervening mechanisms will thus be left open to speculation for future research on the effects of post-hire personal interdependence on performance.

I do not directly address here the conditions that make network effects more or less important. Performance, turnover, and job satisfaction are much more salient in knowledge-intensive industries or jobs—or more generally, in industries that require skilled and highly paid labor, where firms are particularly concerned with attracting and retaining workers. Insofar as that is true, the present research on a specific lower-entry position at the Phone Center provides an especially stringent test of the hypotheses. The workplace that I chose to study does not involve particularly skilled workers, which makes its support of sociological hypotheses all the more remarkable; these are purported to be jobs in which network effects would be less salient. Future research should pay attention to the fact that the industry/job/firm level factors might create variation in the strength of the network effects that I am investigating. This is particularly relevant given the mixed results from past studies on turnover and performance. After all, the relationships between turnover or performance, and network referral may vary by industry, job, and even by location of the organization.

1.6. Discussion and Conclusion

Organizations frequently use referral programs to recruit new employees. Implicit in referral hiring is the assumption that social relationships among existing employees and potential employees will benefit the hiring organization. My paper provides a distinctive and substantive contribution to the literature of economic sociology and labor economics by evaluating the impact of referral ties on employee performance. For the first time in this tradition of research, I have been able to code and carefully analyze unique data on direct measures of employee productivity. As an immediate result, I have been capable of empirically distinguishing among the different theoretical mechanisms by which the hiring of new employees using referrals (as an indicator of pre-existing personal connections) influences employee productivity and turnover over time. The dominant argument in economics posits that personal contact hires perform better than isolated hires because social connections provide difficult to obtain and more realistic information about jobs and candidates. A more sociological explanation, however, emphasizes post-hire social processes that occur among socially connected employees. Employers may still hire referrals at a higher rate simply because personal connections help to *produce* more productive employees in the job. The social support and interactions that occur between referral and referrer can enhance an employee's productivity and attachment to the firm.

My study begins by testing the effectiveness of labor economic models of employee referrals. My findings in the analyses of initial performance suggest that referral programs provide important economic returns to the organization at the very beginning of the job contract: Not only do employees hired via referrals initially perform better than non-referrals but they are also more likely to complete their initial training in the organization. In the long run, however, referral programs do not seem to provide any economic returns: I provide robust evidence that employees hired via referrals do not perform any better than non-referrals over time. In addition, I show that information from a candidate's résumé does not seem to predict her initial productivity, subsequent productivity growth, or tenure on the job.

This study is a significant step forward for the literature on social networks and employment because it illustrates the benefits of including the *post-hire* dynamics of social relations to understand employee productivity over time. I argue that the finding that referrals perform better than non-referrals initially, but not over time, does not provide adequate support for the better-match argument. Instead I show that the explanation for better performance by referrals over non-referrals is sociological given that the initial performance differential seems to be due to the fact that the referrer is possibly around teaching the referral the particulars of the job at the very beginning of the job contract or during the training period or perhaps simply making the workplace a better place to work. Thus, I find that the presence of the referrer seems to be the factor that accounts for any initial performance differential between referrals and non-referrals.

Fig. 5.2 provides a stylized representation of these findings, charting the effects of social interdependence on employee performance growth curves (note that the figure is not a plot of my model parameters, though). Referrals do appear to be better performers than non-referrals right after their training. Any performance advantages of referrals over non-referrals disappear soon after their first month working as a CSR at the Phone Center; I found no significant difference in performance careers between referrals and non-referrals. This pattern of converging productivity for both groups of workers is very much related to the nature of the job selected for this study. Like many lower-entry jobs, working as a phone customer representative is a job with a "hard" performance ceiling. Employees hired via referrals peaked out earlier and were later caught up by

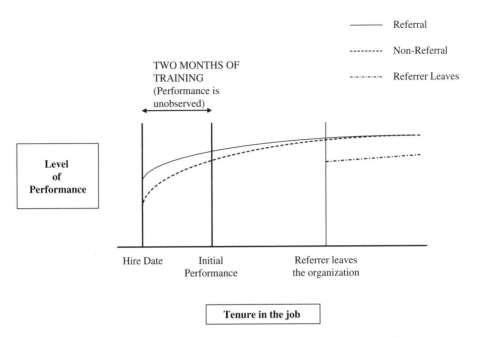

Fig. 5.2 Summary graph of the effects of social interdependence on performance growth curves.

those hired by non-referrals. The performance improvement mostly occurs during the first three months at the job. The performance of the referral, however, is negatively affected by the referrer's departure at any point in time, even when the referral has been in the organization for a while. As shown in this figure, the leaving of the referrer results in a significant predicted parallel downward shift of the performance curve.[83] This finding is of interest because an important predictor of performance evolution over time appears to be the referrer's continued presence in the firm. It is the "breaking of the tie" between referrer and referral that seems to have serious negative implications for the referral employee's productivity over time.

This paper clearly demonstrates that the role of social connections goes beyond the screening and hiring of employees and is an important tool for understanding the dynamics of employee outcomes such as turnover, job satisfaction,

---

[83] There is only an effect if the referrer left. As an anonymous reviewer pointed out, this may be due to the fact that CSRs reach very high-levels performance during the first few months of employment and the only way to go is down. However, I still find that for those referrals whose referrer leaves, this "going down" is more pronounced than for those whose referrer stays (or even for those who were not referrals).

and performance. For this reason, researchers should not restrict their study of employee networks to examining whether employees are connected at a given point in time. The ability to develop prominent personal connections, whether "weak" or "strong," may not be the central determining factor of individual careers (Granovetter, 1973). On the contrary, the main argument in this study is that individual career outcomes may depend more on whether those contacts are present in the right place at the right time. In this sense, I suggest bringing back the original notion of embeddedness (as in Granovetter, 1985) and to conceptualize "embeddedness" as an ongoing set of social relationships whereby employees come to organizations with pre-existing social ties (some of which vanish and some of which remain over time). Once employees get hired, they develop new sets of connections that prevail when they move on to other organizations. Influential and knowledgeable contacts thus depend on one's past mobility; in turn, these contacts influence one's future career moves.

My research indicates that the presence of certain contacts at the workplace (and especially their departure) can influence one's productivity and attachment to the firm. In order to gain a better understanding of the mechanisms by which social relations matter in the hiring and post-hiring of employees in organizations, future work in this field needs to consider the dynamics of these interactions of employees with both pre-existing contacts as well as in new personal relationships developed at the workplace, and should use detailed longitudinal data such as those I have analyzed here. At a more general level, this type of research should continue facilitating necessary dialogue between economic and sociological theories (Baron and Hannan, 1994). I also believe that such work is essential for expanding (and further clarifying) our current knowledge about the impact of social relations on economic behavior and outcomes.

Acknowledgments

I am grateful for the financial support provided by the Social Sciences Research Council (Program of the Corporation as a Social Institution). I have benefited enormously from the extensive and detailed comments of Roberto M. Fernandez, Mark Granovetter, and John W. Meyer. I thank Robert Freeland, Ezra Zuckerman, and Dick Scott for their wonderful suggestions on earlier versions of this paper. I also thank my colleagues in the Management Department at Wharton, especially Mauro F. Guillén, Anne-Marie Knott, Lori Rosenkopf, Nancy Rothbard, Christophe Van den Bulte, and Mark Zbaracki, and all the attendants of the M-square seminar for their comments on earlier drafts. I am also extremely thankful to the entire Fernandez family for all their love and support.

References

Arrow, K.J., 1962, The economic implications of learning by doing, *The Review of Economic Studies* 29 (3), 155–173.

Baron, J.N. and Hannan, M.T., 1994, The impact of economics on contemporary sociology, *Journal of Economic Literature* 32, 1111–1146.

Bartel, A.P. and Borjas, G.J., 1981, Wage growth and job turnover: an empirical analysis. In: *Studies in Labor Markets* (S. Rosen, Ed.), National Bureau of Economic Research, The University of Chicago Press, Chicago, IL, Volume 31, pp. 65–90.

Bassett, G.A., 1967, *Employee Turnover Measurement and Human Resources Accounting*, Personnel and Industrial Relations Services. General Electric, Crotonville.

Bassett, G.A., 1972, Employee turnover measurement and human resources accounting, *Human Resource Management* Fall: 21–30.

Berk, R.A., 1983, An introduction to sample selection bias in sociological data, *American Sociological Review* 48, 386–398.

Breaugh, J.A., 1981, Relationships between recruiting sources and employee performance, absenteeism, and work attitudes, *Academy of Management Journal* 24, 142–147.

Breaugh, J.A. and Mann, R.B., 1984, Recruiting source effects: a test of two alternative explanations, *Journal of Occupational Psychology* 57, 261–267.

Caldwell, D.F. and Spivey, W.A., 1983, The relationship between recruiting source and employee success: an analysis by race, *Personnel Psychology* 36, 67–72.

Castilla, E.J., 2002, Social networks and employee performance, Unpublished Ph.D. thesis, Stanford University.

Corcoran, M., Datcher, L. and Duncan, G.J., 1980, Information and influence networks in labor markets. In: *Five Thousand American Families: Patterns of Economic Progress* (G.J. Duncan and J.N. Morgan, Eds.), Institute for Social Research, University of Michigan, Ann Arbor, MI, Volume VIII, pp. 1–37.

Cox, D.R., 1972, Regression models and life tables (with discussion). *Journal of the Royal Statistical Society, Series B* 34 (2), 187–220.

Cox, D.R., 1975, Partial likelihood, *Biometrika* 62 (2), 269–276.

Dalton, D.R. and Todor, W.D., 1979, Turnover turned over: an expanded and positive perspective, *Academy of Management Review* 4, 225–235.

Datcher, L., 1983., The impact of informal networks on quit behavior, *The Review of Economics and Statistics* 65, 491–495.

Decker, P.J. and Cornelius, E.T., 1979, A note on recruiting sources and job survival rates, *Journal of Applied Psychology* 64, 463–464.

Fernandez, R.M. and Castilla, E.J., 2001, How much is that network worth? Social capital returns for referring prospective hires. In: *Social Capital: Theory and Research* (K. Cook, N. Lin and R. Burt Eds.), Aldine de Gruyter, Hawthorne, NY, pp. 85–104.

Fernandez, R.M., Castilla, E.J. and Moore, P., 2000, Social capital at work: networks and employment at a phone center, *American Journal of Sociology* 105 (5), 1288–1356.

Fernandez, R.M. and Weinberg, N., 1997, Sifting and sorting: personal contacts and hiring in a retail bank, *American Sociological Review* 62, 883–902.

Gannon, M.J., 1971, Source of referral and employee turnover, *Journal of Applied Psychology* 55, 226–228.

Granovetter, M., 1973, The strength of weak ties, *American Journal of Sociology* 78, 1360–1380.

Granovetter, M., 1985, Economic action and social structure: the problem of embeddedness, *American Journal of Sociology* 91, 481–510.

Granovetter, M., 1995, *Getting a Job: A Study of Contacts and Careers*, 2nd ed., Harvard University Press, Cambridge, MA.

Gronau, R., 1974, Wage comparisons: a selectivity bias, *Journal of Political Economy* 82, 1119–1155.

Heckman, J.J., 1976, Life-cycle model of earnings, learning and consumption, *Journal of Political Economy* 84 (4), 11–44.

Heckman, J.J., 1979, Sample selection bias as a specification error, *Econometrica* 45, 153–161.

Jones, S.R.G., 1990, Worker interdependence and output: the Hawthorne studies reevaluated, *American Sociological Review* 55, 176–190.

Jovanovic, B., 1979, Job matching and the theory of turnover, *Journal of Political Economy* 87 (5), 972–990.

Krackhardt, D. and Porter, L.W., 1985, When friends leave: a structural analysis of the relationship between turnover and stayers' attitudes, *Administrative Science Quarterly* 30 (2), 242–261.

Lee, L-F., 1979, Identification and estimation in binary choice models with limited (censored) dependent variables, *Econometrica* 47, 977–996.

Lee, L-F., 1983, Notes and comments: generalized econometric models with selectivity, *Econometrica* 51 (2), 507–513.

Lee, L-F. and Maddala, G.S., 1985, Sequential selection rules and selectivity in discrete choice econometric models, *Econometric Methods and Applications* 2, 311–329.

Lee, L-F., Maddala, G.S. and Trost, R.P., 1980, Asymptotic covariance matrices of two-stage probit and two-stage tobit methods for simultaneous equations models with selectivity, *Econometrica* 48, 491–503.

Lewis, H., 1974, Comments on selectivity biases in wage comparisons, *Journal of Political Economy* 82, 1119–1155.

Liang, K-Y. and Zeger, S.L., 1986, Longitudinal data analysis using generalized linear models, *Biometrika* 73, 13–22.

Lin, N., 1999, Social networks and status attainment, *Annual Review of Sociology* 25, 467–487.

Marsden, P.V. and Hurlbert, J.S., 1988, Social resources and mobility outcomes: a replication and extension, *Social Forces* 66, 1038–1059.

Martin, T.N., Price, J.L. and Mueller, C.W., 1981, Job performance and turnover, *Journal of Applied Psychology* 66, 116–119.

Medoff, J.L. and Abraham, K.G., 1980, Experience, performance, and earnings, *The Quarterly Review of Economics* XCV, 703–736.

Medoff, J.L. and Abraham, K.G., 1981, Are those paid more really more productive? The case of experience, *The Journal of Human Resources* 16, 186–216.

Mobley, W.H., 1980, The uniform guidelines on employee selection procedures: a retrait from reason? *Business and Economic Review* 26, 8–11.

Mobley, W.H., 1982, *Employee Turnover: Causes, Consequences, and Control*, Addison-Wesley Publishing Company, Reading, MA.

Montgomery, J.D., 1991, Social networks and labor-market outcomes: toward an economic analysis, *The American Economic Review* 81, 1408–1418.

Petersen, T., Saporta, I. and Seidel, M-D., 2000, Offering a job: meritocracy and social networks, *American Journal of Sociology* 106 (3), 763–816.

Porter, L.W. and Steers, R.M., 1973, Organizational, work, and personal factors in employee turnover and absenteeism, *Psychological Bulletin* 80, 151–176.

Price, J.L., 1977, *The Study of Turnover*, Iowa State University Press, Ames, IA.

Quaglieri, P.L., 1982, A note on variations in recruiting information obtained through different sources, *Journal of Occupational Psychology* 55, 53–55.

Seybol, J.W., Pavett, C. and Walker, D.D., 1978, Turnover among nurses: it can be managed, *Journal of Nursing Administration* 9, 4–9.

Sicilian, P., 1995, Employer search and worker-firm match quality, *The Quarterly Review of Economics and Finance* 35, 515–532.

Simon, C.J. and Warner, J.T., 1992, Matchmaker, matchmaker: the effect of old boy networks on job match quality, earnings, and tenure, *Journal of Labor Economics* 10 (3), 306–329.

Staw, B.M., 1980, The consequences of turnover, *Journal of Occupational Behavior* 1, 253–273.

Stolzenberg, R.M. and Relles, D.A., 1997, Tools for intuition about sample selection bias and its correction, *American Sociological Review* 62 (3), 494–507.

Swaroff, P., Barclay, L. and Bass, A., 1985, Recruiting sources: another look, *Journal of Applied Psychology* 70 (4), 720–728.

Taylor, S.M. and Schmidt, D.W., 1983, A process oriented investigation of recruitment source effectiveness, *Personnel Psychology* 36, 343–354.

Tuma, N.B., 1976, Rewards, resources, and the rate of mobility: a non-stationary multivariate stochastic model, *American Sociological Review* 41, 338–330.

Ullman, J.C., 1966, Employee referrals: prime tool for recruiting workers, *Personnel* 43, 30–35.

Wanous, J.P., 1978, Realistic job previews: can a procedure to reduce turnover also influence the relationship between abilities and performance, *Personnel Psychology* 31, 249–258.

Wanous, J.P., 1980, *Organizational Entry: Recruitment, Selection, and Socialization of Newcomers*, Addison-Wesley Publishing Company, Reading, MA.

Winship, C. and Mare, R.D., 1992, Models for sample selection bias, *Annual Review of Sociology* 18, 327–350.

2. Institutional Aspects of National Science Activity (Paper 2)

Abstract: In this study, I examine the institutional factors influencing national science activity during the 20th century in the world. In particular, I investigate when and under which conditions countries join any of the unions comprising the ICSU, the primary and oldest international science institution in the world. According to the dynamic analysis of historical data for 166 countries, institutional theories have the greatest power to predict the rate at which nation-states join scientific organizations. The joining rate increases more quickly during the post–World War II era with the rise of the world-system. In addition, as the number of ICSU Unions' memberships of a country increases over time, the hazard rate increases at a decreasing rate. By contrast, development and modernization arguments do not seem to offer significant explanations of the evolution of the ICSU membership. These functional arguments are of importance early in the "science diffusion" process. After 1945, institutional factors better account for worldwide national science activity. Although core countries have higher joining rates than peripheral countries, there is evidence of convergence in the evolution of national joining rates over time. The results also demonstrate substantial invariability in the impact of functional and institutional factors for core and peripheral countries.

2.1. Introduction

The International Council of Scientific Unions (henceforth ICSU) is the oldest and primary existing international science organization committed to international scientific cooperation. Created in 1931 after its predecessor, the International Research Council (1919), ICSU's main goal has been to "promote international scientific activity in the different branches of science and their applications for the benefit of the humanity" (ICSU, 1995: p. 1). ICSU is a non-governmental body with two categories of membership: research councils that are national, multidisciplinary bodies (92 national academies of science), and Scientific Unions, which are international, disciplinary organizations (23 International Scientific Unions). Four objectives justify the existence and functioning of the ICSU. First, it initiates, designs, and coordinates major international, interdisciplinary research programs. Second, the ICSU creates interdisciplinary bodies, which undertake activities and research programs of interest to several member bodies. In addition to these programs, several bodies set up within the ICSU address matters of common concern to all scientists, such as capacity building in science, data, science and technology in developing countries, and ethics and freedom in the conduct of science. Third, the Council also acts as a focus for the exchange of ideas, the communication of scientific information, and the development of scientific standards. Finally, the ICSU also assists in the creation of international and regional networks of scientists with similar interests. Thus, ICSU maintains close working relations with a number of intergovernmental organizations, in particular with UNESCO. Because ICSU is in contact, through its membership, with hundreds of thousands of scientists worldwide, it is being increasingly called upon "to act as the spokesman for the world scientific community and as an adviser in matters ranging from ethics to the environment" (ICSU, 1995: p. 4).

In 1931 there were 38 nation members (193 memberships to the different ICSU scientific unions). The earliest memberships occurred at the beginning of the 20th century. Over 22% of them were created before 1935. Over 75% were created after 1945. Fig. 5.3 shows how the cumulative number of ICSU Unions' membership grew over the 1919–1990 period. The graph clearly demonstrates that the joining of countries to ICSU Unions began early in the century, and steadily increased during the post–II World War era. Thus, in 1991 the ICSU comprised 20 international Scientific Unions and 75 national members, including 8 associates (702 ICSU Unions' memberships in total).

In this study, I examine some of the factors affecting the rate at which nation-states enter into any ICSU Scientific Union over the 1919–1990 period. In other

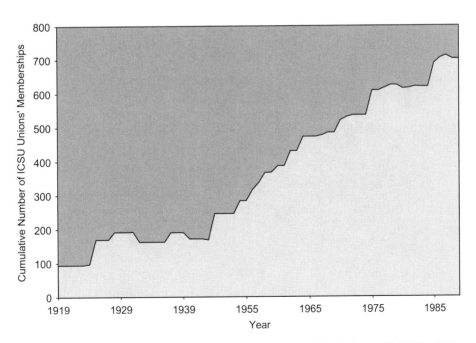

Fig. 5.3　Cumulative number of ICSU Unions' memberships in the world (1919–1990).

words, I investigate when and under which conditions nation-states are more likely to join any of the unions of the ICSU, the primary and oldest international science organization in the world. Using historical and quantitative data, I develop an event history analysis to account for repeating events. I seek to identify and test two major sets of theoretical arguments concerning the expansion of ICSU Unions' memberships over time. First, I want to test variations in the impact of socio-economic development on joining rates across time. Second, I explore the role of worldwide system's linkages and discourse in the process of national adherence to ICSU Unions. Attention is restricted to the effect of (1) the degree of economic openness of a country to the world (i.e., economic linkages) and (2) the degree of rise and linkages to the worldwide system, specifically the expansion of the international discourse and the proliferation of a "scientific consciousness" (institutional linkages). In addition, I control for the effect of previous ICSU Unions' memberships of a country on its joining rate. As the number of ICSU Unions' memberships of a country might increase over time, the hazard rate at which a nation-state joins an additional ICSU Union might increase at a decreasing rate.

2.2.   The International Council of Scientific Unions

The ICSU, according to its founding statutes, is an international non-governmental and non-profit making scientific organization. The principal objectives of ICSU are: (1) to encourage and promote international scientific and technological activity for the benefit and well-being of humanity; (2) to facilitate coordination of the activities of the Scientific Union Members (see Statute 7); (3) to facilitate coordination of the international scientific activities of its National Scientific Members (see Statute 8); (4) to stimulate, design, coordinate, and participate in the implementation of international interdisciplinary scientific programs; (5) to act as a consultative body on scientific issues that have an international dimension; and (6) to engage in any related activities. Each nation-state member has the obligation to support the objectives of ICSU and meet its financial obligations as appropriate. Members may adhere to ICSU in one of two categories (what has been called the "principle of dual membership"): (a) Scientific Union Members and (b) National Scientific Members.

In 1993, the ICSU comprised 84 national members and 20 International Scientific Unions, which "provide a wide spectrum of scientific expertise enabling members to address major international, interdisciplinary issues which none could handle alone" (*Science International* 51, 1993). The scientific unions cover a wide range of disciplines, including all four major branches of basic science: mathematics, physics, chemistry, and biology.

2.3 Hypotheses

The existence and functioning of the ICSU provides an interesting object of analysis because the process by which countries join any of its unions can be used to test several macro sociological theories. In this particular study, I examine the propensity of nation-states to join any ICSU Scientific Union over the 1919–1990 period. The event for the purpose of this paper is a country's entry to any of the ICSU Scientific Unions. This leads to an analysis of repeatable events over time. Based on relevant prior theories and research, I would like to test the following three set of theoretical arguments: (1) development and modernization arguments; (2) institutional arguments; and (3) time period and historical effects.

2.3.1. Development Arguments

According to the development and modernization arguments, more developed countries that have already attained a considerable level of development in science and technology are more likely to join ICSU Scientific Unions, especially, at an earlier stage of diffusion. The following are testable hypotheses regarding development:

Hypothesis 1.1. *Socio-economic development increases the rate at which countries join ICSU Scientific Unions.*

Hypothesis 1.2. *The effects of socio-economic development on joining hazard rates should increase over time.*

Hypothesis 1.3. *More developed countries (core countries) are more likely to join ICSU Unions than peripheral countries.*

However, one should expect certain convergence in the evolution of joining rates over time. In other words, the difference in the likelihood at which developed versus developing countries enter into ICSU Unions should decrease over time.

2.3.2. Institutional Arguments

Following institutional approaches, as the world-system rises and expands after 1945, nation-states tend to behave isomorphically, that is, they tend to be influenced by what other nation-states are doing, as well as by the cultural rules and discourse emerging from communities of sciences and professions and international governmental and non-governmental organizations. According to Meyer (1994) and many other institutionalist theorists (Thomas *et al.*, 1987), the world society is made up of not only nation-states but also of international organizations, particularly in the sciences and professions, which tell nation-states what to do to become legitimate. Institutionalists claim that not only new nation-states whose internal sovereignty has decreased but also others are less able to exercise their authority in a way that they independently choose. The authority of nation-states needs to follow certain international "rules of the game" to have the "approval" of the world-system (i.e., regimes that determine how international exchange has to take place, the adequate form of military actions, and other state obligations). In this sense, membership in the international society means that nation-states in the world both influence and take advantage of the international rules and ideologies concerning what is desirable and undesirable in the relations and actions of states. It can be concluded that today there is not only a higher economic interdependence of countries, but also a greater interdependence in terms of reciprocal legitimization and in terms of dependence on common organizations and rules. The following are a set of hypotheses regarding institutional arguments:

Hypothesis 2.1. *The joining rate increases more quickly during the post–World War II era.*[84]

Hypothesis 2.2. *The degree of linkage of a country to the world-system (as measured by the cumulative number of national memberships in ICSU Unions) should have a positive effect on the rate at which a country joins an ICSU Union.*[85]

Hypothesis 2.3. *As the number of ICSU Unions' memberships of a country increases over time, the hazard rate at which a nation-state joins an additional ICSU Union increases at a decreasing rate.*[86]

Hypothesis 2.4. *The degree of economic linkages to the world-system (as measured by the amount of imports and exports as a percentage of GNP) should have a positive effect on the rate of joining any ICSU Scientific Union.*[87]

In this line of thought, the central argument of institutionalists is that the nation-state system is given worldwide support and legitimacy. Interests and

---

[84] In 1945, the United Nations is created and it is thought to legitimate nation-states and transmit the international rules and recommendations concerning appropriate state structures and actions.

[85] So that as the number of ICSU Unions' memberships of a country increases over time, the joining rate also increases.

[86] In other words, one should expect an inverted U-shaped relationship between the rate at which countries enter into ICSU Scientific Unions and the cumulative number of Unions' membership of a country.

[87] That is, the higher the amount of imports and exports as a percentage of GNP, the more likely a nation-state is to enter into any ICSU Scientific Union.

concerns of the state are increasingly determined by the cultural content of the world polity (i.e., world definitions of the justifications, perspectives, purposes, aims, and policies to be pursued by nation-states). Thus, we witness an increase in the state's responsibilities and obligations along with a decrease in internal sovereignty. Both processes take place simultaneously. On the one hand, the world-system confers increasing powers on states "to control and organize societies politically around the values established in the world political culture" (Meyer, 1980: p. 110). These powers are ideological, cultural, and political reflections of the logic of the world economy.

At the same time, modern nation-states' organizations should be studied as similar social actors under exogenous universal cultural processes. Increasingly, countries are structurally similar and change isomorphically. As world conditions have changed, states have in many respects tended to become more similar despite the internal differences that may have (Grew, 1984).[88] Thus, nation-states do many similar things like ensuring mass education, respecting human rights, taking environmental actions, etc. Indeed, states are expected to be responsible for those actions. Nation-states are driven by pressures and incentives, as well as changes in the world-system. They expand and become more similar in the functions they perform (what Grew calls the "universal standards of the state").

The modern cultural framework comes from an institutionalized system composed of many international organizations and professional associations. Especially in the last two decades, this group of international governmental organizations and international non-governmental organizations, as well as professional groups, has come to define a set of rules and conventions governing state interests at the international level. These organizations have started to outline the new international economic and political order. Multinationals working around the world tend to support these international recipes and principles of management, especially those relating to the definition of people as human capital. In addition, there are the powerful and influential scientific communities of the world. The world-system expands, telling states what to do and not to do. It does not force states to do it because sovereignty still rests at the state level. However, this "global international society" is unlimited in its ability to define actions and decisions at the nation level.

There is an enormous pressure on countries, which do not follow international recommendations. Sanctions can be imposed to them and, most importantly,

---

[88] Activities and operations undertaken by certain leading economies are progressively copied and mastered by other economies. Thus, the United States and Japan have provided models to be copied by other countries throughout the century. However, in the modern period, Meyer and Strang (1993) claim that countries are less likely to copy successes directly rather than to copy them as institutionalized and interpreted by the world-system.

their access to world-system facilities (e.g., participation in international forums and organizations, eligibility for socio-economic aid, and so forth) might consequently be denied. In sum, what all these structures of the world polity produce is a great deal of powerful talk rather than binding authoritative action— i.e., scientific talk, legal talk, non-binding legislation, normative talk, talk about social problems, suggestions, advice, consulting talk, and so on (Brusson, 1989, cited in Meyer, 1994: p. 9).[89] Then, nation-states are supposed to put into decision and action the policies proposed in the international talk.

> **Hypothesis 2.5.** *The rate at which countries enter into ICSU Unions should increase over historical time with the increase in the dimensions of the world-system (as measured by the number of international governmental organizations in the world).*[90]
>
> **Hypothesis 2.6.** *As more states, whether hegemonic or peripheral, join ICSU Scientific Unions, other nation-states imitate the behavior in order to legitimate themselves.*[91]

The purpose of this study is to examine the nation-state adherence to ICSU unions over the 1919–1990 period. Specifically, I examine the effects of development and institutionalization processes on the rate at which a country joins any union using event history data. Event history models can be used to estimate the likelihood that a nation-state at risk of joining a union will join another union at a given time. The unit of analysis for the study is the nation-state (i.e., a sample of 166 nation-states), and the event of joining a union is a recurring event. The dimension of time used in the study is historical time (i.e., 1919–1990). At any given time, a nation may or may not join any of the ICSU unions.

2.4. Description of the Data

The data to be analyzed in this paper are mainly obtained from a cross-national cross-time data set. This set provides information for the period 1815–1990 by single years and by each country. It contains over 250 variables with data on approximately 173 nation-states and colonies. Primary sources include The Arthur

---

[89] A good example is the world's scientific communities which produce a very powerful talk about environmental regulations, about failures and requirements in national development policies, and about the protection of human rights.

[90] The reason is that the international talk on science and development produced by INGOs and the increasing involvement of nation-states in such talk should increase the likelihood of a country joining any of the ICSU Scientific Unions.

[91] In other words, the rate at which countries adhere to ICSU Scientific Unions should increase with the rise in the cumulative number of ICSU Scientific Unions' memberships in existence in the world over time.

Banks Cross-National Time Series Data Archive (1986), Gurr's Polity Persistence and Change Data (1978), The World Bank World Tables (1990), The World Handbook of Social and Political Indicators, various editions of the Yearbook of International Associations, and several UNESCO Statistical Yearbooks.

The sample analyzed here includes 166 countries, after deleting cases with missing information on the dependent variable. The data file analyzed contains multiple records for each of the 166 nation-states—specifically, one for each year. Since the event (i.e., a country's adherence to any of the ICSU Scientific Unions) is repeatable over time, information is available on a total of 6729 years of data for 166 countries.[92] The data begin at the date of independence of each country, or in the case of old countries in 1919 when the International Research Council (the predecessor of the ICSU) was created. The data ends in 1990, the final point at which data were available. There are 713 events, 222 involving core countries and 491 involving peripheral countries. The sample of countries includes 21 core countries and 145 peripheral countries.

The variables have been selected on the basis of (1) relevance to the description of the major factors considered in this analysis; (2) judgment related to the degree of homogeneity sufficient to allow cross-country comparisons; and (3) availability of time-series observations in most of the countries. A detailed description of each variable is provided in the following section.

2.4.1. Measures and Concepts

For the purpose of this study, I assume that an event occurs when a nation-state joins one or more ICSU Unions in a given year.[93] The measure was obtained from the 1995 edition of the Yearbook of the International Council of Scientific Unions published by the ICSU Secretariat in Paris.

The development hypotheses are tested by using *gross national income per capita* as measured by the natural logarithm of GDP per capita for every year. Although this variable is not excessively appropriate for measuring national socio-economic development, it was the best available measure recorded back to the 1919 for the majority of countries.[94] Two other indicators of the effects of

---

[92] Each of the 166 countries is taken as multiple observations over time.

[93] Data were available on the year when a country joined an ICSU Union. Therefore, it is impossible to know the exact date when the event took place (i.e., month and day). In addition, the fact that a country may join one or more unions in a given year should not be seen as different unrelated events. Instead they could be taken as one single event. This is because it is hard to believe that the processes of a country adhering to one or more Unions in a given year are completely independent from each other, and therefore the same explanatory factors may account for the "effort" of joining one, two or more unions in a given year. Nevertheless, I control for the number of ICSU memberships of a given country as an independent variable in the analyses.

[94] Some values are extrapolated using the linear extrapolation facility available in SPSS for Windows.

socio-economic development are dummy variables for least developed countries and new industrializing economies.

Three main indicators are used to operationalize institutional arguments: the cumulative number of ICSU unions' memberships in existence in the world (international-level variable); the cumulative number of national memberships in ICSU unions (national-level variable); and the number of international governmental and non-governmental organizations in the world (international-level variable). All variables are measured from 1919 to 1990.

The *number of ICSU Unions' memberships in existence in the world* is used to test to which extent nation-states are influenced by other nations' behavior relating science (what DiMaggio and Powell have called worldwide mimetic behavior). Values are recorded yearly from 1919 to 1989. Fig. 5.3 displays the cumulative number of memberships in the world, showing extensive activity well after 1945. Well before World War II, the cumulative number of ICSU Unions' memberships was around 180, with core countries holding around 70% of such memberships. By the late 1980s, the number had increased to over 700. The percentage of core countries holding ICSU memberships had decreased to 57%. The ICSU as the primary international science institutions contributes to the creation and maintenance of a world cultural discourse as well as normative pressures in the domains of science and development (see Thomas *et al.*, 1993).

The *cumulative number of national memberships in ICSU Unions* is used to test the degree to which a nation-state is linked to the world-system. Thus, it measures the extent to which each nation-state adheres to the world-level discourse on science. Fig. 5.4 shows the evolution of the average number of national ICSU Unions' memberships per year over the 1919–1990 period. As can be seen, the average number of national ICSU memberships increases from 1.62 in 1919 to 6.28 in 1989 when all countries are considered. The average number of scientific unions per country increases at a rate of almost 3 ICSU Unions' memberships per country per year.

In 1989, core countries belong to almost 20 scientific unions on average. This represents a 65% increase in the average number of memberships by core nations since 1960. However, the average number of scientific unions to which peripheral countries belong in the same year is 4.4, which represents an increase of 120% since 1960.

The pattern of membership growth for core and peripheral countries was the opposite before 1960. Thus, in 1920 the average number of memberships was 3.75 for core countries (20 core countries) and 0.4 for peripheral countries (36 peripheral countries). The average number of memberships increases at a rate of 2 additional memberships per year in the case of core countries versus a rate of almost 6 in the case of peripheral countries.

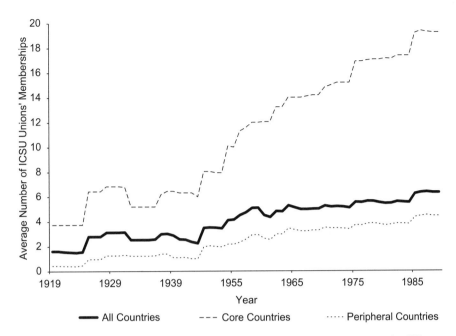

Fig. 5.4 Average number of ICSU Unions' memberships per country per year by different categories of countries (1919–1990).

In addition, there is a trend towards convergence in the average number of national memberships per nation-state. In other words, the number of memberships per country has recently tended to converge among countries. Thus, in the case of number of ICSU memberships for all countries, the coefficient of variation in the national number of memberships is 1.33 in 1920 and 1.27 in 1989. This trend is weak but it is in the direction of convergence. In the case of peripheral countries, the trend in the direction of convergence appears to be very weak—the coefficient of variation changes from 2.31 in 1920 to 1.48 in 1989. This averages out to a 0.5% decrease per year. On the contrary, in the case of core countries, the trend in the direction of convergence is stronger. Thus, the coefficient of variation is 0.56 in 1920 and 0.22 in 1989; this averages out to a 0.87% decrease per year.[95]

---

[95] In this paper, the degree of divergence is measured using the coefficient of variation (i.e., the ratio of the standard deviation to the mean, expressed as a percent) rather than the more common alternatives such as the standard deviation or variance because the coefficient of variation is adjusted for shifts in the mean. The greater the decrease in the coefficient of variation over time, the greater the level of convergence among countries. The mean convergence per year reported in this analysis is calculated as follows:

$$\text{Rate of convergence per year} = [(\text{CV}t_1 - \text{CV}t_2 / \text{CV}t_1] / (t_2 - t_1)$$

where $\text{CV}t_1$ = coefficient of variation at the earlier date, and $\text{CV}t_2$ = coefficient of variation at the later date.

Moreover, the *number of international governmental organizations in the world* gives an idea of the dimensions of the world-system as well as the impact of its discourse on nation-state behavior. These variables are obtained from the Yearbook of International Associations. Values are recorded yearly from 1919 to 1989. The number of IGOs increases exponentially over time, ranging from less than 30 organizations in 1919 to over 1500 by 1990. In the case of INGOs, the number grows enormously from almost 800 in 1920 to over 12000 by 1990. These organizations contribute to the creation and maintenance of a world cultural discourse as well as normative pressures in the domains of science and development (see Thomas *et al.*, 1993).

Economic linkages of countries to the world are measured here by the amount of *imports and exports as a percentage of GNP*. Thus, countries with relatively a high level of imports and exports are more connected to the world economy (what economists called economic openness).

2.5. Exploratory Analysis

Before proceeding to the multivariate methods of event history analysis, it is crucial to undertake exploratory analysis of the longitudinal data. A better understanding of the temporal nature of the data was obtained by estimating integrated hazard rates and hazard rates.[96] These functions are estimated for the full sample of countries. They are also estimated for two subgroups of countries created according to the economic position of a country in the world: core countries and peripheral countries.

The integrated or cumulative hazard rate indicates whether the hazard rate varies with time. Thus, if the hazard rate is constant over time, the integral of this rate over time should yield a straight line. Fig. 5.5 shows a plot of the Nelson–Aalen estimates of the integrated hazard rate at which nation-states adhere to any of the ICSU Scientific Unions versus historical time. The nonlinear relationship between the integrated hazard rate and time suggests that the hazard rate is time dependent, and therefore, a constant rate model may not be appropriate for the purposes of this analysis. Since the curve bends upward, then the likelihood of a country joining at least one of the unions increases over time. As the figure shows, the hazard rate changes relatively constant until around 1950. During the post–II World War era, the hazard rate varies over time. The plot partially supports Hypothesis 2.1 in the sense that the joining rate increases more quickly during the post–II World War era. In 1945, the United Nations is created and that is thought to legitimate nation-states and transmit the "blueprints" concerning state action.

---

[96] The survivor function was not estimated since it does not make sense in the case of repeatable events.

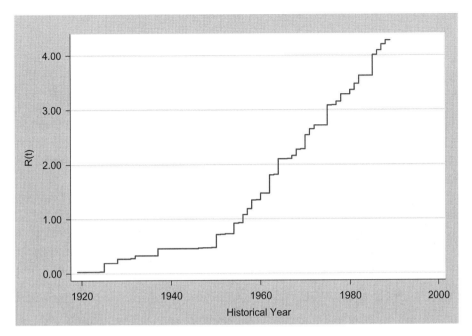

Fig. 5.5 Nelson–Aalen estimates of the integrated hazard rate at which nation-states join at least one ICSU Scientific Union over historical time (1919–1990).
Note: Number of events/countries: 713/166.

Fig. 5.6 presents the integrated hazard rates for core and peripheral countries. Like Fig. 5.5, this plot indicates that the hazard rate is time dependent. The slope of the integrated hazard rate is not constant for either group of countries; however, it varies much more for core countries than for peripheral countries. The slope of the integrated hazard rate for peripheral countries is approximately linear, especially prior to 1950. Although the hazard rate for these countries may not be time dependent, the low level of fluctuation may be in part due to the relatively small number of peripheral countries joining the ICSU Unions, especially prior to the 1945. Furthermore, the plot indicates that the rate of adoption throughout the entire period is higher for core nation-states.

More information is gained from examining the hazard rate itself. The hazard rate is similar to the probability that a nation-state joins any of the ICSU Scientific Unions per unit of time. I estimated the Nelson–Aalen estimates of the hazard rate at which nation-states adhere to any of the ICSU Scientific Unions over the 1919–1990 period for the full sample of countries (estimates are available upon request). Spurts of joinings appear between 1925 and 1930, and later between 1950 and 1977, and finally after 1982. The rate peaks around 1950, when about 51% of all the nation-states at risk (39 out of 76 nation-states) joined at least one ICSU Union. The hazard rate for the second highest peak occurs in 1962

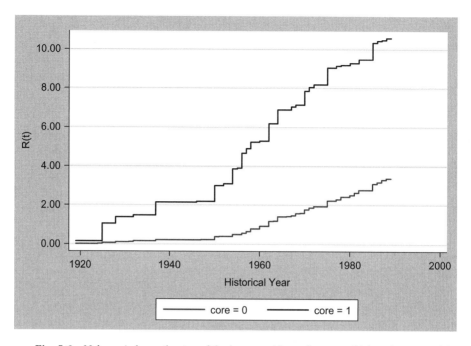

Fig. 5.6 Nelson–Aalen estimates of the integrated hazard rate at which nation-states join at least one ICSU Scientific Union over historical time by category of countries (1919–1990). *Note*: Number of events/countries: 222/21 (core countries); 491/145 (peripheral countries). Core countries (black line); peripheral countries (gray line).

(50% of all 111 nation-states at risk joined at least one ICSU Union). Finally, the third highest peak occurs in 1975, when about 40% of the nation-states at risk (61 out of 142 nation-states) adhere to at least one Union.

Fig. 5.7 shows the smoothed hazard rate over historical time. It shows that the hazard rate for all countries changes non-monotonically over time. The smoothed hazard rates are considerably smaller than the unsmoothed hazard rates at corresponding historical times, as smoothing levels out the "erratic" tendencies of the hazard rate. After smoothing, the hazard rate remains quite constant until 1945 period, when it rises sharply to more than half this rate in 10 years (i.e., from 1950 to 1960). The post-1960 period is characterized by a slight decline in the hazard rate, although it again rises increasingly after 1970, when new countries become independent and in addition, seven new ICSU Unions are created (2 in the 1970s; 2 more in 1982; and 3 more in 1993).[97]

---

[97] In 1993, three new ICSU Unions are created: The International Society of Soil Science (ISSS); the International Brain Research Organization (IBRO); and the International Union of Anthropological and Ethnological Sciences (IUAES). Since this study only covers the 1919–1990 period, these three Unions are excluded from the analysis. Therefore, the maximum number of Scientific Unions analyzed here is 21.

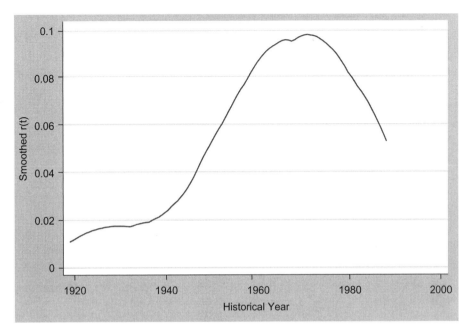

Fig. 5.7 Smoothed estimates of the hazard rate at which nation-states join at least one ICSU Scientific Union versus historical time (1919–1990).
Note: Number of events/countries: 713/166.

Institutional arguments seem to apply on a world-level to the joining of ICSU Scientific Unions, as theory predicts that the highest rates of joining will occur with the rise of the world-system in the post-war era. Thus, the formation of the United Nations in 1945 and the founding of UNESCO soon after contribute to the sharp rise in the hazard rate after 1950. A possible functional explanation of the burst of activity culminating in 1960 could be that this period was characterized by economic expansion and growth just after the World War II. The time-variant nature of the gross national income per capita used in this analysis will help in assessing this suggestion later.

A different picture is obtained when looking at two different categories of countries. Fig. 5.8 presents the hazard rates for core and peripheral countries (again smoothed to ease interpretation and comparison). The overall patterns of the two hazard rates have one similarity, namely that they change non-monotonically over time. On the one side, the hazard rate for core countries rises increasingly until 1960, when it drops slightly in a period of 30 years (i.e., the hazard rate is decreased out to 1% per year over the 1960–1990 period). On the other side, the hazard rate for peripheral countries also experiences one peak around 1960—it remains quite constant until 1945, when it rises sharply to more than half this rate from 1950 to 1960. The post-1960 period is characterized by

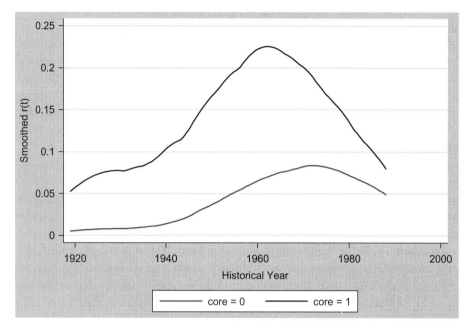

Fig. 5.8 Smoothed estimates of the hazard rate at which nation-states join at least one ICSU Scientific Union over historical time by category of countries (1919–1990).
Note: Number of events/countries: 222/21 (core countries); 491/145 (peripheral countries). Core countries (black line); peripheral countries (gray line).

a small decline in the hazard rate, although it again rises increasingly after 1970. This increase of the hazard rate is due, to some degree, to the fact that 55 peripheral countries become independent after 1960. These recently independent nation-states are the ones that continue entering ICSU Unions at a higher rate in the later period.[98]

Furthermore, the plot in Fig. 5.8 indicates that the rate of adoption throughout the entire period is higher for core nation-states than for peripheral countries. In 1989, core countries belong to almost 20 scientific unions on average. This represents a 65% increase in the average number of memberships by core nations since 1960, and a 215% since 1946. On the contrary, the average number of scientific unions to which peripheral countries belong is 4.4 (an increase of 69% since 1960, and a 300% since 1945). The pattern of membership growth

---

[98] Before 1960, 225 events take place. Forty-eight percent of those events take place in core countries (around 21 core countries) and the rest take place in 43 different peripheral countries (115 events). The opposite occurs during the post-1960 period. Thus, after 1960, 488 events took place. Seventy-seven percent of those events occur in 91 different peripheral countries. Eighty-two percent of those events are joinings of peripheral countries, which become independent before 1960 (i.e., 59 countries). Thirty countries, which become independent after 1960 only account for 69 events in the 1960–1990 time period.

for core and peripheral countries differs before 1960. In 1920, the average number of memberships is 3.75 for core countries (20 countries) and 0.4 for peripheral countries (36 countries). The average number of memberships increases at a rate of 2.23 per decade in the case of core countries (versus a rate of 0.6 per decade in the case of peripheral countries). Since the curves for the two different categories of countries (i.e., core versus peripheral countries) are not equally spaced within the 1919–1990 observational period, the hazard rates for core and peripheral countries are not multiples of one another and, therefore, proportional models do not seem to be appropriate for modeling the phenomena under study. Moreover, the curves for the two groups differ quite dramatically. The differences are greatest prior to 1960, when the joining activity by peripheral countries is very low. The lines only seem to approach each other by the end of the period analyzed, with the rise and expansion of the worldwide system.

2.6. Multivariate Models

Although plots of hazards versus historical time are valuable tools for exploratory analysis and for making basic non-parametric comparisons, this does not let me test any of the theoretical arguments posed here and does not tell much about the effects of some variables on the hazard rate when others are held constant. Both of these goals call for a second stage in the research: specification of a parametric model of the rate at which nation-states join at least one ICSU Union. Event history analysis (Tuma and Hannan, 1979 and 1984) is used to examine some of the internal and external factors affecting the rate at which nation-states enter into any ICSU Scientific Union over the 1919–1990 period. An important advantage of event history analysis is that information on national characteristics and on time-varying covariates can be incorporated in analyses of the time-dependent rate of occurrence of national events. Preliminary exploratory results (available upon request) and the theoretical arguments here proposed suggested that estimating the hazard rate in three periods divided at 1945 and 1965 would allow evaluation of hypotheses in a satisfactory way. For the purpose of this paper, I estimate several piecewise models that allow intercepts and effects of covariates to vary across different historical periods. The values of all covariates except the dummy variable for core countries (e.g., GNP per capita and the cumulative number of national memberships in ICSU Unions) change over time for a given country.

All models are estimated by the method of maximum likelihood. This method is advantageous because it provides estimates that are asymptotically unbiased and efficient. In addition, this method of estimation allows testing nested models using χ^2 statistics (Tuma and Hannan, 1984). The standard errors of parameters in a correctly specified model are asymptotically normally distributed, which allows testing hypotheses that any given parameter in the model differs from some pre-specified value (usually zero). The models are implemented using Tuma's computer program RATE for analyzing event history data (Tuma, 1979).

2.6.1. Selection of Multivariate Models

Neither of the commonly proposed and available parametric specifications for a non-monotonic hazard rate seemed to fit data on ICSU Unions' joinings. I there-fore chose to use piecewise models as a first step for studying the effects of covariates on the hazard rate. Table 5.6 gives the maximum likelihood estimates of three piecewise models of the hazard rate at which nation-states join at least one ICSU Scientific Union in three historical periods (e.g., 1919–1945, 1946–1964, 1965–1990). I begin with the *piecewise exponential model* (PE); results are presented in the first three columns of Table 5.6. This model assumes that the effects of the covariates can vary across different time periods. It expresses the dependence of the hazard rate on covariates explicitly and, there-fore, allows the testing of hypotheses about predicted effects and their variation across historical sub-periods. The PE model of the hazard rate takes the follow-ing functional form:

$$r(t) = \exp(\beta_p' X) \tag{1}$$

where the subscript p denotes a specific historical sub-period (e.g., 1919–1945, 1946–1964, or 1965–1990, respectively), X refers to a set of explanatory and control variables used in the analysis, and β_p are the parameters that measure the effects of X on the rate. The values of all covariates except the dummy variable for core countries (e.g., GNP per capita and the cumulative number of national memberships in ICSU Unions) change over time for a given country. As can be seen in Table 5.6, the PE significantly improves upon a baseline or null model with no covariates and with intercepts that vary across the three historical peri-ods; the probability levels for these tests are well below 0.001.

One inconvenience of the PE model, often overlooked in cross-national com-parative studies, is that it assumes that the covariates have the same effects within each period (i.e., $r(t)$ is a step function). Since I am investigating changing trends over time, with the hypothesis of decreasing positive effects of development and modernization factors and increasing effects of institutional factor on the hazard rate over time, the piecewise exponential framework with period specific effects does not seem to be the most appropriate for the purpose of this paper. In addi-tion, the plots of the hazard rates against historical time shown in the exploratory section clearly suggested that the hazard rate under study is time dependent. Such exploratory results indicated that the data do not meet the assumptions of exponential models because the hazard rates appear to be time dependent.

Table 5.6

Maximum likelihood estimates of models of the hazard rate at which nation-states join at least one ICSU Scientific Union in three historical periods, 1919–1945, 1946–1964, and 1965–1990

Covariates	Piecewise Exponential Model			Piecewise Proportional Gompertz Model			Piecewise Non-Proportional Gompertz Model		
	1919–1945	1946–1964	1965–1990	1919–1945	1946–1964	1965–1990	1919–1945	1946–1964	1965–1990
1. Time-independent vector of covariates (β)									
Intercept	−1.759****	0.026	3.538****	−7.409****	−30.250****	−35.450****	−22.900****	−38.210****	−53.640****
Gross national income per capita (in 000's)	0.626*	0.248**	0.016	1.044**	0.267**	0.063****	−1.653*	0.799	−0.686**
Imports and exports (as % of GNP)	−0.002	−0.001	0.000	−0.002	0.002	0.000	0.000	0.001	0.000
Cumulative number of national memberships in ICSU Scientific Unions	0.594****	0.319****	0.386****	0.567****	0.382****	0.397****	0.642****	0.355****	0.395****
Squared cumulative number of national memberships in ICSU Scientific Unions	−0.072****	−0.017****	−0.017****	−0.071****	−0.020****	−0.018****	−0.079****	−0.018****	−0.018****
Cumulative number of ICSU Unions' memberships in existence in the world	−0.035****	−0.038****	−0.017****	−0.044****	−0.090****	−0.034****	−0.051****	−0.099****	−0.032****
Number of INGOs in the world (in 00's)	0.772****	0.369****	0.037****	3.210****	0.163****	−0.158*	4.134****	4.794****	−0.045*
Core countries (dummy variable)	0.581**	−0.082	−0.052	0.582**	−0.037**	−0.012	0.597	−0.055	0.420
2. Time-dependent vector of covariates (?):									
Intercept				−0.371****	1.346****	1.157****	4.739****	1.598****	1.497****
Core countries (dummy variable)							−0.005	0.006	−0.007
Gross national income per capita (in 000's)							0.245**	−0.013	0.011**
Number of INGOs in the world							−0.079****	−0.007****	−0.002****
Overall rate per year	0.073	0.325	0.193	0.073	0.325	0.193	0.073	0.325	0.193
Number of events	71	270	372	71	270	372	71	270	372
χ^2 statistic									
Relative to the baseline model[a]		588****			1057****			1210****	
Degrees of freedom		21			21			30	
Relative to the piecewise exponential model					470****			622****	
Degrees of freedom					3			12	
Relative to the piecewise proportional Gompertz model								153****	
Degrees of freedom								9	
Number of countries = 166 countries									

[a] The model labeled baseline is a model with no covariates (constants only).

*Significant at the 0.10 level. **Significant at the 0.05 level. ***Significant at the 0.01 level. ****Significant at the 0.001 level.

341

Consequently, I estimate a *piecewise proportional Gompertz model* (PPG) of the hazard rate:

$$r(t) = \exp(\beta_p'X)\,\exp(\mu_p t) = \exp(\beta_p'X + \mu_p t) \tag{2}$$

where μ_p is a constant that changes for each specific historical sub-period p. Note that in addition to the consideration that the covariates can vary across different time periods, this model assumes that the hazard rate also varies over time within each of the sub-periods of analysis—in particular, the model assumes that the rate changes exponentially with time. Thus, the hazard declines, is constant, or rises within each sub-period as μ_p is less than, equal to, or greater than zero. Table 5.6 (columns 4–6) provides the estimates of the PPG modeling the hazard rate at which nation-states join at least one ICSU Scientific Union in three historical periods (e.g., 1919–1945, 1946–1964, and 1965–1990).

In the previous proportional hazards version of the piecewise Gompertz model, one is assuming that the hazard rate depends on a vector of covariates X. The β vector (the so-called time-invariant vector of covariates) describes the effects of heterogeneity of populations in the hazard rate itself. Of course, the β vector is a time-invariant vector within each sub-period p. But since a piecewise version of the Gompertz model is used, the β vector is a time-variant vector of covariates across different sub-periods. The effects of heterogeneity of populations in the time variation in the hazard rate can be controlled directly by using a *piecewise non-proportional Gompertz model* (PNG). To do so, I retain the assumption that the β vector is a function of covariates but now I specify that:

$$\mu = \Omega'Z \tag{3}$$

where Z is a vector of covariates which may or may not overlap with the vector of covariates X. Essentially, the variation in the hazard rate over time depends now on the vector of covariates Z. The functional form for the hazard rate is then:

$$r(t) = \exp(\beta_p'X + \Omega_p'Zt). \tag{4}$$

Note that the vector Ω_p (i.e., the time-dependent vector of covariates) tells the speed at which the ICSU joining rate either falls or rises within each historical sub-period. This piecewise Gompertz model is now non-proportional because time variation in the rate depends on the vector of covariates Z. Columns 7–9 in Table 5.6 report the estimates of the PNG model.

Since the three models—i.e., the PE, the PPG, and the PNG—are nested and can therefore be ordered hierarchically, the likelihood ratio χ^2 tests for significance can be computed to compare simpler with more complicated models.

For the analysis of all 166 countries, the PNG model significantly improves the fit from the previous two models. As can be seen at the bottom of Table 5.6, both the PPG and the PNG models significantly improve upon the PE model whose specification does not allow for changes of the hazard rate over time within each period.[99] The probability levels of the χ^2 statistics are well below 0.001. Thus, a Gompertz model seems to be more appropriate than a constant rate model for this longitudinal and cross-national analysis.[100]

In addition, the likelihood test comparing the PPG model against the PNG model is 153 with 9 degrees of freedom, which also has a p-level well below 0.001. Thus, I can reject the proportional version of the piecewise Gompertz model of the rate at which countries enter into any of the ICSU Scientific Unions with considerable confidence. In substantive terms, the results obtained here suggest two important points. First, the effects of internal and external factors on the ICSU joining rate vary across time periods. Second, the hazard rate is time dependent and, therefore, it changes over time within a historical sub-period in a non-proportional way.[101] This means that certain covariates (in particular, GNP per capita, and the number of INGOs in the world) have a significant effect on the speed of change in the hazard rate within each sub-period.

The results summarized in Table 5.6 led me to prefer the piecewise non-proportional form in order to model the ICSU joining rate because this model fits the data significantly better than the other two. In order to examine national variability in the effects of covariates, I also estimate two piecewise non-proportional model for core countries and for peripheral countries (see Tables 5.7 and 5.8). This analysis would allow me to examine whether the effects of the covariates are different depending on the category of country. Results are discussed in the following section.

---

[99] Table 5.6 shows the χ^2 tests and degrees of freedom for each of the three piecewise models compared to a constant rate model with no exogenous covariates (baseline model). For the piecewise non-proportional Gompertz model, since a χ^2 value of 622 with 12 degrees of freedom is significant ($p < 0.0001$), the inclusion of time-dependent covariates improves the model significantly.

[100] This was previously noted in the exploratory section of this paper when the non-linear relationship between the integrated hazard rate and historical time suggested that the hazard rate is time dependent, and therefore, a constant rate model may not be appropriate for the purpose of this analysis (see Fig. 5.5). Since I am using a relatively large sample of events (i.e., 713), this made it possible to obtain estimates for the PPG and PNG models. I am aware that a relatively small sample of events may make it difficult to estimate the parameters of non-proportional models under a reasonable number of iterations.

[101] This was also suggested in the preliminary exploratory analysis. Fig. 5.8 plotted the smoothed estimates of the hazard rate for core and peripheral countries. Since the curves for the two categories of countries appeared not to be equally spaced within the 1919–1990 observational period, proportional models do not seem to be appropriate for modeling the phenomena under study. The logged hazard rate plots for core and peripheral countries also suggest that the effects of some covariates (or at least this particular covariate) change over time. It could be that the variation over time in the hazard rates is due primarily to the variation over time of covariate effects though.

Table 5.7
Maximum likelihood estimates of the piecewise Gompertz model of the hazard rate at which core countries join at least one ICSU Scientific Union in three historical periods, 1919–1945, 1946–1964, and 1965–1990

Covariates	Core Countries		
	1919–1945	1946–1964	1965–1990
1. Time-independent vector of covariates (β)			
Intercept	−25.740***	−43.650***	−83.920***
Gross national income per capita (in 000's)	−1.817	−0.805	−0.129
Imports and exports (as % of GNP)	0.001	0.001	0.005
Cumulative number of national memberships in ICSU Scientific Unions	0.459*	0.206*	0.823***
Squared cumulative number of national memberships in ICSU Scientific Unions	−0.038	−0.010	−0.036***
Cumulative number of ICSU Unions' memberships in existence in the world	−0.079***	−0.121***	−0.058***
Number of INGOs in the world (in 00's)	4.540***	0.894***	−0.316***
2. Time-dependent vector of covariates (?)			
Intercept	6.353***	1.813***	2.514***
Gross national income per capita (in 000's)	0.171*	0.026	0.004
Number of INGOs in the world (in 00's)	−1.020***	−0.013***	−0.001
Overall rate per year	0.131	0.488	0.293
Number of events	42	100	77
χ^2 statistic			
Non-proportional Gompertz versus baseline model[a]		496***	
Degrees of freedom		24	
Non-proportional Gompertz versus proportional Gompertz		90.7***	
Degrees of freedom		6	
Number of core countries = 21 countries			

[a] The model labeled baseline is a model with no covariates (constants only).
*Significant at the 0.10 level.
**Significant at the 0.05 level.
***Significant at the 0.01 level.

2.7.   Discussion of Results

I estimate the PNG for all countries in three historical periods (e.g., 1919–1945, 1946–1964, and 1965–1990). Table 5.6 (columns 7–9) reports the maximum likelihood estimates of the coefficients for the PNG of the hazard rate at which countries join at least one of the ICSU Unions.

Table 5.8

Maximum likelihood estimates of the piecewise Gompertz model of the hazard rate at which peripheral countries join at least one ICSU Scientific Union in three historical periods, 1919–1945, 1946–1964, and 1965–1990

Covariates	Peripheral Countries		
	1919–1945	1946–1964	1965–1990
1. Time-independent vector of covariates (β)			
Intercept	−20.170***	−34.670***	−49.080***
Gross national income per capita (in 000's)	−2.319	3.344*	−0.605
Imports and exports (as % of GNP)	−0.002	−0.004	0.000
Cumulative number of national memberships in ICSU Scientific Unions	0.847***	0.363***	0.379***
Squared cumulative number of national memberships in ICSU Scientific Unions	−0.133***	−0.019***	−0.017***
Cumulative number of ICSU Unions' memberships in existence in the world	−0.035***	−0.083***	−0.027***
Number of INGOs in the world (in 00's)	3.775***	−0.055***	−0.023
2. Time-dependent vector of covariates (?)			
Intercept	3.578***	1.645***	1.340***
Gross national income per capita (in 000's)	0.377*	−0.070	0.010*
Number of INGOs in the world (in 00's)	−0.624***	0.009***	−0.002***
Overall rate per year	0.004	0.272	0.176
Number of events	29	170	284
χ^2 statistic			
Non-proportional Gompertz versus baseline model[a]		751***	
Degrees of freedom		24	
Non-proportional Gompertz versus proportional Gompertz		70.8***	
Degrees of freedom		6	
Number of peripheral countries = 145 countries			

[a] The model labeled baseline is a model with no covariates (constants only).
*Significant at the 0.10 level.
**Significant at the 0.05 level.
***Significant at the 0.01 level.

The model does not provide support for the hypothesis that socio-economic development has a strong positive effect on the ICSU Unions' joining rate (Hypothesis 1.1). On the contrary, the sign of the coefficients related to the GNP per capita measure is negative in the 1919–1945 and the 1965–1990 period (only significant at the 0.05 level in the later period). This suggests that the relationship between development and joining rates should not be hypothesized in purely internal functional and developmental terms. In addition, the effect of GNP per capita appears to have a statistically insignificant positive effect ($p < 0.05$) on the rate during the 1946–1964 period.

Nevertheless, by looking at the effects of GNP on the hazard rate as estimated by the PPG, one can see that socio-economic development seems to have a strong positive effect on the rate at which countries join ICSU Unions early in the period of analysis. Thus, a $1000 increase in GNP per capita is estimated to increase the joining rate by 184% during the 1919–1945 ($p < 0.05$), other things being equal.[102] The same increase is estimated to increase the rate by 30% in the following period ($p < 0.05$). During the most recent period (i.e., 1965–1990), a $1000 increase in GNP per capita increases the hazard rate by 6% ($p < 0.001$). These results do not confirm Hypothesis 1.2 in the sense that the effect of socio-economic development on the joining rate is not found to increase during the three historical sub-periods. Instead results suggest the opposite trend. Thus, development arguments explaining the national process of joining any ICSU Union might predict the joining of Scientific Unions early in the "science diffusion" process but not after the process is well under way (Tolbert and Zucker, 1983).

More consistent results are found to support institutional arguments in this study. The cumulative number of national memberships in ICSU Scientific Unions, which reflects the extent to which a nation-state is linked to the world-system and its discourse on science, has a strong significant positive effect on the hazard rate at which nation-states enter into any of the ICSU Unions ($p < 0.001$). Thus, in the 1965–1990 period, having one more membership in any of the ICSU Scientific Unions increases the joining rate by almost 50%. Moreover, this degree of national linkage of a country to the world scientific system (as measured by the cumulative number of national memberships in ICSU Unions) increases the hazard rate, which increases at a decreasing rate. This is shown by the negative sign associated to the squared cumulative number of national memberships in ICSU Scientific Unions.[103]

In sum, there is evidence to support Hypotheses 2.2 and 2.3 in the sense that there seems to be an inverted U-shaped relationship between the rate at which countries enter into ICSU Scientific Unions and the cumulative number of Unions' membership of a country. The nature of this association holds consistently throughout the three historical sub-periods.

The number of international governmental organizations in the world, giving an idea of the dimensions of the world-system as well as the impact of its

---

[102] The percentage change in the hazard rate associated with a unit change in GNP per capita (in thousands) is 184% = (exp(1.044)–1) × 100, ceteris paribus.

[103] To test Hypotheses 2.2 and 2.3, I add the cumulative number of national memberships in ICSU Scientific Unions in linear and quadratic form.

discourse on nation-state behavior, also has a strong positive effect at the 0.001 significance level. This supports Hypothesis 2.5. By now looking at the PNG model (columns 7–9 in Table 5.6), an additional 100 INGOs increase the hazard rate by a factor of 62 in the earlier period, and by 120 after 1945 (in the 1946–1965 period), net of other effects.[104] The negative sign of the same effect in the later period suggests that there might be a ceiling in the capacity of additional INGOs to affect nation-state behavior. Thus, after 1965 an additional 100 INGOs do not seem to have much effect on the hazard rate though.

In addition, the number of INGOs in the world (measured in hundreds) was added to the time-dependent vector of covariates (Ω). Surprisingly, the effect of the number of INGOs appears to have a very weak and significant negative effect ($p < 0.001$) on the speed at which the ICSU joining rate changes within each historical sub-period. In other words, the strong positive effect of the number of INGOs on the rate slightly decreases over time within each sub-period. For example, in the 1946–1964 period, the rate slightly decreases every year by 0.7% for each additional 100 INGOs ($p < 0.0001$). The decrease in the effects of the dimensions of the world-system as well as the impact of its discourse on nation-state behavior over time is inappreciable within the latest period of analysis.

This study does not support Hypothesis 2.4 in the sense that the degree of economic linkages to the world-system does not appear to have a significant and consistent effect on the rate of joining any ICSU Scientific Union over time. The amount of imports and exports as a percentage of GNP has neither a significant impact on the hazard rate nor the predicted positive effect on the rate of joining any ICSU Scientific Union.

Contrary to Hypothesis 2.6, the rate does not increase with the rise in the cumulative number of ICSU Scientific Unions' memberships in existence in the world over the 1919–1990 period. According to new institutionalists, mimetic pressures should increase the rate at which countries join ICSU Unions; in other words, as more states, whether hegemonic or peripheral, join ICSU Scientific Unions, other nation-states imitate the behavior in order to legitimate (DiMaggio and Powell, 1983). My analysis does not give support to that argument.

Results in Table 5.6 (columns 7–9) also indicate that the dummy variable for core countries has not significant effect on the hazard rate. This dummy variable most likely does not work because it is highly collinear with other covariates in the analysis (i.e., GNP). As the overall annual rates at which core and peripheral-states enter into any of the ICSU Union clearly indicate, rates not only change across historical times but also vary by category of country. Not surprisingly, the

---

[104] $60 = \exp(4.134)$ and $120 = \exp(4.794)$.

rate per year is higher for core countries than for peripheral countries. Thus, the overall rate for core countries is 0.29 in the 1965–1990 period while the overall rate per year for peripheral countries is 0.18—with rates much higher in the 1945–1965 than in the 1966–1990 period (as presented in Tables 5.7 and 5.8).

Estimates of a simple PPG for all countries where I only add the dummy variable for core countries as a time-invariant covariate confirm that core countries have significantly higher hazard rates than peripheral countries (results are not presented in this article). The model also showed certain convergence in the evolution of joining rates over time. In other words, the difference in the likelihood at which developed versus developing countries enter into ICSU Unions seems to have decreased slowly over time. Thus, being a core country increases the hazard rate of joining ICSU Unions by a factor of almost 3 in the 1919–1945 period, 2 in the 1946–1965, and 1.7 ($p < 0.0001$). This trend is not extremely strong but it is in the direction of convergence as hypothesized by developmental theorists (Hypothesis 1.3).

In order to explore the differences in the effects of covariates for core and peripheral countries, I estimate the PNG separately for each of the two subpopulations of countries in three historical periods (e.g., 1919–1945, 1946–1964, and 1965–1990).[105] Tables 5.7 and 5.8 present the results of this analysis. Table 5.7 reports the maximum likelihood estimates of the coefficients for the PNG of the hazard rate at which core countries join at least one of the ICSU Unions. Table 5.8 presents the estimates of the PNG model for the sample of peripheral countries (145 countries).

In comparison with the results obtained for all countries, there are no changes in the signs of the most significant effects. Only some minor changes occur in the magnitude of such effects for core and peripheral countries. Modernization and development arguments as operationalized by the GNP per capita continue to have a poor explanatory power of the ICSU Union process for both categories of countries. Only in the case of peripheral countries and during the 1946–1965 period, GNP per capita appears to have a significant strong and positive effect on the ICSU joining rate. When GNP per capita is added to the time-dependent vector of covariates (Ω), it only appears to have a significant positive effect ($p < 0.1$) on the speed at which the rate changes within the 1919–1945 sub-period. Thus, a one-unit increase in the national income per capita (expressed in thousands of dollars) raises the hazard rate of core countries by a factor of 1.18 per year within the 1919–1945 sub-period ($p < 0.1$).[106] The insignificant effect of GNP on

---

[105] Fortunately, I could take this approach because I have a relatively large sample of events for core and peripheral countries (number of events/countries: 222/21 (core countries); 491/145 (peripheral countries)). Event data sets on nation-states are not typically large enough to support extensive analysis of this type.
[106] $1.18 = \exp(0.171)$.

the rate after 1945 for both core and peripheral countries once again suggests that development conditions predict the joining of ICSU Scientific Unions at the beginning of the diffusion of science, but not after the process is well under way.

Adding institutional covariates increases the significance of the model under study. As the number of ICSU Unions' memberships for both core and peripheral countries increases over time, the hazard rate increases at a decreasing rate. In the case of external institutional factors, the number of INGOs in the world had a significant positive effect on the hazard rate of core countries up until 1964. After 1964, the number of INGOs has a significant but weak negative effect on the rate. The effect of INGOs is even weaker in the case of peripheral countries. In substantive terms, core countries are responsible for founding Unions early in the period. Therefore, they are the first to join. Once the majority of core countries have joined the existing scientific unions by 1965, peripheral countries follow and copy previous core nation behavior.

2.8. Conclusion

In this study, event history analysis has been applied in order to test some of the factors affecting the rate at which nation-states join any of the ICSU Scientific Unions (a total of 21 Unions in 1990). According to the model here proposed, institutional theories have greater power to predict the rate at which nation-states join any of the Unions over the 1919–1990 period. Thus, the joining rate increases more quickly during the post–World War II era with the rise of the world-system. In addition, the findings in this study suggest that as the number of ICSU Unions' memberships of a country increases over time, the hazard rate at which a nation-state joins an additional ICSU Union increases at a decreasing rate. This inverted U-shaped relationship between the rate and the cumulative number of Unions' memberships of a country holds for both core and peripheral countries. Furthermore, the increasing dimensions of the world-system as well as the impact of its discourse on nation-state behavior (as measured by the number of international governmental organizations in the world) are also proven to have a strong positive effect on the hazard rate for both core and peripheral countries.

Development and modernization arguments do not seem to offer significant explanations of the evolution of the ICSU joining hazard rate over time. The results have shown that more developed countries are more likely to join ICSU Unions than peripheral ones. However, the model does not provide support for the hypothesis that socio-economic development has a strong positive effect on the rate at which countries join ICSU Scientific Unions. Instead I prove that development arguments explaining the national process of joining any ICSU Union might predict the joining of ICSU Scientific Unions early in the "science diffusion" process (i.e., before 1965). After 1945, institutional factors better account for the nature and change of a nation's rate of joining ICSU Scientific Unions.

Finally, results confirm that core countries have significantly higher hazard rates than peripheral countries. There is also evidence of certain convergence in the evolution of joining rates over time. Thus, the difference in the likelihood at which developed versus developing countries enter into ICSU Unions seems to have decreased slowly over time. The results summarized in Tables 5.7 and 5.8 suggest that there is no substantial variability in the effects of the different development and institutional covariates for core and peripheral countries.

An attempt has been made in this paper to introduce the piecewise non-proportional models to the analysis of discrete dependent variables relating national-level events. Whenever the number of events is large enough, such framework might be applied to the study of any time-dependent national process (for example, the hazard rate at which nation-states join any kind of international institutions).

Acknowledgments

I am indebted to John W. Meyer for his guidance and support. I also wish to thank Nancy B. Tuma, Francisco Ramirez, Evan Schofer, and the members of the Stanford Comparative Workshop for their helpful advice and comments on earlier drafts of this paper. This version of the paper was presented at the Annual Meeting of the American Sociological Association (Toronto, August 1997).

References

Brusson, N., 1989, The organization of hypocrisy, Chichester, Wiley, United Kingdom.

DiMaggio, P. and Powell, W., 1983, The iron cage revisited: institutional isomorphism and collective rationality in organizational fields, *American Sociological Review* 48, 147–160.

Grew, R., 1984, The nineteenth century European State. In: *State-making and Social Movements* (C. Bright, S. Harding and Ann Arbor, Eds.), The University of Michigan Press, MI, pp. 83–120.

ICSU – Yearbook of the International Council of Scientific Unions (Paris, France: ICSU Secretariat, 1995).

International Council of Scientific Unions, 1993, *Science International* 51.

Meyer, J.W., 1980, The world polity and the authority of the nation-state. In: *Studies of the Modern World-System* (Albert J.B, Ed.), Academic Press, New York, NY.

Meyer, J.W., 1994, Rationalized environments. In: *Institutional Environments and Organizations: Structural Complexity and Individualism* (W.R. Scott and J.W. Meyer, Eds.), Sage, Thousand Oaks, CA, pp. 28–54.

Strang, D. and Meyer, J., 1993, Institutional conditions for diffusion, *Theory and Society* 22, 487–511.

Thomas, G., Boli, J. and Kim Y.S., 1993, World culture and international non-governmental organization, Paper presented at the *Annual Meeting of the American Sociological Association*, Miami, FL.

Thomas, G.M., Meyer, J.W., Ramirez, F.O. and Boli, J., 1987, *Institutional Structure: Constituting State, Society, and Individual*, Sage, Newberry Park, CA.

Tisdell, C.A., 1981, *Science and Technology Policy: Priorities of Governments*, Chapman & Hall Ltd, London, United Kingdom.

Tolbert, P. and Zucker, L., 1983, Institutional sources of change in the formal structure of organizations: the diffusion of civil service reform, 1880–1935, *Administrative Science Quarterly* 28, 22–39.

Tuma, N.B., 1979, Invoking RATE, 2nd ed., SRI International, Menlo Park, CA.

Tuma, N.B. and Hannan, M.T., 1979, Dynamic analysis of event histories, *American Journal of Sociology* 84 (4), 820–854.

Tuma, N.B. and Hannan, M.T., 1984, *Social Dynamics: Models and Methods*, Academic Press, Orlando, FL.

List of References (with Comments)

The purpose of this list of references is to provide additional sources of information for those who want to continue learning about dynamic methods in the social sciences. I organize the list of references by topics covered in each chapter of this book. Each section identifies further reading to help the reader learn more about the specific areas and techniques covered in this book. In addition to important books and manuals covering both theory and practice, I provide a list of examples of articles published in the top journals in social sciences using the different methodological approaches proposed in this book. Within each section, work cites appear commented, in alphabetical order of author (and within each author by chronological order).

Basic Data Analysis

In Chapter 1, I provide basic guidelines behind *univariate and bivariate statistics*. Many statistical textbooks can be used to learn about basic data analysis. Below I make a few recommendations:

Halmiton, L.C., 1996, *Data Analysis for Social Scientists: A First Course in Applied Statistics*, Wadsworth Inc., Belmont, CA. This book is a wonderful introduction to the basics of data analysis in the social sciences. It is intended to be the main text for a first course in statistics and it covers basic data management and analysis up to an introduction to regression analysis.

Moore, D.S. and McCabe, G.P., 2005, *Introduction to the Practice of Statistics*, 5th ed., W.H. Freeman, New York, NY. Chapters 1 and 2 cover data distributions and relationships.

Rabe-Hesketh, S. and Everitt, B., 2004, *A Handbook of Statistical Analyses Using Stata*, 3rd ed., CRC Press LLC, Boca Raton, FL. This book covers several basic and advanced statistical analyses, with the primary focus on using Stata when analyzing data and interpreting results. Great for new Stata users as well as experienced users who want to have a quick reference guide.

Chapters 2 and 3 assume some basic knowledge of the basic linear regression models and probability models for the study of dichotomous dependent variables. To learn about the *ordinary least square regression model* (OLS), one can also consult the relevant chapters in the following books:

Fox, J., 1997, *Applied Regression Analysis, Linear Models, and Related Methods*, Sage, Thousand Oaks, CA. Chapter 5 covers OLS regression.

Friedrich, R.J., 1982, In defense of multiplicative terms in multiple regression equations, *American Journal of Political Science* 26, 797–833. This article is relevant for those who will be introducing interaction terms in their regression analysis.

Hamilton, L.C., 1991, *Regression with Graphics: A Second Course in Applied Statistics*, Wadsworth Inc., Belmont, CA. Chapters 2 and 3 introduce the basics of multiple OLS regression. Chapters 4, 5, and 6 provide additional information necessary for the evaluation and fitting of the OLS regression model.

Hamilton, L.C., 2006, *Statistics with Stata: Updated for Version 9*, Thompson Brooks/Cole, Belmont, CA. This book is a great introduction to Stata and Stata resources available to the analyses of data. Chapters 6–9 provide an excellent discussion of the linear regression analysis.

Kennedy, P., 1998, *A Guide to Econometrics*, 5th ed., MIT Press, Cambridge, MA, pp. 47–77. Chapter 3 reviews the classical OLS model and Chapter 4 provides a great presentation of interval estimation and hypothesis testing.

For an excellent review of the models for binary outcomes (or dichotomous dependent variables), mainly *logit and probit models*, I recommend reading:

Aldrich, J.H. and Nelson, F.D., 1984, *Linear Probability, Logit and Probit Models*, Sage, Newbury Park, CA. This Sage volume examines the probit and logit models for the analysis of dichotomous dependent variables.

Hamilton, L.C., 1991, *Regression with Graphics: A Second Course in Applied Statistics*, Wadsworth Inc., Belmont, CA. Chapter 7 provides a good practical introduction to the logit regression, including estimation, hypothesis testing, and interpretation of logit results.

Liao, F., 1994, *Interpreting Probability Models: Logit, Probit, and Other Generalized Linear Models*, Sage, Newbury Park, CA. This book presents a variety of probability models used in the social sciences. It begins with a review of the generalized linear model, and covers logit and probit models, ordinal logit and probit models, and multinomial logit models (among many other probability models).

Long, J.S. and Freese, J., 2006, *Regression Models for Categorical Dependent Variables Using Stata*, 2nd ed., Stata Press, College Station, TX. This book is intended to those who will be using Stata for the fitting and interpretation of regression models with categorical dependent variables. Chapter 4 is the one that covers models for binary outcomes and it is the foundation for later chapters covering more advanced models.

Petersen, T., 1985, A comment on presenting results from logit and probit models, *American Sociological Review* 50, 130–131. This is a two-page document helpful to learn how to present results from logit and probit models.

In addition to the basic regression-like models, there are more complications in the OLS and probit/logit models. To learn more about *models for mixed continuous–discrete and polytomous outcomes*, I recommend the following textbooks:

Long, J.S., 1997, *Regression Models for Categorical and Limited Dependent Variables*, Sage, Thousand Oaks, CA. As its title indicates, this book deals with regression models in which the dependent variable is binary, ordinal, nominal, censured, truncated, or counted. They are also referred to as categorical or limited dependent variables. The book begins with a revision of the OLS regression model and an introduction to the maximum-probability estimation methodology. It covers the logit and probit models in the case of dichotomous variables, providing details on each one of the ways in which these models can be interpreted. To learn about models for mixed continuous–discrete and polytomous outcomes, I especially recommend the following chapters: Chapter 5: "Ordinal Outcomes: Ordered Logit and Ordered Probit Analyses;" Chapter 6: "Nominal Outcomes: Multinomial Logit and Related Models;" and Chapter 8: "Count Outcomes: Regression Models for Counts."

Long, J.S. and Freese, J., 2006, *Regression Models for Categorical Dependent Variables Using Stata*, 2nd ed., Stata Press, College Station, TX. Again this book is intended for those who will be using Stata for the fitting and interpretation of regression models with categorical dependent variables. I recommend Chapters 4–8.

Maddala, G.S., 1983, *Limited-dependent and Qualitative Variables in Econometrics*, Econometric Society Monographs 3, Cambridge University Press, Cambridge, United Kingdom. This book covers all methods available to the study of variables that are not continuously observed. It provides many empirical examples throughout.

Longitudinal Data

As discussed, cross-sectional data are recorded at only one point in time. However, it is common these days for cross-sectional data to be collected in a succession of surveys with a new sample of cases each time. Another possibility for longitudinal data occurs when we collect data on the same set of variables for two or more time periods, but for different cases in each period. This is called *repeated cross-sectional data*, and the data for each period may be regarded as a separate cross-section. Because these cases are non-identical in each period, the data have also been described as not-quite-longitudinal data sets. Caution therefore needs to be taken if applying the methods explained in this book. In general, the methods described here will not fit repeated cross-sectional data. For more information about these particular *types of longitudinal data* and the commonly used dynamic models, I recommend consulting:

Dale, A. and Davies, R.B., 1994, *Social Analyzing and Political Change: To Casebook of Methods*, Sage, Thousand Oaks, CA. This book summarizes the most popular statistical methods used for the analysis of change. I highly recommend this book to those without statistical knowledge who are interested in the quantitative methodology in the social sciences. Each chapter begins with a general introduction to a certain method of analysis. It later covers the methodology in detail and presents an example that uses such methodology. At the end of each chapter the advantages and problems of each methodology are emphasized, as well as the statistical packages available. The book is the result of the collaboration between a sociologist and a statistician. For the case of the longitudinal analysis of variables of quantitative nature, see Chapter 4 "The Analysis of Pooled Cross-Sectional Data" by John Micklewright; Chapter 5 "Analyzing Change Over Time Using LISREL" by John Bynner; and Chapter 9 "Time-Series Techniques for Repeated Cross-Section Data" by David Sanders and Hugh Ward. For the case of variables of qualitative nature, see the excellent revision of the event history analysis in Chapter 7 by Nancy B. Tuma.

Menard, S.W., 2002, *Longitudinal Research*, 2nd ed., Sage, Newbury Park, CA. This Sage volume provides advice on major issues involved in longitudinal research. It covers topics such as research design strategies, methods of data collection, and how longitudinal and cross-sectional research compares in terms of consistency and accuracy of results. Chapter 3 provides a good discussion of the different designs for longitudinal data collection, including repeated cross-sectional data designs as well as revolving panel designs.

Applications of Basic Data Analysis in the Social Sciences

There are many examples of empirical papers using basic data analysis. Here I provide a list of examples published in top journals in the social sciences (mostly sociology and management journals). The interested reader can easily search for more articles of their interest online. Two websites I recommend are the Scholarly Journal Archive at www.jstor.org and Google Scholar at scholar.google.com. To read examples of OLS regression, have a look at:

Finkelstein, S. and Hambrick, D.C., 1990, Top management-team tenure and organizational outcomes: the moderating role of managerial discretion, *Administrative Science Quarterly* 35 (3), 484–503.

Land, K.C., Davis, W.R. and Blau, J.R., 1994, Organizing the boys of summer: the evolution of U.S. minor-league baseball, 1883–1990, *American Journal of Sociology* 100 (3), 781–813.

Pelled, L.H., Eisenhardt, K.M. and Xin, K.R., 1999, Exploring the black box: an analysis of work group diversity, conflict, and performance, *Administrative Science Quarterly* 44 (1), 1–28.

Seidel, M.-D.L., Polzer, J.T. and Stewart, K.J., 2000, Friends in high places: the effects of social networks on discrimination in salary negotiations, *Administrative Science Quarterly* 45 (1), 1–24.

Swaminathan, A., 1995, The proliferation of specialist organizations in the American wine industry, 1941–1990, *Administrative Science Quarterly* 40 (4), 653–680.

Thornton, A., Axinn, W.G. and Hill, D.H., 1992, Reciprocal effects of religiosity, cohabitation, and marriage, *American Journal of Sociology* 98 (3), 628–651.

To read examples of *models for binary outcomes*, have a look at some of the articles listed below:

Berk, R.A., Bridges, W.P. and Shih, A., 1981, Does IQ really matter? A study of the use of IQ scores for the tracking of the mentally retarded, *American Sociological Review* 46 (1), 58–71.

Burt, R.S., 2003, Structural holes and good ideas, *American Journal of Sociology* 110 (2), 349–399.

Greeley, A.M. and Hout, M., 1999, Faith, hope, and charity Americans' increasing belief in life after death: religious competition and acculturation, *American Sociological Review* 64 (6), 813–835.

Greve, H.R., 1998, Performance, aspirations, and risky organizational change, *Administrative Science Quarterly* 43 (1), 58–86.

Ingram, P. and Simons, T., 1995, Institutional and resource dependence determinants of responsiveness to work-family issues, *Academy of Management Journal* 38 (5), 1466–1482.

Paternoster, R., Bushway, S., Brame, R. and Apel, R., 2003, The effect of teenage employment on delinquency and problem behaviors, *Social Forces* 82 (1), 297–335.

Petersen, T., Saporta, I. and Seidel, M.-D.L., 2000, Offering a job: meritocracy and social networks, *American Journal of Sociology* 106 (3), 763–816.

Wade, J., O'Reilly III, C.A. and Chandratat, I., 1990, Golden parachutes: CEOs and the exercise of social influence, *Administrative Science Quarterly* 35 (4), 587–603.

To read examples of *models for mixed continuous discrete and polytomous outcomes*, have a look at the articles listed below:

Barkema, H.G. and Vermeulen, F., 1998, International expansion through start up or acquisition: a learning perspective, *Academy of Management Journal* 41 (1), 7–26.

DiPrete, T.A., 1987, Horizontal and vertical mobility in organizations, *Administrative Science Quarterly* 32 (3), 422–444.

DiPrete, T.A., 1990, Adding covariates to log-linear models for the study of social mobility, *American Sociological Review* 55 (5), 757–773.

Gulati, R. and Singh, H., 1998, The architecture of cooperation: managing coordination costs and appropriation concerns in strategic alliances, *Administrative Science Quarterly* 43 (4), 781–814.

Kane, J. and Spizman, L.M., 1994, Race, financial aid awards and college attendance: parents and geography matter, *American Journal of Economics and Sociology* 53(1), 85–97.

Logan, J.A., 1996, Opportunity and choice in socially structured labor markets, *American Journal of Sociology* 102 (1), 114–160.

Lucas, S.R., 2001, Effectively maintained inequality: education transitions, track mobility, and social background effects, *American Journal of Sociology* 106 (6), 1642–1690.

McNamara, G. and Bromiley, P., 1997, Decision making in an organizational setting: cognitive and organizational influences on risk assessment in commercial lending, *Academy of Management Journal* 40 (5), 1063–1088.

Stolzenberg, R.M., 2001, It's about time and gender: spousal employment and health, *American Journal of Sociology* 107 (1), 61–100.

Yamaguchi, K., 2000, Multinomial logit latent-class regression models: an analysis of the predictors of gender-role attitudes among Japanese women, *American Journal of Sociology* 105 (6), 1702–1740.

Longitudinal Analyses of Quantitative Variables

In this section, I identify additional readings for the topics covered in Chapter 2 of this book. To continue learning about *models for the analysis of panel data*, there are many good references out there; below I include some of these references:

Arminger, G., Clogg, C.C. and Sobel, M.E., 1995, *Handbook of Statistical Modeling for the Social and Behavioral Sciences*, Plenum Press, New York, NY. This manual is intended for social scientists. Chapter 7 contains a theoretical and practical treatment of the analysis of longitudinal data for continuous variables. It is written by Cheng Hsiau, whose classical book on the study of panel data is also mentioned in this bibliography. In Chapter 8, the techniques available for the analysis of data of qualitative nature are studied. Event history analysis is covered in Chapter 9.

Greene, W.H., 1997, *Econometric Analysis*, 5th ed., Prentice Hall, Upper Saddle River, NJ. Here I especially recommend Chapter 13 on "Models for Panel Data." In practice, one can also read chapter 17 of the LIMDEP manual (which was written by Greene as well). The chapter is titled "Linear Models with Panel Data." To learn more about the program LIMDEP and its regression commands, I recommend looking at the LIMDEP manual as well:

Greene, W.H., 1995, *LIMDEP: Version 7.0. Users' Manual*, Econometric Software, Inc., Bellport, New York, NY. LIMDEP is a statistical program developed by Greene. It began being a statistical program to estimate probit models, that is, LIMited variable DEPendent models. The last versions of LIMDEP allow to estimate most of the more popular and novel models.

Halaby, C., 2004, Panel models in sociological research: theory into practice, *Annual Review of Sociology* 30, 507–544. This is a nice review of the principles and applications behind panel models used in sociological empirical studies.

Hsiao, C., 1986, *Analysis of Panel Data*, Cambridge University Press, Cambridge, United Kingdom. A good manual written to learn about the analysis of panel data.

Kmenta, J., 1986, *Elements of Econometrics*, MacMillan, New York, NY. In this manual, basic econometric theory is reviewed, including simple regression, violation of the basic assumptions of the simple linear regression, estimation with incomplete data, and multiple regression analysis. It also includes a chapter on the formulation and estimation of special models like models with dependent variables of qualitative nature. Chapter 12 reviews models with cross-section time series data. This book is difficult to read for those without a rigorous methodological and technical background.

Markus, G.B., 1979, *Analyzing Panel Data*, Sage, Newbury Park, CA. This green book discusses an array of techniques for the analysis of data collected on the same units of analysis at two or more points in time.

Menard, S.W., 2002, *Longitudinal Research*, 2nd ed., Sage, Newbury Park, CA. This Sage volume provides advice on major issues involved in longitudinal research. It covers topics such as research design strategies, methods of data collection, and how longitudinal and cross-sectional research compares in terms of consistency and accuracy of results. Chapter 4 covers the most important issues behind longitudinal research, including topics such as panel attrition, repeated measuring, and panel conditioning. Chapter 5 provides a discussion of the different types of longitudinal causal models including cross-sectional cross-time data, time series, short-time series with many sections or cases, and method for long-time series with many sections.

Sayrs, L.W., 1989, *Pooled Time Series Analysis*, Sage, Newbury Park, CA. This Sage volume is an introduction to the statistical models to analyze time series and cross-sectional data.

Stata Corporation, 2005, *Stata Release 9.0: Cross-Sectional Time-Series* (Manual), Stata Press, College Station, TX. This manual is useful to learn about the cross-sectional time series estimation and post-estimation commands in Stata. I especially recommend getting started with the commands `xt`, `xtsum`, and `xttab`, before learning about `xtreg`, `xtgee`, `xtgls`, or other more advanced commands for the study of change over time.

For a thorough introduction to the *GLS and GEE estimation*, consult Maddala (1983), and Greene (1997) referenced earlier in this section. GEE is a variation of the generalized least square method of estimation (GLS). Information on the GLS method of estimation can be found in Greene (1997) Chapter 16 on "Models That Use Both Cross-Section and Time-Series Data." The GEE model provides a richer description of the within-group correlation structure for the panels than standard GLS. For a comprehensive introduction to GEE in the estimation of GLS models, see:

Liang, K.-Y. and Zeger, S.L., 1986, Longitudinal data analysis using generalized linear models, *Biometrika* 73, 13–22.

Zeger, S.L., Liang, K.-Y. and Albert, P.S., 1988, Models for longitudinal data: a generalized estimating equation approach, *Biometrics* 44, 1049–1060.

To learn more about *models for time series data* or *ARIMA models*, consult any advanced book on econometrics. Classic books covering these topics are:

Box, G. and Jenkins, G.M., 1976, *Time Analysis Series: Forecasting and Control*, Holden-Day Inc., San Francisco, CA. This is the classic manual on dynamic models in which the Box–Jenkins methodology was introduced for the study of time series. Examples are included using data of economic interest.

Kendall, M. and Ord, J.K., 1973, *Time Series*, 3rd ed., Oxford University Press, New York, NY.

Mills, T.C., 1991, *Time Series Techniques for Economists*, 2nd ed., Cambridge University Press, Cambridge, United Kingdom.

For a few great introductions to the time series analysis, the following books are highly recommended:

Cromwell, J.B., Labys, W.C. and Terraza, M., 1993, *Univariate Tests for Time Series Models*, Sage, Newbury Park, CA. This book explores how to test for stationarity, normality, independence, linearity, model order, and properties of the residual process in the case of time series data. The book includes examples that illustrate each concept and test and provides advice on how to perform the tests using different software packages.

Cromwell, J.B., Labys, W.C., Hannan, M.J. and Terraza, M., 1994, *Multivariate Tests for Time Series Models*, Sage, Newbury Park, CA. This is recommended for multivariate analyses of time series data. The book helps readers to identify the appropriate multivariate time series model to use. It requires some working knowledge of time series data (such as the knowledge provided in Cromwell, Labys, and Terraza, 1993).

Ostrom, C.W., 1978, *Time Analysis Series: Regression Techniques*, Sage, Beverly Hills, CA. This book introduces readers to the analysis of temporary series by means of regression models.

Stata Corporation, 2005, *Stata Release 9.0: Time Series* (Manual), Stata Press, College Station, TX. This manual is useful to learn commands for time series data in Stata. The time series commands are grouped into two different sections. The first section, "Univariate Time Series" includes all procedures for univariate time series. The "Multivariate Time Series" groups together all different techniques designed for the estimation and post-estimation commands related to time series.

Yaffee, R.A. (with M. McGee), 2000, *Introduction to Time Series Analysis and Forecasting: With Applications of SAS and SPSS*, Academic Press, San Diego, CA. This book provides the basics behind time series analysis, with applications using SAS and SPSS.

To learn more about *structural equations modeling (SEM)*, I recommend Bollen (1989) cited below. For more information about SEM, I recommend consulting manuals such as the ones by Greene (1997) or Kennedy (1998), among many other available books also cited earlier in this list of references, to deepen one's knowledge of structural equations. In my opinion, the best comprehensive manual about structural equations is still the seminal work by Bollen (1989):

Bollen, K.A., 1989, *Structural Equations with Latent Variable*, John Wiley, New York, NY. This book offers a great theoretical overview of simultaneous equations that are used in dynamic models of regression to analyze longitudinal data where the dependent variable is continuous.

SmallWaters Corporation, 1997, *Amos User's Guide. Version 6*, SmallWaters Corporation, Chicago, IL. AMOS is a specific program used for the estimation of structural equations and needs to be purchased independently (even though it is a program module that is easily added to SPSS). I believe that today this is the easiest statistical package available to estimate structural equation models. It offers a useful graphical interface for the presentation and estimation of the model parameters. This program is quite powerful; the user can specify, estimate, assess, and present any model by drawing an intuitive path diagram with the hypothesized relationships among variables.

There are additional models that can be estimated in order to analyze a longitudinal dependent variable measured as a continuous or ordinal variable. Many methodology books talk about these models as *multilevel models for the analysis of change* in general, because they allow researchers to address within-case and between-case questions about change simultaneously. Many of these multilevel models specify two different sub-models. The first one describes how each case changes over time and the second describes how these changes differ across cases. Many more sophisticated models for change also assume that changes in the dependent variable are discontinuous or non-linear. All these models are beyond the introductory scope of this book. Nevertheless, the reader can learn more in:

Bryk, A.S. and Raudenbush, S.W., 1987, Applications of hierarchical linear models to assessing change, *Psychological Bulletin* 101, 147–158.

Bryk, A.S. and Raudenbush, S.W., 2002, *Hierarchical Linear Models: Applications and Data Analysis eMethods*, 2nd ed., Sage, Thousand Oaks, CA.

Rabe-Hesketh, S. and Skrondal, A., 2005, *Multilevel and Longitudinal Modeling Using Stata*, Stata Press, College Station, TX.

Rogosa, D., Brandt, D. and Zimowski, M., 1982, A growth curve approach to the measurement of change, *Psychological Bulletin* 90, 726–748.

Rogosa, D. and Willett, J.D., 1985, Understanding correlates of change by modeling individual differences in growth, *Psychometrika* 50, 203–228.

Singer, J.D. and Willett, J.B., 2003, *Applied Longitudinal Data Analysis: Modeling Change and Event Occurrence*, Oxford University Press, New York, NY.

Applications of Longitudinal Analysis of Quantitative Variables

Below I provide a list of articles published in top journals in the social sciences using the longitudinal analysis of quantitative variables (mostly sociology and management journals). Again, the interested reader can easily search online for additional articles of their interest. Two sites I highly recommend are the Scholarly Journal Archive at www.jstor.org or the more recently developed site, Google Scholar at scholar.google.com:

Baron, J.N., Hannan, M.T. and Burton, M.D., 2001, Labor pains: change in organizational models and employee turnover in young, high-tech firms, *American Journal of Sociology* 106 (4), 960–1012.

Berk, R.A., Rauma, D., Messinger, S.L. and Cooley, T.F., 1981, A test of the stability of punishment hypothesis, *American Sociological Review* 46, 805–828.

Castilla, E.J., 2004, Organizing health care: a comparative analysis of national institutions and inequality over time, *International Sociology* 19 (4), 403–435.

Castilla, E.J., 2005, Social networks and employee performance in a call center, *American Journal of Sociology* 110 (5), 1243–1283.

Cohen, L.E. and Felson, M., 1979, Social change and crime rate trends: a routine activities approach, *American Sociological Review* 44, 588–607.

Doreian, P. and Hummon, N., 1976, *Modeling Social Processes*, Elsevier, Amsterdam, Netherlands.

Hicks, A.M., 1994, The social democratic corporatist model of economic performance in the short- and medium-run perspective. In: *The Comparative Political Economy of the Welfare State* (T. Janoski and A.M. Hicks, Eds.), Cambridge University Press, New York, NY, pp. 189–217.

Hsiao, C., Appelbe, T.W. and Dineen, C.R., 1993, A general framework for panel data models with an application to Canadian customer-dialed long distance telephone service, *Journal of Econometrics* 59, 1–2.

Land, K.C. and Felson, M., 1976, A general framework for building dynamic macro social indicator models: including an analysis of changes in crime rates and police expenditures, *American Journal of Sociology* 82, 565–604.

Lazarsfeld, P.F., 1940, Panel studies, *Public Opinion Quarterly* 4, 122–128.

Nelson, J.F., 1981, Multiple victimization in American cities: a statistical analysis of rare events, *American Journal Sociology* 85, 871–891.

O'Connell, P.J., 1994, National variation in the fortunes of labor: a pooled and cross-sectional analysis of the impact of economic crisis in the advanced capitalist nations. In: *The Comparative Political Economy of the Welfare State* (T. Janoski and A.M. Hicks, Eds.), Cambridge University Press, New York, NY, pp. 218–242.

Shorter, E. and Tilly, C., 1970, The shape of strikes in France, *Comparative Studies of Society and History* 13, 60–86.

Snyder, D. and Tilly, C., 1972, Hardship and collective violence in France, 1830–1960, *American Sociological Review* 37, 520–532.

Wolpin, K.I., 1978, An economic analysis of crime and punishment in England and Wales, 1894–1967, *The Journal of Political Economy* 86 (5), 815–840.

Event History Analysis

In this section, I identify additional readings for the topics covered in Chapter 3 of this book. To learn the basic concepts behind the analysis of events, consult:

Cleves, M., Gould, W. and Gutierrez, R., 2004, *An Introduction to Survival Analysis Using Stata*, revised edition, Stata Press, College Station, TX. Chapter 1 is a great practical introduction to the analysis of events in Stata; the chapter is titled "The Problem of Survival Analysis." This book provides an in-depth coverage of Stata's routines for event history analysis.

Singer, J.D. and Willett J.B., 2003, *Applied Longitudinal Data Analysis: Modeling Change and Event Occurrence*, Oxford University Press, New York, NY. Chapter 9 titled "A Framework for Investigating Event Occurrence" provides a good introduction to the analysis of events.

Stata Corporation, 2005, *Stata Release 9.0: Survival Analysis and Epidemiological Tables* (Manual), Stata Press, College Station, TX. I especially recommend looking at the entries covering the commands `stset`, `stsum`, `stdes`, `stvary` as basic EHA commands in Stata. I also recommend exploring the commands `stfill`, `stjoin`, `stsplit`, and `stgen`.

To learn more about the *exploratory analysis of event history data*, consult:

Carroll, G.R., 1983, Dynamic analysis of discrete dependent variables: a didactic essay, *Quality and Quantity* 17, 425–460.

Cleves, M., Gould, W. and Gutierrez, R., 2004. Cited earlier in this section. Here I recommend Chapter 2 titled "Describing the Distribution of Failure Times."

Singer, J.D. and Willett, J.B., 2003. Cited earlier in this section. Here I recommend the reading of Chapter 10: "Describing Discrete-Time Event Occurrence Data" and Chapter 13: "Describing Continuous-Time Event Occurrence Data."

To learn more about the *explanatory or multivariate analysis of event history data*, consult:

Cleves, M., Gould, W. and Gutierrez, R., 2004. Cited earlier in this section. Here I recommend Chapter 3: "Hazard Models." Also Chapter 12: "Parametric Models" and Chapter 13: "A Survey of Parametric Regression Models in Stata." There are sections on accelerated failure time models.

Singer, J.D. and Willett, J.B., 2003. Cited earlier in this section. Chapter 11: "Fitting Basic Discrete-Time Hazard Models" and Chapter 14: "Fitting Cox Regression Models."

Other guides and chapters to *event history analysis* for researchers and students in the social sciences (also covering more advanced models within EHA) are:

Allison, P.D., 1984, *Event History Analysis: For Regression Longitudinal Event Data*, Sage, Beverly Hills, CA. This book is a brief and yet great introduction to the models of regression for analysis of events. In practice, the reader can also read:

Allison, P.D., 1995, *Survival Analysis Using the SAS System: To Practical Guide*, SAS Inc. Institute, Cary, North Carolina. This practical book is meant to learn how to estimate event history models using SAS. The book presents a good description of the six main procedures that SAS contains to perform exploratory and explanatory analyses of events. The commands are the following: To produce tables of life or graphs of the survival function (PROC LIFETEST); to estimate regression models with continuous data in the time (PROC LIFEREG); to estimate regression models using the method of Cox (partial PROC PHREG); to estimate several exponential models by time intervals (with GENMOD), among other options.

Box-Steffensmeier, J.M. and Jones, B.S., 2004, *Event History Modeling: A Guide for Social Scientists*, Cambridge University Press, New York, NY. This is a nice and accessible guide to event history analysis. It presents the principles of event history analysis with examples using Stata and S-Plus.

Blossfeld, H.-P. and Rohwer, G., 1995, *Techniques of Event History Modeling: New Approaches to Causal Analysis*, Lawrence Erlbaum, Mahwah, NJ. This book provides a comprehensive introduction to event history analysis. In addition the book is a manual of the computer program TDA (Transition Data Analysis) with practical examples. TDA is one of the statistical programs available for the analysis of longitudinal data. This book is especially good to get familiar with examples and practical applications. The appendix of the book includes a great introduction to TDA.

Cox, D.R. and Oakes, D., 1984, *Analysis of Survival Data*, Chapman and Hall, London, United Kingdom. This is a classic manual in the analysis of event history data.

Hannan, M. and Tuma, N.B., 1984, Approaches to the censoring problem in the analysis of event histories. In: *Sociological Methodology* (S. Leinhardt, Ed.), Jossey-Bass, San Francisco, CA, pp. 1–60. This article is helpful to learn about the different available methodological solutions to the problem of censoring.

Lancaster, T., 1990, *The Analysis of Transition Data*, Cambridge University Press, New York, NY. This is also a great manual to learn about event history analysis. It is especially useful to learn more about the common specifications of the function of time $q(t)$.

Mayer, K.U. and Tuma, N.B., 1990, *Event History Analysis in Life Course Research*, The University of Wisconsin Press, Madison, WI. This book is an excellent compilation of examples and applications of event history analysis in life course studies. It includes empirical studies on work histories and careers, migration, fecundity, and marriages, among many. In the third section of this book, there is a presentation on some of methodological topics most discussed in event history analysis.

Petersen, T., 1995, Analysis of event histories. In: *Handbook of Statistical Modeling for the Social and Behavioral Sciences* (G. Arminger, C.C. Clogg and M.E. Sobel, Eds.), Plenum Press, New York, NY.

Tuma, N.B., Hannan, M. and Groeneveld, L.P., 1979, Dynamic analysis of event histories, *American Journal of Sociology* 84, 820–854.

Tuma, N.B. and Hannan, M., 1978, *Social Dynamics: Models and Methods*, Academic Press, Orlando, FL. This is an advanced text on dynamic regression models for data of quantitative and qualitative nature. This book is now classic in the study of event history analysis.

Strang, D., 1994, Introduction to event history methods. In: *The Comparative Political Economy of the Welfare State* (T. Janoski and A.M. Hicks, Eds.), Cambridge University Press, New York, NY, pp. 245–253.

Wu, L.L., 1990, Simple graphical goodness-of-fit tests for hazard rate models. In: *Event History Analysis in Life Course Research* (K.U. Mayer and N.B. Tuma, Eds.), University of Wisconsin Press, Madison, WI, pp. 184–199.

Wu, L.L., 2003, Event history models for life course analysis. In: *Handbook of the Life Course* (J.T. Mortimer and M.J. Shanahan, Eds.), Kluwer Academic/Plenum, New York, NY.

Yamaguchi, K., 1991, *Event History Analysis*, Sage, Newbury Park, CA. This is another famous monograph on the models, methods, and applications of event history analysis.

Applications of Event History Analysis

The following list provides a combination of classic and more recent studies using the techniques behind event history analysis. Again, the interested reader can easily search for additional articles of their interest online:

Barnett, W.P., 1990, The organizational ecology of a technical system, *Administrative Science Quarterly* 35, 31–60.

Bienen, H.S. and Van Den Walle, N., 1989, *Of Time and Power*, Stanford University Press, Stanford, CA.

Drass, K.A., 1986, The effect of gender identity on conversation, *Social Psychology Quarterly* 49, 294–301.

Edelman, L.B., 1990, Legal environments and organizational governance: the expansion of due process in the American workplace, *American Journal of Sociology* 96 (6), 1401–1440.

Felmlee, D.H., Sprecher, S. and Bassin, E., 1990, The dissolution of intimate relationships: a hazard model, *Social Psychology Quarterly* 53 (1), 13–30.

Fernandez, R.M., Castilla, E.J. and Moore, P., 2000, Social capital at work: networks and employment at a phone center, *American Journal of Sociology* 105 (5), 1288–1356.

Fernandez, R.M. and Castilla, E.J., 2001, How much is that network worth? Social capital in employee referral networks. In: *Social Capital: Theory and Research* (K. Cook, N. Lin and R.S. Burt, Eds.), Aldine-deGruyter, Chicago, IL, pp. 85–104.

Flinn, C. and Heckmann, J., 1983, Are unemployment and out of the labor force behaviorally distinct labor force states? *Journal of Labor Economics* 1, 28–42.

Griffin, W.A. and Gardner, W., 1989, Analysis of behavioral durations in observational studies of social interaction, *Psychological Bulletin* 106 (3), 497–502.

Halliday, T.C., Powell, M. and Granfors, M., 1987, Minimalist organizations: vital events in state bar associations, 1970–1930, *American Sociological Review* 52 (4), 456–471.

Hannan, M.T., 1989, Macrosociological applications of event history analysis: state transitions and event recurrences, *Quality and Quantity* 23, 351–383.

Hannan, M. and Carroll, G., 1981, The dynamics of formal political structure: an event-history analysis, *American Sociological Review* 46, 19–35.

Hannan, M. and Carroll, G., 1989, Density delay in the evolution of organizational populations: a model and five empirical tests, *Administrative Science Quarterly* 34, 4111–4130.

Hobcraft, J. and Murphy, M., 1986, Demographic event history analysis: a selective review, *Population Index* 52 (1), 3–27.

Hujer, R. and Schneider, H., 1990, Unemployment duration as a function of individual characteristics and economic trends. In: *Event History Analysis in Life Course Research* (K.U. Mayer and N.B. Tuma, Eds.), University of Wisconsin Press, Madison, WI, pp. 113–129.

Kalev, A., Dobbin, F. and Kelly, E., 2006, Best practices or best guesses? Diversity management and the remediation of inequality, *American Sociological Review* 71, 589–917.

Knoke, D., 1982, The spread of municipal reform: temporal, spatial, and social dynamics, *American Journal of Sociology* 87, 1314–1339.

Marsden, P.V. and Podolny, J., 1990, Dynamic analysis of network diffusion. In: *Social Networks Through Time* (H. Flap and J. Weesie, Eds.), Rijksuniversiteit Utrecht, Utrecht, the Netherlands.

Mayer, K.U. and Carroll, G.R., 1990, Jobs and classes: structural constraints on career mobility. In: *Event History Analysis in Life Course Research* (K.U. Mayer and N.B. Tuma, Eds.), University of Wisconsin Press, Madison, WI, pp. 23–52.

Olzak, S., 1989, Labor unrest, immigration, and ethnic conflict: urban America, 1880–1915, *American Journal of Sociology* 94, 1303–1333.

Pavalko, E.K., 1989, State timing of policy adoption workmen's compensation in the United States, 1909–1929, *American Journal of Sociology* 95, 592–615.

Robinson, D. and Smith-Lovin, L., 1990, Timing of interruptions in group discussions, *Advances in Group Processes* 7, 45–73.

Soysal, Y.N. and Strang, D., 1989, Construction of the first mass educational systems in nineteenth century Europe, *Sociology of Education* 62, 277–288.

Strang, D., 1990, From dependency to sovereignty: an event history analysis of decolonization, *American Sociological Review* 55, 846–860.

Sutton, J.R., 1988, *Stubborn Children: Controlling Delinquency in the United States, 1640–1982*, University of California Press, Berkeley, CA.

Tolbert, P. and Zucker, L., 1983, Institutional sources of change in the formal structure of organizations: the diffusion of civil service reform, *Administrative Science Quarterly* 28, 22–39.

Usui, C., 1994, Welfare state development in a world system context: event history analysis of first social insurance legislation among 60 countries, 1880–1960. In: *The Comparative Political Economy of the Welfare State* (T. Janoski and A.M. Hicks, Eds.), Cambridge University Press, New York, NY, pp. 254–277.

Wu, L.L., 1996, Effects of family instability, income, and income instability on the risk of a premarital birth, *American Sociological Review* 61 (3), 386–406.

On Writing and Publishing Research Papers

If the reader has not done so by now, I recommend using the classic famous little writing manual: William Strunk Jr. and E. B. White, 2000, *The Elements of Style*, 4th ed., Allyn and Bacon, Boston, MA. This slim volume gives more specific advice on elementary rules of usage and the principles of composition, along with important suggestions for developing one's writing.

The University of Chicago Press has a series of publications that can be very helpful to improve on writing, editing, and publishing. Below are a few examples of these books:

Booth, W.C. and Colomb, G.G., 2003, *The Craft of Research*, The University of Chicago Press, 2nd ed., Chicago, IL.

Becker, H.S., 1986, *Writing for Social Scientists: How to Start and Finish Your Thesis, Book, or Article*, The University of Chicago Press, Chicago, IL.

Powell, W.W., 1985, *Getting into Print*, The University of Chicago Press, Chicago, IL.

Turabian, K.L., 2007, *A Manual for Writers of Research Papers, Theses, and Dissertations*, 7th ed., The University of Chicago Press, Chicago, IL.

Williams, J.M., 1990, *Style: Towards Clarity and Grace*, 6th ed., The University of Chicago Press, Chicago, IL.

The University of Chicago Press Chicago Guide to Preparing Electronic Manuscripts, 1987, Prepared by the Staff of the University of Chicago Press.

There are a couple of great good books on social research and inquiry that I highly recommend:

King, G., Keohane, R.O. and Verba, S., 1994, *Designing Social Inquiry: Scientific Inference in Qualitative Research*, Princeton University Press, Princeton, NJ. The goal of this book is to be practical when designing research that will produce valid inferences about social phenomena.

Ragin, C.C., 1994, *Constructing Social Research*, Pine Forge Press, Thousand Oaks, CA. This book offers an excellent broad, integrative overview of social research. As Ragin states in his preface, the book "answers the question: 'What is social research?' with diverse examples that illustrate current thinking about broad issues in social science methodology" (p. xi).

Stinchcombe, A.L., 2005, *The Logic of Social Research*, The University of Chicago Press, Chicago, IL. This book introduces students to the logic of the many methods available to study social causation.

I also recommend reading the classic books on the visual presentation of data by Edward Tufte:

Cleveland, W.S., 1985, *The Elements of Graphing Data*, Wadsworth Press, Belmont, CA.

Tufte, E.R., 1983, *The Visual Display of Quantitative Information*, Graphics Press, Cheshire, CT.

Tufte, E.R., 1990, *Envisioning Information*, Graphics Press, Cheshire, CT.

Tufte, E.R., 1997, *Visual Explanations*, Graphics Press, Cheshire, CT.

To Learn About Statistical Software Programs

Stata, SPSS, and TDA were primarily used to perform most of the dynamic analyses in this book. However, there are a wide variety of statistical software programs and resources available for the analysis of longitudinal data sets. The following websites provide information about the software as well as publications that can help the reader continue learning about these methods. The list is not meant to be exhaustive but rather to give the reader an idea of the most commonly-used statistical software programs for the analysis of longitudinal data sets:

AMOS (http://www.spss.com/amos/index.html)

LIMDEP (http://www.limdep.com)

SAS (http://www.sas.com)

S-Plus (http://www.insightful.com/products/splus)
SPSS (http://www.spss.com)
Stata (http://www.stata.com)

There are two software programs specifically developed for the analysis of event history data:

RATE.   To learn more about this program, consult: Tuma, N.B., 1979, *Invoking RATE*, 2nd ed., SRI International, Menlo Park, CA.

TDA.   To learn more about this program, consult: Blossfeld, H.-P. and Rohwer, G., 1995, *Techniques of Event History Modeling: New Approaches to Causal Analysis*, Lawrence Erlbaum, Mahwah, NJ.

        Go to www.lrz-muenchen.de/~wlm/tdaframe.htm for a complete introduction to this powerful program offering access to the latest developments in event history and transition data analysis.

Finally, there are two useful software programs to help you convert an original data set from one program format to a different one:

Stat Transfer (http://www.stattransfer.com)
DBMSCopy (http://www.dataflux.com/Product-Services/Products/dbms.asp).

Other Chapter References

Akaike, H., 1973, Information theory and an extension of the maximum likelihood principle. In: *Second International Symposium on Information Theory* (B. Petrov and F. Csake, Eds.), Ajademiai Kiado, Budapest.

Akaike, H., 1974, A new look at the statistical model identification, *IEEE Transactions on Automatic Control* 19 (6), 716–723.

Arellano, M. and Bond, S., 1991, Some tests of specification for panel data: Monte Carlo evidence and an application to employment equations, *The Review of Economic Studies* 58, 277–297.

Arrow, K.J., 1962, The economic implications of learning by doing, *The Review of Economic Studies* 29 (3), 155–173.

Auster, R., Levenson, I. and Saracheck, D., 1969, The production of health: an exploratory study, *Journal of Human Resources* 4, 411–436.

Castilla, E.J., 2003, Networks of venture capital firms in Silicon Valley, *International Journal of Technology Management* 25 (1/2), 113–135. Copyright © 2003 by Inderscience Publishers.

Castilla, E.J., 2004, Organizing health care: a comparative analysis of national institutions and inequality over time, *International Sociology* 19 (4), 403–435. Copyright © 2004 by Sage Publications.

Castilla, E.J., 2005, Social networks and employee performance in a call center, *American Journal of Sociology* 110 (5), 1243–1283. Copyright © 2005 by The University of Chicago Press.

Cox, D.R., 1972, Regression models and life tables (with discussion), *Journal of the Royal Statistical Society, ser. B* 34 (2), 187–220.

Cox, D.R., 1975, Partial likelihood, *Biometrika* 62 (2), 269–276.

Drukker, D.M., 2003, Testing for serial correlation in linear panel-data models, *Stata Journal* 3, 168–177.

Endnote, 1997, The Thomson Corporation (http://www.endnote.com/)

Florida, R.L. and Kenney, M., 1987, Venture capital and high technology entrepreneurship, *Journal of Business Venturing* 3, 301–319.

Fuchs, V.R., 1979, Economics, health, and post-industrial society, *Milbank Memorial Fund Quarterly/Health and Society* 57 (2), 153–182.

Hofstede, G., 1980, *Culture's Consequences: International Differences in Work-Related Values*, Sage, London, UK.

Kraemer, K., Dedrick, J., Hwang, C.-Y., Tu, T.-C. and Yap, C.-S., 1996, Entrepreneurship, flexibility and policy coordination: Taiwan's computer industry, *The Information Society* 12, 215–249.

Mathews, J.A., 1997, A Silicon Valley of the East: creating Taiwan's semiconductor industry, *California Management Review* 39 (4), 26–54.

Nohria, N., 1992, Information and search in the creation of new business ventures: the case of the 128 Venture Group. In: *Networks and Organizations: Structure, Form, and Action* (N. Nohria and R. Eccles, Eds.), Harvard Business School Press, Boston, MA, pp. 240–261.

OECD, 1995, OECD Health Data 1995, OECD, Paris, France.

Richardson, D.C. and Richardson, J.S., 1992, The Kinmage: a tool for scientific communication, *Protein Science* 1, 3–9.

Saxenian, A., 1990, Regional networks and the resurgence of Silicon Valley, *California Management Review* 33, 89–112.

Saxenian, A., 1994, *Regional Advantage: Culture and Competitions in Silicon Valley and Route 128*, Harvard University Press, Cambridge, MA.

Sorensen, J., 1999, *STPIECE: Stata module to estimate piecewise-constant hazard rate models.* EconPapers at http://econpapers.repec.org/

Swamy, P., 1970, Efficient inference in a random coefficient regression model, *Econometrica* 38, 311–323.

Triandis, H.C., 1995, *Individualism and Collectivism*, Westview Press, Boulder, CO.

United Nations Development Programme (UNDP), Several Years, *Human Development Report,* UNDP, New York, NY (http://hdr.undp.org/)

Williamson, J.B. and Fleming, J.J., 1977, Convergence theory and the social welfare sector: a cross-national analysis, *International Journal of Comparative Sociology* 18 (3–4), 242–253.

Wooldridge, J.M., 2002, *Econometric Analysis of Cross Section and Panel Data*, MIT Press, Cambridge, MA.

Index

Index of Program Commands and Options in Stata

Index of Program Commands and Options in TDA